High Pressure Engineering

CHEMICAL AND PROCESS ENGINEERING SERIES
Series Editor: I. L. Hepner

High Pressure Engineering

W. R. D. Manning M.A.(Cantab.), C.Eng., F.I.Mech.E., M.I.Chem.E.

S. Labrow M.Sc., C.Eng., F.I.Mech.E., V.D.I.

LEONARD HILL—LONDON
1974

An Intertext Publisher

A Leonard Hill Book
Chemical and Process Engineering Series
Series Editor: I. L. Hepner
First published in Great Britain in 1971 by

Leonard Hill Books
A Division of International Textbook Co. Ltd.
24 Market Square,
Aylesbury, Bucks HP20 1TL

This edition published 1974

ISBN 0 249 44132 2

Filmset by Photoprint Plates Limited, Rayleigh, Essex and printed in Great Britain by Butler and Tanner Limited, Frome and London

Contents

Preface

High pressure, as a tool in the improvement of chemical processes and in various metallurgical developments, is finding increasing industrial application, and—on a much smaller scale, but at much greater intensities—it is to be found in many laboratories. Not only is its importance in chemical and physical research continuing, but various other sciences are calling in its aid; geology and biochemistry are two examples.

Our object in this book is to explain the fundamental principles upon which these developments are based and to illustrate them with various laboratory and industrial applications. We hope in this way to assist students who may be attracted by the potentialities of the subject, and those who may be considering its use in various research projects; also we have tried to show some of the larger scale equipment, and to indicate the economics of pressure plants and their possible advantages, as well as the problems of safe operation and the protection of personnel from the results of accidents directly related to high pressure equipment. It may be worth noting, at this early stage, that high pressure working is by no means as expensive or as dangerous as is often feared.

The problem of units needs a brief note of explanation. It is complicated at the time of writing by the need to fall into line with the international system now coming into force which recommends that both pressures and stresses should be expressed in newtons per metre^2 (pascals), abbreviated N/m^2. As however this is an inconveniently small unit for most practical purposes "bars" and "kilobars" are being used increasingly, and in what follows the latter have been adopted, although in a number of instances the corresponding values in pounds force per square inch (here abbreviated lbf/in^2) have been added in brackets, for the sake of comparison with quoted work. An extensive *Conversion Table* is given on p. xv, from which the reader should have little difficulty in obtaining the corresponding values of any items he may require.

It is a fortunate fact that the "bar", as now generally understood,† is quite close to the standard (760 mm of mercury) "atmosphere", being in fact less than $1\frac{1}{2}\%$ smaller, and only 1·9% larger than the kilogram (force)/cm^2 which has always been extensively used on the Continent of Europe. In these preliminary pages we shall talk in terms of atmospheres since these are still (at the time of writing) the most widely understood and used of pressure units, although within the order of precision used in these general and historical notes the atmosphere and the bar can be regarded as the same.

Scope of the Book

The discussion of high pressure raises the question of what is meant by "high". As we are concerned here with engineering considerations, it will be convenient to define high pressure as that which requires some appreciable modifications to normal practice. Although, of course, the limits vary over the range of equipment considered, this seems to have fewer objections than any other method of classification; and one must remember that the special procedures introduced by one generation of engineers become the accepted practice of the next, and whatever method was adopted would be out of date before very long.

Generally speaking the lower limits of our classification are lowest for pumps and compressors because of the more difficult problems encountered, e.g. fatigue. With the design and construction of containers, however, the level can be raised considerably, largely because of the great mass of past experience, much of it gained by the gun designers in the arsenals of the "Great Powers", as we shall note in the next section.

The problems to be considered fall more or less into three main categories, namely those of containing, of generating, and of controlling high pressures. In Part I (Chapters 2 to 7) therefore we discuss the design and construction of vessels, pipes, and other containers, and the most suitable materials to use and the appropriate processes for their preparation.

Part II (Chapters 8 to 12) deals with the problems of compression and the design of machines in which this can be carried out. This section contains a brief survey of the thermodynamic principles involved, and emphasises the importance (especially for the benefit of university students) of realising that real gases at high pressures depart so much from the "ideal" gas laws that these are of little value in dealing with the practical problems which arise.

In Part III (Chapters 13 to 16) we consider the design and construction of the various types of control valve, and of other ancillary pieces of equipment needed to make a plant, either for commercial production, or for small-scale research studies. The problems of determining the pressure and of suitable types of measuring instrument are then discussed together with methods of calibration, maintenance, etc. The measurement of other variables, such as temperature and flow, is also briefly considered.

Finally, in Chapter 17, we consider the problems of safety, and the methods by which protection can be provided for operatives and other personnel in the neighbourhood.

In considering pressure containers emphasis has been laid on those made from forgings or built up by various compounding methods from thin plate or strip, and no attempt has been made to deal with thick-walled vessels fabricated from plate rolled to shape and seam welded. This is a subject which would require a book to itself at least as large as this one, and it is to a large extent covered by various national codes, for instance B.S. 1515. In some instances the equipment we describe can certainly be made by this technique, and the detailed design can then be safely left in the hands of the more experienced manufacturers.

We have also omitted consideration of the techniques of producing and controlling short duration pressures such as those produced in shock tubes, or by hollow charge explosives. There is of course no suggestion that these are unimportant; on the contrary much valuable work is going on in various quarters with developments of this kind, but they have different applications and are therefore not within the scope of this book.

The authors would like to express their thanks to all those who have so kindly supplied photographs and illustrations.

List of Figures

Figure | Page

CONVERSION TABLE

UNIT	Bar	Newton/m² (pascal)	Dyne/cm²	†Standard atmosphere (760 mm Hg)	‡kg(force)/cm²	kg(force)/mm²	lb(force)/in² (p.s.i.)	Tons(long)/in²	Kilobars (kb)
Bar	1·0	10^5	10^6	0·9869	1·0197	$1{\cdot}0197 \times 10^{-2}$	14·50(4)	$6{\cdot}475 \times 10^{-3}$	10^{-3}
Newton/m² (pascals)	10^{-5}	1·0	10	$9{\cdot}869 \times 10^{-6}$	$1{\cdot}0197 \times 10^{-5}$	$1{\cdot}0197 \times 10^{-7}$	$1{\cdot}450 \times 10^{-4}$	$6{\cdot}475 \times 10^{-8}$	10^{-8}
Dyne/cm²	10^{-6}	10^{-1}	1·0	$9{\cdot}869 \times 10^{-7}$	$1{\cdot}0197 \times 10^{-6}$	$1{\cdot}0197 \times 10^{-8}$	$1{\cdot}450 \times 10^{-5}$	$6{\cdot}475 \times 10^{-9}$	10^{-9}
Standard atmosphere (760 mm Hg)	1·0133	$1{\cdot}0133 \times 10^5$	$1{\cdot}0133 \times 10^6$	1·0	1·0332	$1{\cdot}0332 \times 10^{-2}$	14·696	$6{\cdot}565 \times 10^{-3}$	$1{\cdot}0133 \times 10^{-3}$
Kilogram (force)/cm²	0·9807	$9{\cdot}807 \times 10^4$	$9{\cdot}807 \times 10^5$	0·9679	1·0	10^{-2}	14·22	$6{\cdot}349 \times 10^{-3}$	$9{\cdot}807 \times 10^{-4}$
Kilogram (force)/mm²	98·07	$9{\cdot}807 \times 10^6$	$9{\cdot}807 \times 10^7$	96·79	10^2	1·0	$1{\cdot}422 \times 10^3$	0·6349	$9{\cdot}807 \times 10^{-2}$
lb(force)/in² (p.s.i.)	$6{\cdot}895 \times 10^{-2}$	$6{\cdot}895 \times 10^3$	$6{\cdot}895 \times 10^4$	$6{\cdot}805 \times 10^{-2}$	$7{\cdot}031 \times 10^{-2}$	$7{\cdot}031 \times 10^{-4}$	1·0	$4{\cdot}464 \times 10^{-3}$	$6{\cdot}895 \times 10^{-5}$
Ton (long)/in²	154·4	$1{\cdot}544 \times 10^7$	$1{\cdot}544 \times 10^8$	152·4	157·5	1·575	$2{\cdot}240 \times 10^3$	1·0	0·1544
Kilobars (kb)	1,000	10^8	10^9	$9{\cdot}869 \times 10^2$	$1{\cdot}0197 \times 10^3$	10·197	$1{\cdot}450 \times 10^4$	6·475	1·0

To convert a pressure given in a particular unit to its corresponding value in the unit required, find the given unit in the left-hand column and then move horizontally across the table to the column headed by the required unit. The figure appearing there is the factor by which the given quantity has to be multiplied to bring it to the required unit.

For example, if given a pressure as 2,865 lb (force)/in² and requiring to know its corresponding value in standard (760 mm Hg) atmospheres, we find the factor to be $6{\cdot}805 \times 10^{-2}$, and thus the required figure is $2{,}865 \times 6{\cdot}805 \times 10^{-2}$, or 194·96 atm.

† This is also known as the "physical atmosphere".

‡ This is also known as the "technical atmosphere". The kilogram (force) is also known as the 'Kilopond', especially in Germany; Readers should also note that some U.S. writers use the term 'kilopound' to denote 1,000 lb which can easily be confused with 'kilopond'.

1 Historical

The desire to compress gases and to pump liquids has existed for a very long time. Probably the earliest record we have is in an Egyptian tomb painting of about the sixteenth century B.C. which shows two "blowers" used to supply air to a furnace for melting metals. These were similar to our ordinary domestic bellows, consisting of a pair of leather bags attached to a frame on the ground which was worked by the hands and feet of the operators, and as there were two of these the resulting flow was no doubt fairly steady, although the pressure can have been only slightly above atmospheric. In Roman times larger and more elaborate blowers were introduced having metal valves and other features which foreshadowed the reciprocating compressor of today.

Some of the latter were used to pump water, probably by a device of the air-lift type, although there is clear evidence that the Romans were never able to produce a satisfactory pump. Their appreciation of the requirements of good water supplies for their cities must have emphasised this need, but they evidently failed to satisfy it and had to fall back on huge civil engineering projects such as the superb aqueduct of Pont du Gard, near Avignon in Southern France. No doubt the biggest impediment to mechanical developments of all kinds was the lack of materials capable of resisting tensile stresses, and a whole millenium passed by before any real progress was to be made in what we now know as mechanical engineering. However, Leonardo da Vinci, towards the end of the fifteenth century A.D. described various designs of valves for pumps and compressors, some of which bear a striking resemblance to types in use today. For example, he shows a multi-flap valve for dealing with large flows, in which a group of small flap valves (to reduce inertia forces) were assembled in a housing in such a way as to give a very favourable flow pattern. We do not know if any of these were ever made and tested, but the lack of suitable materials makes this rather unlikely. Leonardo also took a great interest in various weapons of war, but there does not seem to be any record of his contributing to the design of guns, which were the first real high pressure vessels.

The earliest guns had been made by packing together a ring of wrought iron bars round a cylindrical core, rather like the staves of a barrel, and holding them together by smith's welds along their sides. These were then reinforced by forcing or shrinking on a series of forged hoops. The practical difficulties of carrying out this sort of operation without the assistance of power hammers must indeed have been extreme, and one wonders to what extent the "staves" were really capable of resisting the tangential forces tending to separate them. Nevertheless, guns of formidable size were made in this way (the famous "Mons Meg" now preserved in Edinburgh Castle has a bore of $19\frac{1}{2}$ in) and they were used extensively in the warfare of the period. We can but admire the incredible skill of those early smiths, and if history tells us little of the guns that burst on firing we know that many were capable of a considerable effect on the morale of the enemy.

Guns built up in this way eventually gave place to castings, first in bronze at the end of the fifteenth century and then in iron about fifty years later.

The basis upon which these were designed is not known, but they were tested by firing with propellent charges considerably in excess (sometimes as much as 50%) of the designed charge, which was usually one half the weight of the projectile. Surprising ranges were achieved by these guns; for instance, in 1651 during operations in the isles of Scilly, which had remained in Royalist hands throughout the Civil War, Cromwell's forces landed guns on the island of Tresco from which they were able to control the harbour of St. Mary's, the principal port of the islands. This must have involved a range of at least 2 kilometres and chamber pressures in the guns of 100 atmospheres or more. We know nothing of the life of these guns and they can hardly have been very accurate at such ranges. On the other hand the tensile stress involved must have been alarmingly high for cast iron.

It was not for nearly another two centuries that the problem of determining the stresses in thick-walled cylinders of elastic material was finally solved, by Lamé and Clapeyron[1.1] in France. Their paper was published in 1833 and contains the solution (known by the name of Lamé) which we still use.

High pressure chemical studies probably began when Papin, a Huguenot refugee working in London in 1681, made a bronze digester for extracting marrow from bones[1.2], but there was very little further interest in this for many years. At the end of the eighteenth century, however, Jacob Perkins[1.3], first in America and later in England, measured the compressibility of water and also the effect of pressure on melting phenomena. His results were far from accurate, but he worked up to 100 atmospheres, using a gun as his pressure vessel. Later, with an improved and strengthened vessel, he claimed to have reached about 2,000 atmospheres. Various other scientists reported working in what we would now call the kilobar range, but it was not until Cailletet and Amagat appeared on the scene that the technique was placed on a scientific footing and the pressures were measured with reasonable accuracy. These two men worked in Paris and published their work between the years 1869 and 1894. Although most of it was published separately, they appear to have collaborated in the design of apparatus, and there is no doubt that theirs was a major contribution, both to high pressure technique, and to the measurement of high pressure physical effects. It is to be presumed that they knew of Lamé's work although neither of them appears to have acknowledged this. In fact few details are given of their experimental methods,† but their results especially those of Amagat are quite astonishingly accurate, even up to 3,000 atmospheres.

By 1890 autoclaves for chemical reactions on an industrial scale were becoming quite common, although the operating pressures remained below 100 atmospheres and the sizes seldom reached 500 litres. They were mostly of cast iron which was doubtless one of the main limitations of pressure. Working at high temperatures also introduced sealing problems which were probably troublesome at that time. On the other hand, Andrews[(1.5)] had given an account of his successful liquefaction of gases at low temperatures and this focused much interest on gases at pressures up to several hundred atmospheres.

This work on liquefaction, leading to the commercial separation of oxygen from the air, became an important stimulus to modern high pressure engineering, especially in the design and development of compressors, but – before turning to consider this further – mention should be made of the work of W. Spring[(1.6)] in Belgium at the end of the nineteenth century. He was one of the first geologists to become interested in high pressure, chiefly in an attempt to throw light on the formation of minerals in the early stages of the Earth's existence, and his technique is interesting for two reasons: first, he used tool steel plungers and cylinders to contain and produce his pressures, and secondly he applied these pressures direct to powdered solids. He claimed to have worked at over 6,000 atmospheres and he certainly obtained some surprising and reproducible results (such as the direct combination of iron and sulphur to iron sulphide at room temperature). On the other hand there seems to be some doubt about the uniformity of his pressure distribution, which was probably by no means hydrostatic. But his ideas were the forerunners of some of the techniques used in diamond synthesis, and his experiments also have a bearing on the isostatic compacting operations which are an essential part of modern powder technologies.

Returning to the subject of pressure work in the gas phase, Linde saw the importance of the liquefaction and fractionation of air and other mixed gases, and in 1895 he built the first continuous compressor for service at 200 atmospheres with an air liquefier attached. He used a 2-stage machine with horizontally opposed cylinders which, with their intercoolers, were wholly immersed in water[(1.7)].

Linde's compressors which were developed rapidly in the early years of the present century undoubtedly paved the way for Haber and Bosch when they came to build the first synthetic ammonia plant in 1913[(1.8)]. Many variations of this process followed, the highest pressure being 1,000 atmospheres in the Claude process of 1919, while others more recently have worked at pressures as low as 120 atmospheres. In the case of the former it seems that the materials then available were hardly sufficient for such conditions and most of Claude's licensees

† For a review of the work of Cailletet and Amagat reference should be made to Bridgman's work[(1.4)] which also describes other early work in this field.

were content to lower the maximum pressure to around 800 atmospheres. In recent years the advent of rotary compressors to work at pressures of 200 atmospheres and more with very large throughputs has, in some instances, led to a lowering of the working pressure to about that originally used by Haber, and it is clear that many other factors, some local, can affect the economically optimum pressure for this reaction, but it would seem that in general this is likely to lie between 300 and 450 atmospheres. The more recent developments have been in the building of larger plants, and compressor units requiring 12,000 horsepower are in operation; certain American manufacturers are prepared to consider even larger machines (up to 15,000 hp).

The importance of synthetic ammonia as a means of "fixing" atmospheric nitrogen and its consequently enormous importance in increasing crop yields throughout the world, has tended to overshadow other important high pressure chemical processes, such as the synthesis of methanol and urea. For the most part these use similar engineering techniques to those of the ammonia processes, and the pioneer studies of Bergius in Germany, and the full scale plants of I. G. Farben and of Imperial Chemical Industries Ltd. [1.9] at Billingham fall into this category.

While these great advances in high pressure industrial chemistry were taking place, considerable research activity was also in progress. Imperial Chemical Industries were active in this in their Northwich (Cheshire) laboratories, and in the course of a comprehensive survey at pressures up to 3,000 atmospheres (with occasional scouting experiments up to 12,000 atmospheres) the solid polymer of ethylene† was discovered (in March 1933), and this was quickly developed on a plant scale in view of its great military importance for submarine cables, Radar, and other wartime applications. The first I.C.I. plant began production in September 1939 at a pressure of about 1,500 atmospheres.

This pressure was considerably in excess of that produced by any commercially available gas compressor at the time, and various expedients had to be adopted to deal with the situation. For instance I.C.I. used a reciprocating lute of mercury in a U-tube to transmit the pressure from an oil pump to the gas, and the Du Pont Co. in America liquefied the ethylene and compressed it with a modified oil pump. Both of these expedients introduced many extra problems, although both succeeded in maintaining the required supply of the product. By about 1950, however, most of the major compressor manufacturers on the European Continent and in the U.S.A. (though not, unfortunately, in Britain) had found out how to raise the delivery pressure of their conventional machines to meet these demands, which have now increased still further (to at least 3,000 atmospheres).

The development of reaction vessels for these pressures was also a considerable problem, although the gun designers and manufacturers had already shown what could be done in the way of resisting internal pressure for very short intervals of time. Small-scale research had also begun to throw more light on the strength of containers, and – in the event – the vessels designed by I.C.I. in 1940

† This was given the generic name "Polythene" in Britain, but in the U.S. it was usually known as "Polyethylene". The recent developments in other olefine polymers, notably polypropylene, seems to make the latter preferable in spite of its extra syllables.

gave very little trouble, and many are still being manufactured to what is virtually the same design.

The next major development using high pressures to be reported was the synthesis of diamond in 1954[1.10] by the General Electric Company of America, although it was subsequently revealed by the Swedish company, A.S.E.A., that they had actually achieved this 12 months earlier.[1.11] The important fact in this work is that very much higher pressures were required—at least 50,000 atmospheres with simultaneous high temperatures, well above 1,000°C. Published details of these processes (which are now in commercial operation by at least five organisations) are somewhat scanty, as might be expected, but enough is known of their main features to enable us to discuss the general background, which we do in Chapter 16, where it is explained that the process is probably possible only because of some rather fortunate physical properties in graphite and diamond, and—as we shall show—there are reasons for thinking that this is getting near the limit of true hydrostatic pressure which can be contained in conventional equipment.

The metallurgical industries have become interested in high pressure techniques in recent years. Static pressure for pre-shaping of components to be made from powder, both metal and ceramic, has been practised for 30 years or more, but more recently the pressure levels have been greatly increased and 10,000 atmospheres is considered desirable for certain applications.[1.12] Another development in this field is that of hydrostatic extrusion whereby materials which are difficult or impossible to work by conventional extrusion methods can be thus processed. In some instances the pressures required are very high because the chamber into which the extrudate is delivered also has to be at high pressure, and the pressure in the main chamber has to be sufficiently high for the difference to be large enough to force the material through the die.[1.13]

So far in this brief review no mention has been made of P. W. Bridgman. The reason for this is that he has contributed more than anyone else to developments in nearly all applications of high pressure, and it is hardly possible to cite each contribution. Beginning in 1909 until his death in 1961 he continued his pioneering work, most of which has been collected in his book *The Physics of High Pressure*, first published in 1931[1.4] and going through several editions until 1958. This was supplemented in 1952 by his *Studies in Large Plastic Flow and Fracture*[1.14] and lastly his *Collected Experimental Papers* were republished in 1964.[1.15] The first of these books also contains a much more detailed historical survey than that presented here.

REFERENCES

1.1. LAMÉ, G. and CLAPEYRON, B. P. E., *Mém. présentés par Divers Savants,* **4,** Paris 1833.
1.2. TONGUE, H., *The Design and Construction of High Pressure Chemical Plant,* Chapman & Hall, London 1959, 2.
1.3. PERKINS, JACOB, *Phil. Trans. Roy. Soc.,* **110,** 324, 1820.
1.4. BRIDGMAN, P. W., *The Physics of High Pressure,* Bell, London 1931, 6.
1.5. ANDREWS, T., *Brit. Ass. Rep.,* 76, 1861.
1.6. SPRING, W., *Zeit. Phys. Chem.,* **1,** 227, 1887; see also Ref. 1.4, 6.
1.7. LINDE, C. VON, Publication of Linde Eismaschinen A. G., 1964.
1.8. HABER, F. and BOSCH, C., *British Patent* No. 14023, 1910.

1.9. SMITH, F. E., *Proc. Inst. Mech. Eng.,* **133,** 139, 1936.
1.10. BUNDY, F. R. *et al., Nature,* **176,** 51, 1955.
1.11. LIANDER, H. and LUNDBLAD, E., *Ark. Kemi,* **16,** 139, 1960.
1.12. ROLFE, B. W., Conference on High Pressure Engineering, London, Sept. 1967, Paper No. 18, *Proc. Inst. Mech. Eng.,* **182,** Pt 3C, 239.
1.13. PUGH, H. LL. D., *Bulleid Memorial Lectures,* Nottingham University 1965, Lecture III.
1.14. BRIDGMAN, P. W., *Studies of Large Plastic Flow and Fracture,* McGraw-Hill, New York and London 1952.
1.15. BRIDGMAN, P. W., *Collected Experimental Papers,* Harvard University Press, Cambridge, Mass. 1964.

PART I

STRESSES AND MATERIALS

Notation Used in Part I

A	Constant of integration
a	Area of cross-section (of wound-on strip, or of cylinders)
B	Constant of integration
C	Constant used in empirical relations; also to simplify calculations, etc.
c	Specific heat
D	Diameter
E	Young's Modulus
G	Shear (rigidity) Modulus
J	Joint factor
K	Overall diameter (or radius) ratio of cylinder or sphere
K_h	Thermal conductivity (Chapter 5)
K_{1c}	'Stress intensity factor' in special brittleness tests
k	Diameter or radius ratio of part of cylinder or sphere
L	Length
n	Number of components (in composite cylinder); also positive integer greater than 1 (in Chapter 3, eq. 3.33)
P	Pull required in winding on wire or strip for reinforcement
p	Pressure
R	Contact pressure between surfaces
r	Radius (of sphere or cylinder)
S	Ratio of working pressure to tensile proof stress $= p_W/\sigma_T^*$
T	Torque in torsion test records; also temperature in Chapter 5
t	Wall thickness of cylinder; also time
U	Stress function
u	Radial shift
V	Strain energy
v	Radial shift
w	Shrinkage allowance
X	Product of k_m and k_{m+1}
Y	Ratio of wall thickness to internal diameter in cylinders $= t/D_i$
Asterisk(*)	Limiting (proof stress) value of stress; corresponding pressure
α	Coefficient of linear expansion; also angle in strained section
β	Angle in strained section
γ	Shear strain
ε	Direct strain
θ	Angular position; also angle of twist in torsion test
Θ	Temperature difference between surfaces
λ	Constant $= u_o^2 + 2u_o r_o$
ν	Poisson's Ratio
ρ	Density
σ	Direct stress
τ	Shear stress

Suffixes applied to symbols for stresses, strains, pressures, etc.

1, 2 and 3	Directions in general three-dimensional systems
θ, z and r	Tangential, longitudinal and radial directions
θ–r	In tangential–radial plane (applied to shear stress and strain)
al	Allowable
c	Stress to cause a given creep (1% in 10^5 hours unless otherwise stated)
$c+15$	Ditto at temperature 15°C higher than operating temperature
hy	Hydrostatic (applied to pressure and the strains thus produced)
L	Relating to the liner in liner–mantle combinations
M	Relating to the mantle in liner–mantle combinations; also to the Maxwell (von Mises) hypothesis with stress function *U*
oct	In the octahedral plane
R	Residual (working pressure removed) stress conditions
Ru	Stress to cause rupture in high temperature creep (in 10^5 hours unless otherwise stated)
s	In strip or wire windings
sh	Shear (applied to strain energy)
T	In tensile test
tot	Total (applied to strain energy)
W	Working (applied to stress conditions)

Suffixes applied to dimensions, etc.

c	Critical, as applied to transition between elastic and plastic; also relating to core in reinforced cylinders
i and *o*	Internal and external
s	In strip or wire reinforcement

$1, 2 \ldots . m-1, m+1 \ldots . n$ Number of component in multi-component cylinder

2 The Simple Cylinder under Static Loading

2.1 General Remarks

The cylindrical shape is used more than any other in high pressure work; virtually every pressure vessel has at least some part which is cylindrical in section, and almost all pipes are of that shape. Fortunately it is one which can now be completely solved, both for elastic and for both partly and fully overstrained conditions. The first was dealt with more than 100 years ago by Lamé and Clapeyron[2.1] in France, and the analysis known usually by the name of the former has been proved by the matching of the theoretical and measured strains in many tests. Partial overstrain (autofrettage) is much used in gun design and owes its solution mainly to Macrae.[2.2]

In this chapter consideration is given to the simple cylinder, i.e. one which is made in one piece, and to the conditions which cause overstrain, and to its remaining strength after overstrain has begun. The practical use of partially overstrained cylinders, now almost universally known as Autofrettage, is also dealt with in some detail.

2.2 The General Case

The only assumptions made here are that the material is homogeneous, that it is free from internal stress before the pressures are applied, and that it is geometrically perfect, i.e. that its axis is straight and that the cylindrical surfaces are circular, concentric and of constant diameter.

The principal directions of stress and strain must, by symmetry, be the tangential, the axial, and the radial. The stresses are denoted by σ and the strains by ε, each with the appropriate suffix: θ for tangential, z for axial, and r for radial. Shear stresses and strains are denoted by τ and γ respectively, and usually two suffixes are added so as to define the planes to which they refer; thus ${}_{\theta}\tau_{r}$ represents the shear stress in the cross-section at right-angles to the axis.

The applied pressures, p_i at the inside and p_o at the outside, are assumed to be uniform over the surfaces, both inside and out, and—except where otherwise stated—over the ends as well. They are regarded as positive, but as the direct stresses, when compressive, are regarded as negative these latter will at the surfaces have the same numerical values as the pressures but the opposite sign. Thus:

$$p_o = -(\sigma_r)_o \text{ and } p_i = -(\sigma_r)_i \tag{2.1}$$

Consider now the static equilibrium about the diameter of the piece of cylinder of length L shown in Fig. 2.1; this is given by:

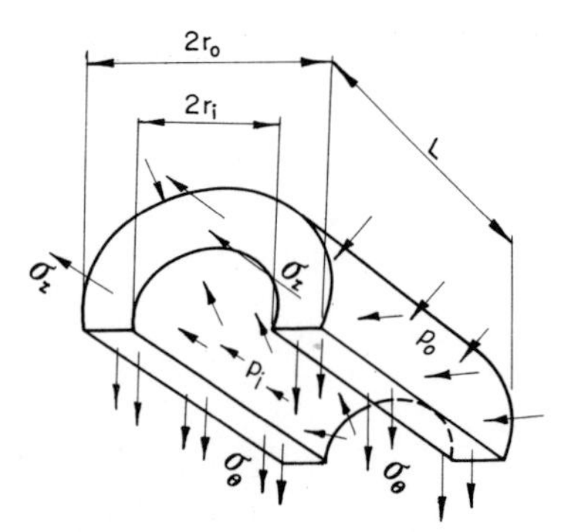

Figure 2.1 Static equilibrium of cylinder

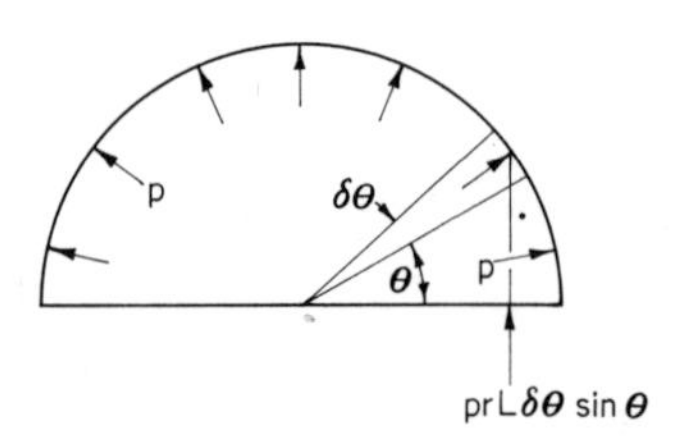

Figure 2.2 Diagram to illustrate resultant of pressure on cylindrical surface

$$2Lp_o r_o + 2L\int_{r_i}^{r_o} \sigma_\theta \, dr = 2Lp_i r_i \tag{2.2}$$

r being any radius and r_o and r_i the radii at the respective surfaces.† Substitution from eq. (2.1) leads to:

$$\int_{r_i}^{r_o} \sigma_\theta \, dr = (\sigma_r)_o r_o - (\sigma_r)_i r_i = \left[r\sigma_r \right]_{r=r_i}^{r=r_o}$$

which, on differentiation with respect to r, gives:

$$\sigma_\theta = \frac{d}{dr}(r\sigma_r) \tag{2.3}$$

This is more conveniently written:

$$\sigma_\theta - \sigma_r = r\frac{d\sigma_r}{dr} \tag{2.4}$$

† The fact that the resultant force normal to a diametral plane of a pressure p acting on a cylindrical surface of radius r and of length L is $2prL$ is not always clear. It can however be easily demonstrated as follows.

Consider a small strip of the surface subtending an angle $\delta\theta$ and running parallel with the axis, and let the position of this make an angle θ with the diametral plane concerned (see Fig. 2.2).

The radial force on this will be $prL\delta\theta$, and the component of this force normal to the diametral plane will be $prL\delta\theta \sin\theta$. Consequently the resultant taken over the whole semi-cylindrical surface will be:

$$prL\int_0^\pi \sin\theta \, d\theta = -prL[\cos\theta]_0^\pi = 2prL$$

and since the tangential and radial planes are those of principal stress:

$$ {}_{\theta}\tau_r = \tfrac{1}{2}(\sigma_\theta - \sigma_r) \tag{2.5} $$

Substitution into eq. (2.4) gives:

$$ 2\,{}_{\theta}\tau_r = r\frac{d\sigma_r}{dr} $$

which on integration gives:

$$ [\sigma_r]_{r=r_i}^{r=r_o} = 2\int_{r_i}^{r_o} ({}_{\theta}\tau_r/r)dr \tag{2.6} $$

or, by substituting from eq. (2.1) for the values of σ_r at the surfaces, we get:

$$ p_i - p_o = 2\int_{r_i}^{r_o} {}_{\theta}\tau_r \; d\ln r \tag{2.7} $$

It will be noted that the derivation of this equation is quite general in that no connection between stress and strain has been assumed. Its importance lies in the fact that, if we can discover the way in which the shear stress varies across the wall of a cylinder, then by eq. (2.7) we can determine the pressure difference at the surfaces which would be required to cause it. As will be shown in Section 2.8 it can be used to follow the course of progressive overstrain in a cylinder, where the strains may be so great as to cause considerable changes in the values of r_o and r_i. In such cases eq. (2.7) has to be written:

$$ p_i - p_o = 2\int_{r_i'}^{r_o'} {}_{\theta}\tau_r \,.\, d\ln r \tag{2.7a} $$

where the dashes indicate the strained values of r_o and r_i.†

2.3 The Simple Elastic Cylinder

Since the principal directions of stress are the tangential, the axial, and the

† Equation (2.7) can be solved directly if the shear stress does not vary with the radius, i.e. with the strain. This is the case of true plasticity, and it is approached by the behaviour of some of the modern low alloy steels when they have passed the yield point. If the constant value of the shear stress is τ^*, then the integration gives:

$$ p_i - p_o = \tau^* \ln K^2 \tag{2.7b} $$

where K is the diameter or radius ratio r_o/r_i.
In practice there is nearly always some strain-hardening, although the effect is often quite small. In a cylinder this means that ${}_{\theta}\tau_r$ increases somewhat across the wall towards the bore; on the other hand, the straining causes r_i to increase more than r_o and consequently:

$$ \frac{r_o'}{r_i'} < \frac{r_o}{r_i} $$

and therefore the straining causes K to decrease. This effect tends to offset that due to the strain-hardening, so that—as we shall see in Section 2.10—many problems can be solved with sufficient accuracy for practical purposes by means of eq. (2.7b).

radial, it follows that stress strain relations can be written down for each direction thus:

$$E\varepsilon_\theta = \sigma_\theta - \nu(\sigma_z + \sigma_r) \qquad (2.8)$$

$$E\varepsilon_z = \sigma_z - \nu(\sigma_r + \sigma_\theta) \qquad (2.9)$$

$$E\varepsilon_r = \sigma_r - \nu(\sigma_\theta + \sigma_z) \qquad (2.10)$$

where E is Young's Modulus and ν Poisson's Ratio.

Now it is an observed fact that elastic cylinders of uniform material and section strain symmetrically and without axial bending, and therefore it is a reasonable presumption that plane sections remain plane. In other words the values of ε_z cannot vary with r, or:

$$\frac{d\varepsilon_z}{dr} = 0 \qquad (2.11)$$

It is possible also to express ε_θ and ε_r in terms of a single parameter, known as the radial shift, generally denoted u, and defined as the radial distance moved by a point initially at radius r before the pressures are applied. As can be seen from

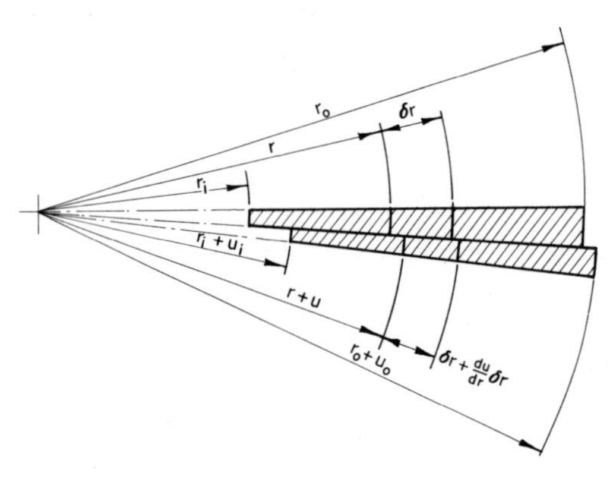

Figure 2.3 Correlation of strains in transverse plane of cylinder with radial shift

Fig. 2.3, the length of the circumference through such a point increases from $2\pi r$ to $2\pi(r+u)$ and consequently:

$$\varepsilon_\theta = \frac{2\pi(r+u) - 2\pi r}{2\pi r} = \frac{u}{r} \qquad (2.12)$$

Similarly, in the radial direction, a small length δr increases to:

$$\delta r + \frac{du}{dr}\delta r,$$

and therefore:

$$\varepsilon_r = \frac{\delta r + \dfrac{du}{dr}\,\delta r - \delta r}{\delta r} = \frac{du}{dr} \tag{2.13}$$

We thus have, in equations (2.8) to (2.13) inclusive, together with eq. (2.4), seven equations containing seven unknowns, and their solution is comparatively simple. If we first substitute for ε_θ and ε_r in terms of u in equations (2.8) and (2.10), and subtract the former from the latter and then substitute from eq. (2.4), we obtain:

$$E\frac{d}{dr}\left(\frac{u}{r}\right) = -(1+\nu)\frac{d\sigma_r}{dr} \tag{2.14}$$

Differentiation of eq. (2.8) with respect to r with the appropriate substitution for ε_θ then gives the same left-hand side as in eq. (2.14), but we then have a term containing σ_z on the right-hand side. This can be eliminated by differentiating eq. (2.9) with respect to r and substituting equation (2.11), with the result that:

$$E\frac{d}{dr}\left(\frac{u}{r}\right) = \frac{d\sigma_\theta}{dr} - \nu^2\frac{d\sigma_r}{dr} - \nu^2\frac{d\sigma_\theta}{dr} - \nu\frac{d\sigma_r}{dr} \tag{2.15}$$

Subtracting eq. (2.14) from this and dividing through by the factor $1-\nu^2$ then gives:

$$\frac{d\sigma_\theta}{dr} + \frac{d\sigma_r}{dr} = 0 \tag{2.16}$$

Integrating this gives:

$$\sigma_\theta + \sigma_r = 2A \tag{2.17}$$

where $2A$ is the constant of integration. This combined with eq. (2.4) leads to:

$$2\sigma_r + r\frac{d\sigma_r}{dr} = 2A \tag{2.18}$$

which can be written:

$$\frac{1}{r}\,\frac{d}{dr}(r^2\sigma_r) = 2A$$

which, on further integration, leads to:

$$r^2\sigma_r = r^2A + B$$

where B is a further constant of integration.

Thus:

$$\sigma_r = A + B/r^2 \tag{2.19}$$

and correspondingly, from eq. (2.17):

$$\sigma_\theta = A - B/r^2 \tag{2.20}$$

The evaluation of the constants is obtained by substituting the boundary conditions, i.e. $\sigma_r = -p_i$ at $r = r_i$, and $\sigma_r = -p_o$ at $r = r_o$; whence:

$$A = \frac{-p_o K^2 + p_i}{K^2 - 1} \tag{2.21}$$

and

$$B = \frac{(p_o - p_i) K^2 r_i^2}{K^2 - 1} \tag{2.22}$$

where $K = r_o/r_i$, which is usually referred to as the diameter (or radius) ratio.

To evaluate σ_z it is necessary to know whether or not the whole of the end forces are balanced by the axial stresses, but it is evident from eq. (2.9), (2.11) and (2.17) that σ_z does not vary with r, i.e. that it is constant across the section.

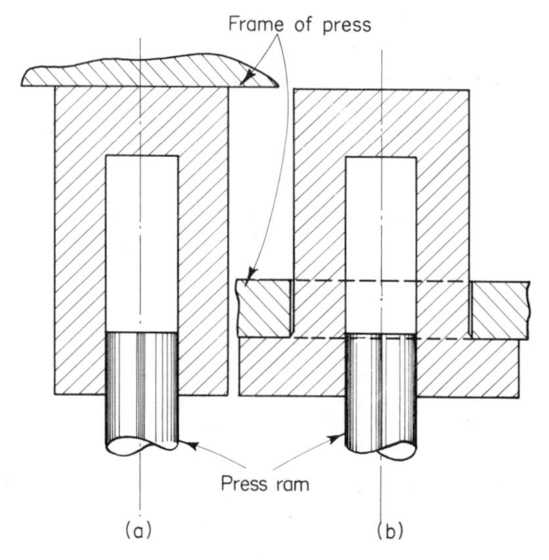

Figure 2.4 Types of hydraulic press showing effect on longitudinal stress in press cylinder

The two commonest conditions are (i) where the walls take all the end load due to the pressure, and (ii) where this load is taken wholly by external means so that the axial stresses are then negligible. Fig. 2.4 exemplifies these two conditions for two typical arrangements of hydraulic press, that at *a* corresponding to (ii) above, and that at *b* to (i). For this case the axial force is clearly $p_i \times \pi r_i^2$ and if this is uniformly distributed over the section the mean stress will be:

$$\frac{p_i \times \pi r_i^2}{\pi r_o^2 - \pi r_i^2}$$

or in the general case where there is also an external pressure:

$$\sigma_z = \frac{-p_o K^2 + p_i}{K^2 - 1}$$

Thus the complete solution in this case of the elastic cylinder wholly supporting its own end loads is:

$$\sigma_\theta = \frac{-p_o K^2 + p_i}{K^2 - 1} - \frac{(p_o - p_i)K^2}{K^2 - 1}\left(\frac{r_i}{r}\right)^2 \qquad (2.23)$$

$$\sigma_z = \frac{-p_o K^2 + p_i}{K^2 - 1} \qquad (2.24)$$

$$\sigma_r = \frac{-p_o K^2 + p_i}{K^2 - 1} + \frac{(p_o - p_i)K^2}{K^2 - 1}\left(\frac{r_i}{r}\right)^2 \qquad (2.25)$$

Also

$$_\theta\tau_r = -\frac{(p_o - p_i)K^2}{K^2 - 1}\left(\frac{r_i}{r}\right)^2 \qquad (2.26)$$

and

$$\varepsilon_\theta = \frac{-p_o K^2 + p_i}{K^2 - 1}(1 - 2\nu) - \frac{(p_o - p_i)K^2}{K^2 - 1}\left(\frac{r_i}{r}\right)^2 (1 + \nu) \qquad (2.27)$$

This is the complete solution of the problem, but for the simpler case which is more often met in practice, namely the cylinder subjected only to internal pressure, and for the location $r = r_i$ where the stresses are highest:

$$(\sigma_\theta)_{max} = p_i \frac{K^2 + 1}{K^2 - 1} \qquad (2.28)$$

and

$$(_\theta\tau_r)_{max} = p_i \frac{K^2}{K^2 - 1} \qquad (2.29)$$

These are usually known as the Lamé formulae.

As a check on the validity of such calculations an actual test may be cited in which a pressure of 34,000 lbf/in^2 (2,350 bars) was applied to a portion of a reactor of $11\frac{3}{4}$ in external diameter and $7\frac{1}{2}$ in bore, with an overall length of 6 ft 8 in, made from an alloy steel with an ultimate strength of 123,000 lbf/in^2 (8,480 bars). It had a value for Young's Modulus of $2{\cdot}00 \times 10^6$ bars and for Poisson's Ratio of 0·28.

If u_o is the radial shift of the outside surface, it is clear that the change in outside diameter should be given by $2r_o(\varepsilon_\theta)_o$ or $D_o \, . \, (\varepsilon_\theta)_o$, and thus with no external pressure and for $r = r_o$ eq. (2.27) simplifies to give:

$$\Delta D_o = \frac{p_i D_o}{E(K^2 - 1)}(2 - \nu) \qquad (2.30)$$

Also the change in overall length will be $L\varepsilon_z$, or:

$$\Delta L = \frac{p_i L}{E(K^2-1)}(1-2\nu) \tag{2.31}$$

Putting in these figures gives values for ΔD_o and ΔL, which are compared with actual measured values in Table 2.1.

Table 2.1

	Calculated	*Measured*
Expansion of external diameter (in)	0·0163	0·0170
Increase in overall length (in)	0·0283	0·0275

The fact that the cylinder actually swelled about 4% more than the theory predicted may have been due to some errors in the elastic constants used, or to some slight departure from elasticity in this test, which carried the material at the bore very nearly to its elastic limit and could therefore have exceeded it where, for instance, heat treatment stresses had been locked up in the metal. More surprising at first sight is the fact that the lengthwise stretch came out at about 3% less than the calculated value. On the other hand, this is much more sensitive to variations in the value of ν, and an increase to 0·306 would bring the observed and calculated values of the stretch to exactly the same figure.

In spite of these discrepancies, however, these test results show that the Lamé equations do provide a good approximation, and no other existing formula is as good. Their derivation depends, as we have seen, only on the theory of elasticity and there is therefore every reason to expect a satisfactory correlation; calculations relating to elastic conditions in cylinders will in consequence be based on this theory in the subsequent pages.

2.4 Stress Distribution in Elastic Cylinders

Reference to equations (2.23), (2.24) and (2.25) shows three important facts:

(i) The stress system consists of a pure shear equal to

$$\frac{(p_o-p_i)K^2}{K^2-1}\left(\frac{r_i}{r}\right)^2$$

with a superimposed hydrostatic stress equal to

$$\frac{-p_oK^2+p_i}{K^2-1}$$

(ii) The shear stress has its maximum value at the inner surfaces.

(iii) The maximum values of all the stresses are dependent only on the ratio of the diameters and not on their actual size.

Fig. 2.5 shows for cylinders of 3 : 1 diameter ratio the stress distributions resulting from three systems of applied pressure. In the first (Fig. 2.5a) there is an

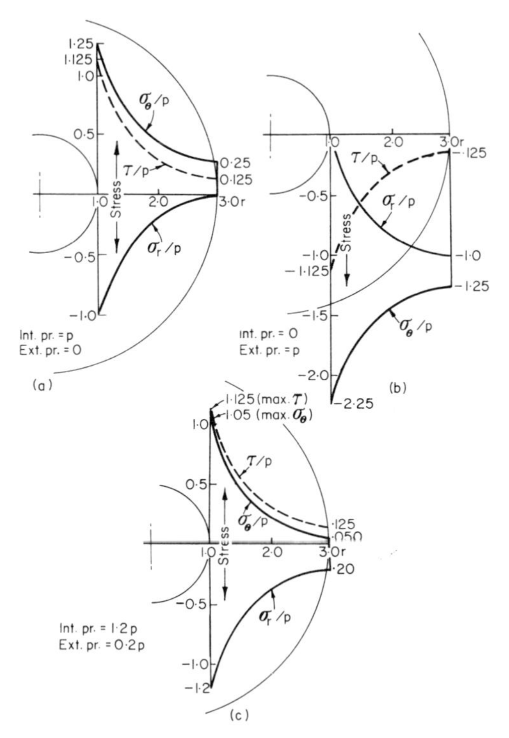

Figure 2.5 Distribution of stresses in cylinders sections: (a) internal pressure only; (b) external pressure only; (c) pressures on both surfaces

internal pressure of p and no external pressure; in Fig. 2.5b the situation is reversed with p at the outside and no internal pressure, while in Fig. 2.5c there is an internal pressure of $1{\cdot}2p$ and an external pressure of $0{\cdot}2p$. Thus the difference in pressure is numerically the same in each case and consequently the shear stress has numerically the same value at all corresponding points (although the sign is reversed in Fig. 2.5b) in conformity with eq. (2.26). The sharpness of the peaks and the general unevenness of all the stresses in the transverse plane are very noteworthy; the material of the wall is clearly loaded in a most uneconomic manner.

2.5 Criteria of Elastic Breakdown

As we have seen in the foregoing section, the stresses in an elastic cylinder rise to sharp peaks at the inner surface, and it is evident therefore that this will be the location at which overstrain first appears. It is also likely, having regard to the importance of shearing stresses in such a system, that whatever criterion is found to be most suitable will be based on shear.

The reliable published information relating to overstraining tests with cylinders is scanty, which is not altogether surprising when the experimental difficulties are taken into account. Usually the strain measurements have to be taken at the outside surface, and the plot of pressure against change in outside diameter normally shows only a very gradual departure from linearity as the pressure is increased. This is only to be expected, since—except for material in which there is a pronounced yield with a secondary yield stress well below the primary—the yielding in the cylinder begins in the innermost layers when the stress there no longer rises quite in direct proportion to the strain; but there will then still remain a great thickness of elastic metal outside these layers and this will tend to mask the transition, and in many of the heat-treated alloy steels most suitable for high pressure work this transition—even in a tensile test—is not very clearly marked.

The situation with materials showing a well marked secondary yield is in some respects easier to judge, although the overstrain is usually accompanied by gross and unsymmetrical deformation with the section going out of the circular.

In what follows two sets of experiments will be examined; the first due to Cook[2.3] is for a low carbon steel (Table 2.2), and the second by Crossland and Bones[2.4] for an alloy steel with an ultimate strength of approximately 8,275 bars (120,000 lbf/in^2), (Table 2.3) p. 23.

Table 2.2

Diameter ratio	*Pressure at onset of yield (bars)*	*Maximum direct stress (bars)*	*Maximum shear stress (bars)*
1·167	538	3,510	2,020
1·168	522	3,395	1,960
1·50	1,081	2,805	1,945
2·0	1,514	2,525	2,020
3·0	1,835	2,295	2,065
4·0	1,945	2,205	2,075

As was shown in the preceding section, the mechanism of elastic breakdown is likely to be a matter primarily of shear stresses, and a glance at the third and fourth columns of Table 2.2 shows that, whereas the maximum direct stress at the point of overstrain progressively decreases and shows a variation of more than 25% from the mean, the shear stress only varies by about 3% on each side of its mean.

Now the primary yield of a tensile specimen of the same material was 3,600 bars (52,300 lbf/in^2). Thus, if the maximum shear stress in each case is the criterion,† overstrain in the cylinders would have been expected when the shear stress reached one half of the tensile yield stress, i.e. at 1,803 bars, compared with a mean for all the cylinder specimens of 2,015 bars, or a discrepancy of some 10%. In view of this it may be desirable to look at some other possible criteria for the

† Usually known as the Tresca criterion.

correlation of elastic breakdown in tensile specimens and thick cylinders. The best known of these is certainly that originally suggested by Clerk Maxwell in 1856[2.5], but subsequently rediscovered by von Mises in 1913[2.6] and now usually called after the latter. It is perhaps easiest to regard this criterion as a simple symmetrical function of *all three* principal stresses, i.e.

$$(\sigma_1-\sigma_2)^2+(\sigma_2-\sigma_3)^2+(\sigma_3-\sigma_1)^2$$

but for convenience it is usual to write each term as one half the difference between the stresses. If one then inserts a further factor of $\frac{1}{2}$ in front of the function and takes its square root, the result when applied to a simple tensile test comes out at one-half the value of the tensile stress, which is of course the same as the maximum shear stress at the breakdown of elasticity. Thus if U_M be the value of the Maxwell (von Mises) function:

$$U_M = \sqrt{\left[\frac{1}{2}\left\{\left(\frac{\sigma_1-\sigma_2}{2}\right)^2+\left(\frac{\sigma_2-\sigma_3}{2}\right)^2+\left(\frac{\sigma_3-\sigma_1}{2}\right)^2\right\}\right]} \tag{2.32}$$

its maximum value for the cylinder (where $\sigma_1 = \sigma_\theta$, $\sigma_2 = \sigma_z$ and $\sigma_3 = \sigma_r$) is, from equations (2.23), (2.24) and (2.25), given by:

$$U_M = (p_o-p_i)\frac{\sqrt{3}}{2}\frac{K^2}{K^2-1} \tag{2.33}$$

and thus, if σ_T^* be the limiting elastic stress in the tensile test, overstrain in the cylinder will occur when:

$$p_o-p_i = \frac{\sigma_T^*(K^2-1)}{\sqrt{3}\,.\,K^2} \tag{2.34}$$

In the Tresca relation, the corresponding criterion is:

$$p_o-p_i = \frac{\sigma_T^*(K^2-1)}{2K^2} \tag{2.35}$$

Thus, when referred to tensile data, the Maxwell criterion expects the elastic range in the cylinder wall to persist to a higher value of the applied pressure difference (in the ratio 2 to $\sqrt{3}$), and referring to Cook's results (Table 2.2) this means that the cylinders would be expected to overstrain when the maximum pressure difference sets up a shear stress in the walls equal to the tensile limit divided by $\sqrt{3}$, or to 2,080 bars. As we have seen, this is about 3% above the mean value in the 4th column of Table 2.2. In view of the experimental difficulty of judging the actual beginning of non-elastic conditions in a cylinder wall, this is probably as close as we are likely to get with a simple criterion of this kind.

It is worth noting, while discussing this criterion, that the shear stress on the octahedral plane, τ_{oct} is given by:

$$\tau_{oct} = \tfrac{1}{3}\sqrt{\{(\sigma_1-\sigma_2)^2+(\sigma_2-\sigma_3)^2+(\sigma_3-\sigma_1)^2\}} \tag{2.36}$$

whence the function U_M given in eq. (2.32) differs by a constant factor of about 1·06 from this particular shear stress. For this reason this is sometimes referred to as the *Octahedral Shear Stress* criterion. It is also called the *Shear Strain Energy* criterion, because the strain energy given to a piece of material by a shear stress τ causing in it a shear strain γ is given by $\tau^2/2G$, where G is the shear modulus, and so for a three-dimensional stress system the total shear strain energy V_{sh} is given by:

$$V_{sh} = \frac{1}{2G}\left\{\left(\frac{\sigma_1-\sigma_2}{2}\right)^2+\left(\frac{\sigma_2-\sigma_3}{2}\right)^2+\left(\frac{\sigma_3-\sigma_1}{2}\right)^2\right\} \tag{2.37}$$

so that:

$$V_{sh} = U_M^2/G \tag{2.38}$$

Before leaving this question of the appropriate criteria it may be useful to consider various other hypotheses which have been put forward from time to time. The best known of these are the ones due to Rankine and St. Venant, the former stating that overstrain in any system will occur when the maximum principal stress reaches a critical value, and the latter when the maximum principal strain reaches a critical value. In addition it may be worth considering the total strain energy of the system, since several authorities, for instance Beltrami and Haigh, have in the past preferred this quantity as the correlating hypothesis.[2.7] To these might also be added a symmetrical function of the principal direct strains, suggested by Rôs and Eichinger[2.8], and also the maximum shear strain, but further examination will show that the former corresponds with the Maxwell criterion, and the latter with that of Tresca, i.e. the maximum shear stress theory.

To test the St. Venant and the total strain energy theories we need to have an accurate value for Poisson's Ratio, and it is therefore preferable to use the experimental data of Crossland and Bones[2.4], where this quantity was determined from specimens cut from the same bar of material. It will also simplify the comparison if we consider the case of cylinders loaded only with internal pressure (i.e. where p_o is zero) as in these experimental results.

Considering first the Rankine hypothesis, the direct stress in the tangential direction is evidently the greatest, with the value:

$$p_i \cdot \frac{K^2+1}{K^2-1}$$

at the inner surface, and thus, if p_i^* is the internal pressure at the onset of overstrain:

$$p_i^* = \frac{K^2-1}{K^2+1}\sigma_T^* \tag{2.39}$$

For the St. Venant criterion the maximum positive strain evidently occurs in the tangential direction at the inner surface, and this has to be compared with the strain at the elastic limit in the tensile test, i.e.

$$\frac{1}{E}[(\sigma_\theta)_i - \nu\{(\sigma_z)_i + (\sigma_r)_i\}] = \frac{\sigma_T^*}{E}$$

which, on substituting for the stresses in terms of p_i^* and K, gives:

$$p_i^* = \frac{\sigma_T^*(K^2-1)}{(K^2+1)+\nu(K^2-2)} \tag{2.40}$$

Finally, there is the total strain energy criterion; this involves rather more algebra although it is fairly easily dealt with. Thus the total strain energy in the tangential direction is:

$$\tfrac{1}{2}\sigma_\theta \varepsilon_\theta$$

and consequently, if V_{tot} be the total strain energy:

$$V_{tot} = \frac{1}{2E}\{(\sigma_\theta^2+\sigma_z^2+\sigma_r^2)-2\nu(\sigma_\theta\sigma_z+\sigma_z\sigma_r+\sigma_r\sigma_\theta)\}$$

for the cylinder, and

$$V_{tot} = \frac{(\sigma_T^*)^2}{2E}$$

for the tensile specimen.
Substituting for σ_θ, etc. and equating these two relations:

$$p_i^* = \frac{\sigma_T^*(K^2-1)}{\sqrt{\{(2K^4+3)+2\nu(K^4-3)\}}} \tag{2.41}$$

We can now compare these results with those of Crossland and Bones for an alloy steel containing 0·30% C; 0·24% Si; 0·012% S; 0·013% P; 0·66% Mn; 2·57% Ni; 0·58% Cr; 0·60% Mo, which had been oil hardened and tempered to an ultimate tensile strength of 8,450 bars (122,500 lbf/in^2). The yield stress occurred at 7,450 bars (108,000 lbf/in^2), and it is recorded that there was no evidence of primary/secondary yielding. Unfortunately no Proof Stress values are given.

In Table 2.3 are given the observed values of pressure at which the cylinder specimens departed from true elastic behaviour and the corresponding values at which this would be expected according to the various criteria. In these

Table 2.3

All pressures given in bars

Diameter ratio of specimen	*Actual internal pressure at start of overstrain*	*Pressure at start of overstrain according to hypothesis of*				
		Rankine	*St. Venant*	*Tresca (also max. shear strain)*	*Maxwell (also Rôs and Eichinger)*	*Haigh (total strain energy)*
1·2	1,260	1,340	1,435	1,140	1,315	1,270
1·4	1,965	2,455	2,425	1,825	2,105	2,140
1·6	2,535	3,270	3,110	2,270	2,620	2,725
1·8	2,840	3,870	3,630	2,580	2,970	3,145
2·0	3,090	4,470	4,015	2,795	3,220	3,435

calculations σ_T^* has been taken as 7,450 bars (i.e. 108,000 lbf/in^2) and ν as 0·283.

It is thus seen that the Maxwell (or von Mises) criterion is the most satisfactory, with a maximum error of about 7%. Since the practical determination of this limit in a cylinder is very difficult, and unlikely to be more accurate than say 2% or 3%, it seems fair to conclude that this criterion is sufficient for most practical purposes when dealing with ductile materials and when tensile data only are available. The problem where brittle materials are concerned is discussed later.

Crossland and Bones also carried out some careful torsion tests on specimens cut from the same bars as the test cylinders, from which they found the torsion elastic limit to be 60,500 lbf/in^2 (4,170 bars) shear stress. If we accept that this is directly comparable with the cylinder stress (we have seen in equations (2.33), (2.24) and (2.25) that the latter is a pure shear with a superimposed hydrostatic stress, while torsion in cylindrical specimens is a pure shear only), then it would be reasonable to expect overstrain in either when the shear stress reached this value.† Table 2.4 shows the pressures measured as initiating overstrain compared with those required to bring the shear stress in the walls to 4,170 bars.

Table 2.4

All pressures in bars

Diameter ratio	*Actual pressure at onset of overstrain*	*Calculated pressure at onset of overstrain (from torsion data)*	*Difference %*
1·2	1,260	1,275	1·2
1·4	1,965	2,045	4·1
1·6	2,535	2,540	0·2
1·8	2,840	2,880	1·4
2·0	3,090	3,125	1·1

This shows that, with the exception of the 1·4 diameter ratio specimen, the agreement is to about 1%, and examination of the paper by Crossland and Bones from which these are taken suggests there may be some error in this one particular measured departure from elasticity. If therefore one ignores this, the remaining four tests in Table 2.3 show differences between the values calculated from tensile data according to the Maxwell (von Mises) criterion of about 4% above the measured values, which is much the same as we saw with Cook's experiments with low carbon steel shown in Table 2.2.

A reasonable conclusion from this is that, in spite of this discrepancy of 3% or 4%, the Maxwell (von Mises) relation is the most satisfactory of the various

† It should be noted that in torsion the principal stresses are σ_s, 0, and $-\sigma_s$ from which the Maxwell function U_M, eq. (2.32), has the value $\sqrt{3}\ \sigma_s/2$, where σ_s is numerically the same as the shear stress τ. As we see in eq. (2.33) by substitution from eq. (2.24), the critical pressure in each cylinder should be that which brings the shear stress up to the above value of 4,170 bars.

relatively simple criteria of correlation which we have discussed here. On the other hand, a much closer correlation is obtained with torsion data, and it is strongly recommended that, for highly stressed cylinders, a torsion test should be specified when ordering the material, and the design based on the results thereof.

2.6 Brittle Materials

Mention was made on p. 24 of the different behaviour of cylinders made of brittle material. There is however very little data to go on in this, but the general belief is that failure occurs substantially in accordance with the Rankine (maximum direct tensile stress) theory, and the work of Cook and Robertson[2.9] on grey cast iron certainly supports this.

More recently such materials as concrete and cemented tungsten carbide have been considered for certain parts of pressure vessels – for example, the latter is now extensively used for liners in the high pressure cylinders of reciprocating compressors – but in nearly all cases it has been found necessary to ensure that external support is sufficient to prevent tensile stresses developing in the brittle material. This does not mean that the latter cannot carry a useful pressure difference (as we shall see when considering built up cylinders, Section 3.3), but special design and operating precautions are then essential.

The main uncertainty in this correlation is the variation from specimen to specimen in even the ultimate tensile stress, although there are some indications that sometimes materials which show virtually no ductility in tensile testing may deform quite appreciably as the walls of a thick cylinder. For instance, Bridgman[2.10] cites the case of a "tool steel" cylindrical specimen which he burst, but not before its bore had expanded to twice its initial value. Unfortunately he gives no details of the material, although he does reproduce a photograph of the fracture. In view therefore of the scarcity of reliable data and the discouraging nature of what there is we must conclude that, except for special requirements, brittle materials should not be used for containing pressure.

2.7 Graphical Representation of Elastic Cylinders

There are certain aspects of the theory of elastic cylinders (and indeed of over-strained cylinders also, as we shall see later) which facilitate the preparation of charts to aid the designer.

The stress system, as represented by equations (2.23), (2.24) and (2.25), consists of a pure shear with a superimposed hydrostatic stress, as has already been pointed out. This latter will be tensile, zero, or compressive, depending on whether $p_o K^2$ is numerically less than, equal to, or greater than p_i. We have also seen in Section 2.5 that the shear stress is virtually the only factor that needs to be taken into consideration by the designer if his cylinder is to remain elastic in its working life. It is also clear, from eq. (2.26), that the shear stress depends only on the difference of the applied pressures and not on their absolute magnitudes. From these arguments it can reasonably be accepted that superimposing an additional hydrostatic stress on such a system will not alter the correlation between the limit of its elastic range and the difference between the applied pressures at the walls.

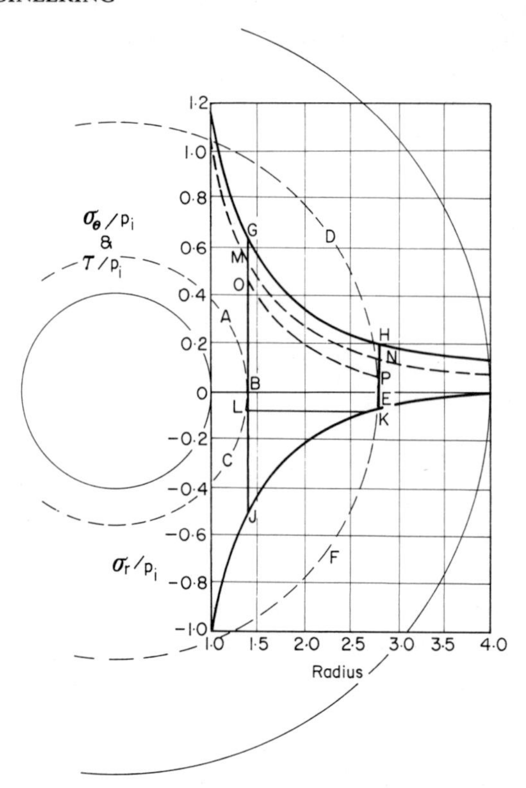

Figure 2.6 Correlation of stresses in cylinder with concentric portion of larger cylinder

To illustrate this further consider Fig. 2.6, in which the stress distribution is plotted for a cylinder in which $r_o = 4 \times r_i$, the stresses being as ratios of the applied pressure p_i at the bore, p_o being zero. As will be seen, at the inner surface, $\sigma_\theta/p_i = 1{\cdot}133$; ${}_\theta\tau_r/p_i = 1{\cdot}067$; and $\sigma_r/p_i = -1{\cdot}0$.†

Also in the diagram are two broken line circles ABC and DEF, having radii of 1·4 and 2·8 times r_i respectively. Drawing vertical lines through B and E to cut the σ_r curve in J and K gives the radial stresses at these points as $-0{\cdot}477p_i$ and $-0{\cdot}069p_i$ respectively. Thus this zone (between the circles ABC and DEF) will behave, so far as the stresses are concerned, as if it was a 2 : 1 diameter ratio cylinder with an internal pressure of $0{\cdot}477p_i$ and an external pressure of $0{\cdot}069p_i$. One could say in fact that an element of the material within this zone would not know whether it was part of a 4 : 1 cylinder carrying an internal pressure of p_i and no external pressure, or whether it was part of a 2 : 1 cylinder carrying an internal pressure of $0{\cdot}477p_i$ and an external pressure of $0{\cdot}069p_i$; and this in turn could be regarded as an internal pressure of $0{\cdot}408p_i$ accompanied by a hydrostatic pressure of $0{\cdot}069p_i$. The values of the shear stress will be unchanged as represented by the broken line MN referred to the base line BE. Similarly the

† For the sake of simplicity the suffixes used with τ will henceforth be omitted except where doubt might arise as to the actual plane of the shear stress.

tangential and radial stresses will be represented by the curves GH and JK referred to the base line BE. If now we consider a 2 : 1 cylinder carrying an internal pressure of $0{\cdot}408p_i$ and no external pressure, there will be no change in the shear stress distribution because the only change has been to remove a hydrostatic stress. On the other hand the radial stress will be reduced by $0{\cdot}069p_i$ at every point and the tangential stress increased by the same amount. In fact the relevant curves are still GH and JK, but now referred to LK as base instead of BE. To complete the diagram for the 2 : 1 cylinder it is only necessary to draw in a new shear curve OP, parallel to MN but displaced through a distance BL, so that it now refers to LK as its zero base line.

Consideration of this shows that, at any rate where the internal pressure is greater than the external, it should be possible to regard any actual elastic cylinder as a part of an infinite cylinder just overstraining at its bore, adjusted by an appropriate hydrostatic pressure. Thus we assume K to be infinite, p_i to have the value which just raises the maximum shear stress to its limiting elastic value τ^*, p_o to be zero, and r_i unity. Substituting these into equations (2.23) to (2.27) inclusive gives:

$$\sigma_\theta = p_i/r^2;\ \sigma_z = 0;\ \sigma_r = -p_i/r^2;\ \tau = p_i/r^2 \tag{2.42}$$

and

$$\varepsilon_\theta = \frac{(1+\nu)p_i}{Er^2} \tag{2.43}$$

Since p_i is constant and τ has its maximum value when $r = r_i$ (or, in this case, when $r = 1$), it is evident that:

$$p_i = \tau^* \tag{2.44}$$

and the above equations can now be rewritten:

$$\sigma_\theta = \tau^*/r^2;\ \sigma_z = 0;\ \sigma_r = -\tau^*/r^2;\ \tau = \tau^*/r^2 \tag{2.45}$$

and

$$\varepsilon_\theta = \frac{(1+\nu)\tau^*}{Er^2} \tag{2.46}$$

Now, the chart based on this consists of the curves for σ_θ/τ^*, σ_r/τ^*, τ/τ^* plotted as ordinates on a linear scale against r plotted as abscissa on a $\log_{10}$ scale. All the curves are the same except that for σ_r/τ^*, which, being negative, is a mirror image of the other two, which are coincident. Such a chart is shown in Fig. 2.7. It will be appreciated that this is of *universal application* because we are dealing with stress ratios referred to the limiting shear stress (to be determined preferably by torsion testing) and because of the logarithmic scale for radius. Thus a vertical intercept represents a stress or pressure difference and a horizontal intercept a radius ratio.

The method of using such a chart is best illustrated by means of an example which refers to the actual test quoted in Section 2.3.

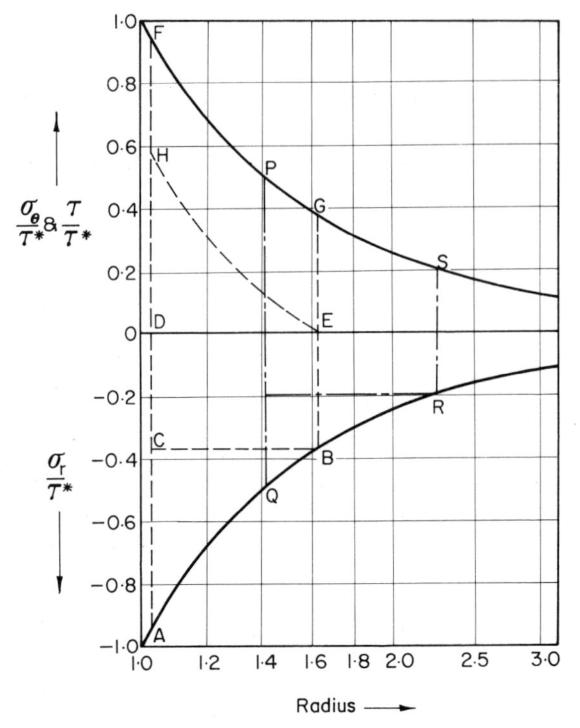

Figure 2.7 Chart for solution of elastic cylinder problems

Here the applied internal pressure was 2,350 bars, and the external and internal diameters $11\frac{3}{4}$ in and $7\frac{1}{2}$ in respectively. The torsional elastic limit was not measured, but the material had a tensile yield stress of 105,000 lbf/in^2 and, as it was to the same specification (B.S.S. En 25) as that used by Crossland and Bones[(2.4)] we can assume a similar value, i.e. 4,170 bars, for τ^*. Thus the ratio of internal pressure to τ^* is:

$$2{,}350/4{,}170 \text{ or } 0{\cdot}564$$

and the radius (or diameter) ratio is 1·567.

We need therefore to find two points on the σ_r curve whose horizontal intercept is 1·567 and whose vertical intercept is 0·564. The easiest way to do this is to take a rectangular card and mark off (to the appropriate scale) a distance equal to 1·567 from the top left-hand corner along the top edge, and a corresponding vertical distance equal to 0·564 down the left-hand edge from the top; the card is then moved about over the chart, keeping its axes parallel with those of the chart, until both these points lie on the σ_r curve. These positions are marked at A and B in Fig. 2.7. If then we draw vertical lines from these points to cut the zero stress line in D and E and the upper curve in F and G respectively, and if we also draw a horizontal through B to cut the line ADF in C, we have the complete stress diagram for a cylinder of this diameter ratio when subjected to an internal pressure of approximately $0{\cdot}950\tau^*$ and a simultaneous external pressure of $0{\cdot}386\tau^*$.

At this point it should be noted that the chart required could hardly be simpler, being curves for $1/r^2$ and $-1/r^2$ on log linear coordinates. In Fig. 2.7 this has naturally had to be reduced below the scale needed to read off these distances to the third decimal place, although it should still be quite large enough to indicate the procedure.

Returning to Fig. 2.7, AB is the curve of radial stress and FG that of both tangential stress and shear stress (all referred to the base line DE), the last two being coincident because by the nature of the construction:

$$p_o K^2 = p_i$$

and therefore the hydrostatic stress term in equations (2.23), (2.24) and (2.25) is zero.

To correct for the fact that, in the actual test, there was no hydrostatic pressure, we need only move the base line from DE to CB in the diagram, and the tangential and radial stresses are given by FG and AB as before. As, however, changes in hydrostatic stress do not affect the shear, we must draw in a new shear stress curve parallel to FG but displaced downwards by the same vertical distance as between DE and CB. In other words we draw a curve parallel to FG and passing through E on the zero stress line of the chart. If a complete stress diagram is then needed, it is only necessary to multiply the ordinates by the value of τ^* to bring them to absolute magnitudes and replot them on a linear radius scale, marked from $3\frac{3}{4}$ in at r_i to $5\frac{7}{8}$ in at r_o. This has been done in Fig. 2.8.

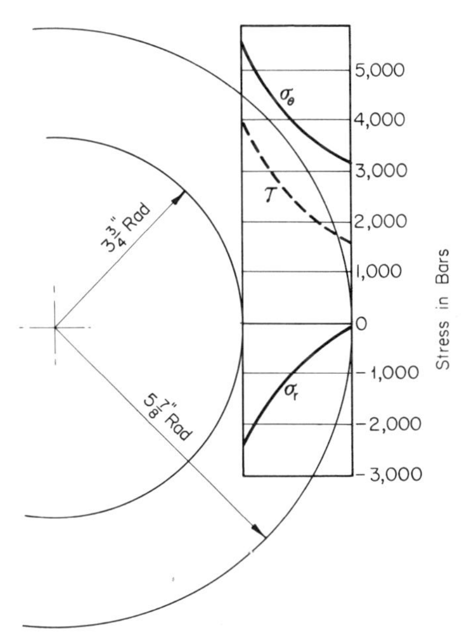

Figure 2.8 Stresses in example

The chart shows at a glance how near the reactor was to overstrain in this test. The same stress curves could of course have been very simply obtained by merely plotting them from equations (2.23), (2.25) and (2.26), but the chart method provides an easy means of seeing if the elastic conditions are likely to be exceeded.

We come now to the question of strains. The point G represents, when multiplied by $(1+\nu)/E$, the value of ε_θ at the outer surface when the internal pressure of $0{\cdot}564\tau^*$ is accompanied by a hydrostatic pressure of $0{\cdot}386\tau^*$. The intercept GE is, of course, equal to BE and has therefore this same value of $0{\cdot}386\tau^*$. Hence the value of ε_θ, given that $E = 2{\cdot}00 \times 10^6$ and $v = 0{\cdot}28$, will be:

$$\frac{0{\cdot}386 \times 4{\cdot}170 \times 1{\cdot}28}{2{\cdot}0 \times 10^3} = 1{\cdot}030 \times 10^{-3}$$

Now the effect of the hydrostatic pressure, which is present in the circumstances represented by the diagram but was not present in the test, will be to reduce the strains; thus, to find the extent of this reduction, we must calculate the magnitude of the strain caused by such a hydrostatic stress and *add* it to the value given above.

Going back to eq. (2.8), it is clear that if all three principal stresses are the same (as they will be for hydrostatic conditions), then:

$$E(\varepsilon_\theta)_{hy} = p_{hy}(1-2v) \tag{2.47}$$

where $(\varepsilon_\theta)_{hy}$ and p_{hy} are respectively the tangential strain and the hydrostatic pressure which causes it. Thus:

$$(\varepsilon_\theta)_{hy} = \frac{0{\cdot}386 \times 4{\cdot}170(1-2 \times 0{\cdot}28)}{2{\cdot}0 \times 10^3} = 0{\cdot}354 \times 10^{-3}$$

Thus, the value of ε_θ in the actual test would have been $(1{\cdot}030 + 0{\cdot}354) \times 10^{-3} = 1{\cdot}384 \times 10^{-3}$, which on an external diameter of $11\frac{3}{4}$ in would mean a swelling of 0·0163 in, which is, of course, the same as that obtained by direct calculation and given in Table 2.1.

Finally, regarding the longitudinal strain, this will be zero in the condition represented on the diagram, so the only actual longitudinal strain will be that due to the hydrostatic pressure, i.e. $0{\cdot}354 \times 10^{-3}$ in/in. Thus on a length of 6 ft 8 in (80 in) this will be:

$$80 \times 0{\cdot}354 \times 10^{-3} = 0{\cdot}0283 \text{ in}$$

One can also see very easily the safe working pressures which could be set for this reactor for various safety factors based on the limiting elastic pressure. If, for instance, a factor of 2 is accepted, then τ/τ^* must equal 0·5. Thus, we can draw in the point P on the shear stress curve of Fig. 2.7, and drop a perpendicular to cut the radial stress curve in Q. Moving thence to the right a distance equal to 1·567, the corresponding positions R and S can be marked as shown. The vertical intercept between Q and R now represents the allowable working pressure for this safety factor. As will be seen, even on the small scale of Fig. 2.7, this is about 0·3, so the safe working pressure will be about $0{\cdot}3 \times \tau^*$ or say 1,250 bars.

2.8 The Effects of Overstrain

The deliberate use of overstrained material in engineering service is a comparatively recent development, but probably its earliest application was in gun barrels, which—according to Jacob[2.11]—was suggested by a French artillery designer named Malaval as long ago as 1906. The procedure, now generally known as *autofrettage*, was gradually developed in the period between the two

World Wars so that, although hardly used at all in the first, it became standard practice with all the belligerents in the second. It is now also a most important factor in the design of very high pressure industrial equipment.

Before discussing the basis of the problem it is important to consider how a cylinder behaves as overstrain is developed within it, and this varies considerably with different classes of material. Fig. 2.9 shows schematic load-extension diagrams of the sort obtained in tensile tests for (*a*) low carbon steels, (*b*) heat-treated alloy steels, and (*c*) softened austenitic stainless steels and many non-ferrous metals.

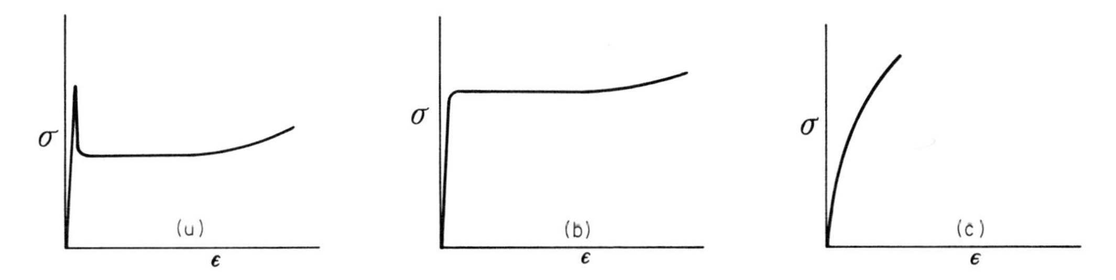

Figure 2.9 Typical stress-strain characteristics of different materials

With (*a*) when an element in a cylinder yields it ceases altogether to take its share of the load, thereby throwing more on to those adjacent to it, especially in the axial direction, and, as a result, a narrow belt of overstrain runs quickly along the cylinder lengthwise. This in turn throws more load on the material radially behind it and causes overstrain to advance through the wall on a narrow front until it reaches the outside, whereupon a similar phenomenon occurs at a position diametrically opposite, often causing the section to depart markedly from the circular. This was noted by Cook[2.3] and also by Crossland and Bones[2.4], and it has been beautifully demonstrated with the aid of strain figures by Steele and Young[2.12], who showed that the process of overstrain in deep wedges on a narrow front would go on repeating itself in different positions until the whole wall was overstrained, by which time the section would have once again become truly circular.

The second class of materials does not have a secondary yield at all and even slight overstrain usually brings with it some hardening. In consequence the element that first overstrains is given sufficient extra strength to carry on and thus gradually the whole inner surface yields. This puts a bigger load on the material behind so that the overstraining proceeds symmetrically outwards from the centre and at any intermediate stage the transition front between overstrained and still elastic material is very nearly a concentric circle.

With the third type this strain-hardening effect is much more marked. In general however these materials are seldom of high strength and only come into these considerations in special circumstances.

The essential requirement of autofrettage is that it should be a symmetrical process about the centre in all its stages, and evidently therefore we cannot use it satisfactorily with low carbon steels. This leaves us the main group of high tensile, low alloy, steels, which are fortunately the most suitable for high pressure service for various other reasons.

As we have seen in Fig. 2.7, according to this theory the material of an infinitely thick-walled cylinder is subjected to finite stresses by finite internal pressure. In that diagram it is supposed that the pressure is just sufficient to cause the bore layers to be on the point of overstraining. If therefore we could increase this internal pressure somewhat, these inner layers would be overstrained. We can therefore imagine another infinite cylinder with a bore radius less than unity and stressed so that all the material for which $r<1$ is overstrained, while all for which $r>1$ is still elastic. Evidently this latter part will be identical with Fig. 2.7. Our problem then is to draw similar curves, i.e. curves for σ_θ, σ_r, and τ, in the region of overstrain.

Here reference must be made back to eq. (2.7a) in Section 2.2. This shows that if the variation of τ across the section could be found it would be easy to evaluate the variation of σ_r with radius. One possibility at once comes to mind, since many of the class (*b*) materials discussed above have an appreciable range of nearly constant shear stress straining, and if we neglect the change in the radii due to the straining, i.e. use eq. (2.7), the problem can be solved, since the shear curve becomes a horizontal straight line. The tangential stress has then to be obtained for each particular problem from eq. (2.5) rearranged thus:

$$\sigma_\theta = 2\tau - \sigma_r \tag{2.48}$$

If however the situation is such that this straining at constant shear stress, i.e. true plasticity, can be assumed, a simple analytical solution is available as described in Section 2.10; we are more concerned here however with the provision of a chart which takes account of strain hardening as well as of the change in radii due to straining, which reduces the radial thickness of any given band of the wall. As we shall see, this can be done fairly simply on the basis of three assumptions, which appear by the results to be justified. These are:

(i) That deformation in the region of overstrain occurs without change in density.

(ii) That there is no strain in the axial direction (or that it is negligible in comparison with that in the other principal directions).

(iii) That the relationship between true shear stress and true shear strain is the same in the transverse plane of a cylinder as it is in a torsion test.

The first two of these assumptions indicate that overstraining in a cylinder is a constant area process and therefore it is merely a matter of simple geometry to find the shear strain at any radius in terms of (say) the change in radius at the bore. The third then allows us to determine the corresponding shear stress distribution from which, by integration, we can obtain values of the radial stress.

Various complications arise in practice, but these do not present any insurmountable difficulties. For a more detailed account of the preparation of this part of the chart, see the Appendix to this chapter.

It must be appreciated that the overstrain chart is no longer of universal application; the overstrain characteristics of different materials cannot be specified by a simple parameter as is done by Young's Modulus within the elastic range. On the other hand alloy steels of similar metallurgical characteristics, heat-treated to comparable strengths, give – for the most part – charts which are sufficiently similar to enable one to be used for a whole group of materials; and, as we shall see later (p. 39), this can be carried still further by certain assumptions and, although we cannot offer complete universality as in the elastic case, we can go some distance in that direction.

Franklin and Morrison[2.13] carried out a very careful study of the overstrain characteristics of an alloy steel (B.S.S. En 25) which is very similar to that used by Crossland and Bones[2.4] and also to the material of the reactor of which the elastic strains during test were discussed in Section 2.3. Franklin and Morrison, however, determined the shear stress strain characteristics by torsion tests on thin tubes up to shear strains of the order of 0·015. Unfortunately, for the preparation of a chart of general utility, we need a rather larger range of overstrain, but the torsion curves reported by Crossland and Bones agree well with

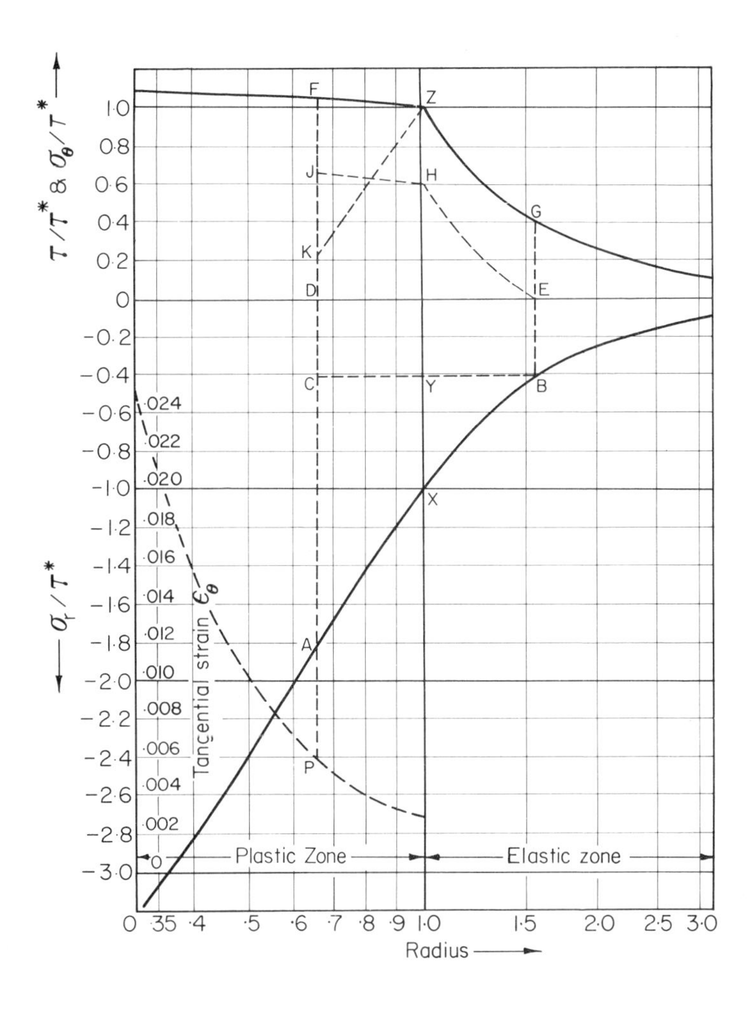

Figure 2.10 Chart extended to cover partial overstrain in nickel-chrome-molybdenum steel (B.S.En 25; 8,500 bar U.T.S.)

those of Franklin and Morrison† so that the two sets of results together enable the necessary chart to be constructed, as has been done in Fig. 2.10. The latter authors also reported a systematic study of the effects of overstrain on small specimens, and the use of the chart will be explained by taking one such experimental investigation as an example.

In Fig. 2.10 the upper curve represents the values of the shear stress in terms of the shear yield stress τ^*, which has the value of 4,095 bars; in the elastic region it also represents the tangential stress as the ratio σ_θ/τ^*, as has been explained in Section 2.7. The lower curve represents the corresponding ratio for the radial stress, i.e. σ_r/τ^*. The broken line curve in the plastic region represents the value of ε_θ at each radius; as will be seen it has the value of 0·00257 at $r = 1{\cdot}0$, which is the limiting elastic strain, or τ^*/G. Franklin and Morrison give the following values for the elastic constants, determined in the course of their experiments:

$$E = 2{\cdot}038 \times 10^6 \text{ bars}$$
$$G = 0{\cdot}795 \times 10^6 \text{ bars}$$
$$\nu = 0{\cdot}282$$

In one experiment these authors applied an internal pressure of 38 tonf/in^2 (5,870 bars) to a specimen of 0·750 in bore by 1·80 in outside diameter. As an example we will follow this experiment by means of the chart and estimate the proportion of the total wall material that is overstrained, the inside and outside surface strains, and the stress distribution while this pressure is acting.

The diameter ratio is 1·80/0·750 = 2·40, and the ratio of the applied pressure to the shear yield stress is:

$$5{,}870/4{,}095 = 1{\cdot}43$$

Now, to solve this problem, we take a rectangular card and mark off along the top edge the distance representing 2·40 (on the scale of the chart) from the left-hand top corner; then, down the left-hand side, we mark off a distance corresponding to 1·43. The card is then moved about over the chart with its sides parallel to the chart's axes until these two points fall on the curve of σ_r/τ^*. In Fig. 2.10 these occur at A and B, and the diagram can be completed by drawing a vertical line through A to cut the tangential strain line, the zero stress line, and the shear stress curve in P, D and F respectively. Similarly, we draw a vertical through B to cut the zero stress line and shear curve at E and G respectively, and a horizontal through B to cut the vertical through A at C and to cut the vertical line representing $r = 1{\cdot}0$ at Y.

Then the part represented by CY is overstrained, and that by YB remains elastic. CY represents a ratio of 1·51 approximately, so the radius of the transition circle will be $1{\cdot}51 \times 0{\cdot}750/2 = 0{\cdot}568$ in. In the actual experiment this transition

† These authors also showed how the extent of the permanent set left when the applied pressure was removed could be calculated from the unloading and reversed torsion records. A simpler approximation, which may however be in error by about 20%, is to regard the unloading of the overstrained cylinder as what it would be in a truly elastic cylinder; hence the term "induced elasticity" occasionally applied in such cases.

was determined by observing the release of elastic strain as the overstrained zone was bored out, and the critical radius was found to be 0·59 in. This discrepancy of about 4% could well have been due to variations in hardness of the material, which—in spite of very careful selection—was found to be less consistent in this respect than could have been desired. Also the accuracy of this method of measurement may have been no better than the order of this difference, which in any case is hardly significant in most design work.

The area of overstrained metal in the section, according to this estimate, was:

$$\pi(0{\cdot}568^2 - 0{\cdot}375^2)$$

and so the proportion of overstrained metal, assuming that the radius of the transition does not vary along the axis of the cylinder, was:

$$\frac{\pi(0{\cdot}568^2 - 0{\cdot}375^2)}{\pi(0{\cdot}900^2 - 0{\cdot}375^2)} \quad \text{or } 27\%$$

It should be noted that in this experiment the transition occurred very nearly at the geometric mean radius. This has been suggested as a basis for design, but—as is shown in Section 2.9—it would leave very little margin of safety.

In regard to strains, the point P represents the value of the tangential strain (0·0059) at the inner surface of the cylinder when carrying a pressure difference represented by AC, together with a hydrostatic pressure represented by BE or $0{\cdot}40\tau^*$. This latter will produce a negative strain in all directions equal to:

$$\frac{0{\cdot}40 \times 4{,}095(1 - 2 \times 0{\cdot}282)}{2{\cdot}038 \times 10^6}$$

or 0·00035 in/in.

Thus the total tangential strain at the bore will be the sum of these two, i.e. $0{\cdot}00590 + 0{\cdot}00035 = 0{\cdot}00625$, and the change in bore diameter will be obtained by multiplying it by this strain, i.e. $0{\cdot}75 \times 0{\cdot}00625 = 0{\cdot}0047$ in.

At the outer surface, since conditions are elastic, we can derive the strain from the point B in accordance with eq. (2.46), again making allowance for the effect of the hydrostatic pressure. The value of ε_θ given by point B is:

$$\frac{0{\cdot}40 \times 4{,}095 \times (1 + 0{\cdot}282)}{2{\cdot}038 \times 10^6} = 0{\cdot}00103 \text{ in/in}$$

and we must again add on the hydrostatic strain to give a total of $0{\cdot}00103 + 0{\cdot}00035 = 0{\cdot}00138$, which—on a diameter of 1·80 in—represents a swelling of 0·0025 in. This corresponds almost exactly with the amount measured by Franklin and Morrison.

Finally, to obtain the stress distribution across the section, the radial stress is given directly by the curve AXB relative to the base line CYB. This is, of course, plotted on a logarithmic scale for the radius. Similarly the curve GZ gives the tangential stress in the elastic region, again referred to the base CYB. The shear stress is represented by the line GZF, but this is with respect to the original zero stress line DE, and to make it comparable with the other stress curves it needs to be shifted vertically downwards by the distance BE, which represents the hydrostatic stress which was not in fact acting. Thus we draw the parallel curve EHJ to represent the shear stress with respect to the base line CYB. Finally, the tangential stress in the overstrain region is found from the relation:

$$\sigma_\theta = 2\tau - \sigma_r$$

which gives the line ZK; thus GZK is the curve of tangential stress about the base line CYB.

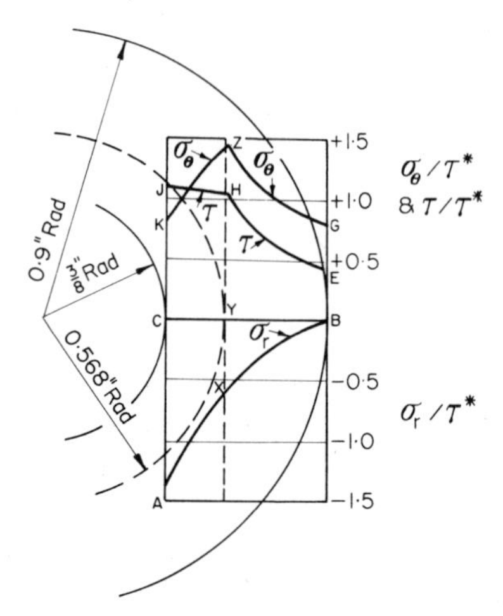

Figure 2.11 Stresses in example (overstrained)

These curves have been replotted on a linear radius scale in Fig. 2.11. It will be seen that the tangential stress passes through a maximum at the transition radius and then falls toward the inner surface, and it is evident that with thicker walls this could actually become negative, i.e. compressive, in the inner layers.

It will also be noticed on the chart, Fig. 2.10, that the σ_r/τ^* curve becomes much straighter in the overstrain region; it actually goes through a point of inflexion, so that the vertical intercept on a card representing a constant value of the radius ratio will, when the card is moved over the chart with the right-hand radius ratio mark kept on the σ_r/τ^* curve, pass through a maximum. This will then represent the Ultimate Bursting Pressure for a cylinder of the material for which the chart has been prepared, and of the radius ratio represented on the top edge of the card. This quantity will be discussed further in the next section.

2.9 Design Considerations with Autofrettage

The usual practice in engineering design is still, where possible, to relate the design to the elastic range of the material (see, for instance, B.S. 1515), but this is evidently impossible when we are dealing with autofrettage, since overstrain has already taken place. Instead the relation must be with the final breakdown, and this needs some discussion.

If one is going to base a design on some fraction of the pressure which causes final breakdown, it is important that the latter should be known accurately. If for instance a fictitiously high value of this pressure were used, then the resulting design would err on the side of danger; and the experimental problem is such that tests to destruction are very liable to produce results which are in fact too high. The idea was therefore introduced (2.14) of defining the *Ultimate Bursting Pressure* as the *lowest* pressure which would ultimately cause failure if maintained for a long enough time. In practice one finds that the quicker a bursting test is carried out, the higher the pressure tends to be. Thus, the smaller the specimen, given the same pressure-generating equipment, the quicker will be the test and consequently the greater the danger of error.

It must also be recognised that the number of published results of carefully controlled bursting tests is very small. Apart from that of Crossland and Bones(2.4) and the much earlier work of Cook and Robertson(2.9) the only other papers known to the authors at the time of writing, which can be recommended from this point of view, are those of Faupel and Furbeck(2.15) and Wellinger and Uebing.(2.16)

For this reason the preferred practice is to derive a relation between ultimate bursting pressure and initial diameter ratio for a given material (by the method explained in the Appendix to this chapter) and to check this with one or two actual tests. The theoretical basis of this analysis presumes that there is no scale factor, and therefore that the U.B.P., like the limiting elastic pressure, depends only on the diameter ratio and not on the absolute dimensions. The experimental justification for this is small, although the authors know of no work which actually contradicts it†. Crossland and Bones, in replying to the discussion of their paper (loc. cit.), reported a series of tests to destruction of cylinders made from a bar of 0·17% C steel, all with the same diameter ratio of 2·0, but varying from $\frac{1}{2}$ in to 3 in in the bore. The bursting pressures varied about $\pm 1{\cdot}2\%$ above and below the mean value of about 2,700 bars, a reasonably satisfactory correlation.

Assuming that this and the other assumptions are therefore justified, and that

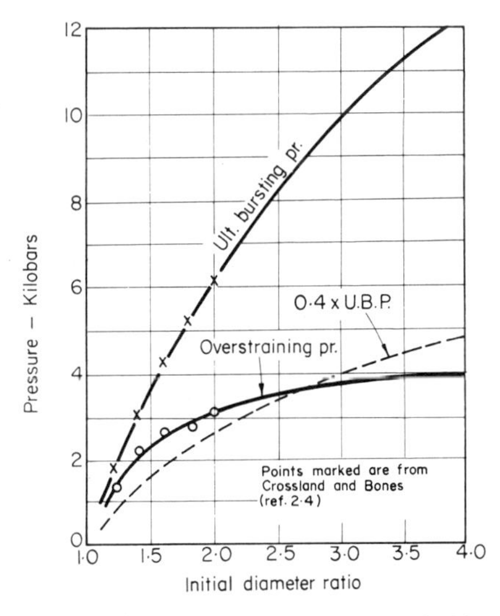

Figure 2.12 Curves showing relations with initial diameters of ultimate bursting pressures and pressures to cause initial overstrain in cylinders of B.S.En 25 steel: *Crossland & Bones, Ref* 2.4

† Langenburg (2.20) reported that in autofrettage tests the same pressure produced the same proportion of overstrain in cylinders of the same material and same diameter ratio irrespective of absolute size.

the U.B.P. values thus arrived at are acceptable, it is useful to plot them against diameter ratios, and also to plot on the same diagram the limiting elastic pressures. This has been done in Fig. 2.12 for the material used by the experimenters of References 2.4 and 2.13. It is seen at once that there is a considerable range between these two pressure limits, and that this range increases rapidly with the diameter ratio. It is true that the U.B.P. curve here has had to be derived theoretically for ratios above 2·0, although for the thinner specimens the agreement between theory and experiment is seen to be very good.

Also in this diagram is plotted a broken line representing 40% of the U.B.P., and it is suggested that for many applications this might form a useful design rule. Normally it will mean that only cylinders whose diameter ratio exceeds about 2·8 will be overstrained at the working pressure. On the other hand, it is normal practice to apply a test pressure higher than the designed working pressure. A reasonable figure for the test would be 33% above the working pressure, and Fig. 2.12 then shows that, on that basis, any cylinder to work at 3,000 bars or more will be overstrained in the course of the test.

For certain special applications, particularly in research work where extra safety precautions can be adopted, much lower margins can be tolerated. For the example quoted above of the small specimen of $\frac{3}{4}$ in bore by 1·80 in outside diameter, the U.B.P. is seen from Fig. 2.12 to be approximately 7,650 bars, whereas the pressure which overstrained it to approximately the geometric mean radius (according to Franklin and Morrison's experimental work) was 5,870 bars. Thus, to work at that pressure would mean accepting a factor of safety on ultimate rupture of no more than 1·3, i.e. an overpressure of 30% would probably burst the cylinder.

A more reasonable procedure for most requirements would be to regard the pressure which overstrained the material out to the geometric mean radius as the "test" pressure, and then to allow $\frac{2}{3}$ of it as the working pressure, i.e. about 3,900 bars. The safety factor would then be nearly 2. In fact, with a slightly more conservative interpretation we might with advantage limit the working pressure to 3,800 bars, and thus obtain a reasonably reliable safety factor of 2 to 1.

As we shall see later (Chapter 6) the use of autofrettage has a number of advantages, particularly where repetitions or alterations of load are to be expected and where, in consequence, fatigue troubles could arise. It can also prove effective in smoothing out localised peaks of stress, caused for instance by changes of section, side connections, or other irregularities; the effect of the overstrain is to spread the stress more evenly over the surrounding material. This could also reduce the danger of brittle failure, and the avoidance of this is very important from the safety point of view.

The estimation of the required overstraining pressure and of the dimensional changes that it will cause is a simple matter if the appropriate chart is available, as we have seen in Section 2.8. If however we do not have such a chart the analytical procedure described in Section 2.10 will probably suffice unless the proposed diameter ratio is very large, e.g. greater than say 3, when it becomes necessary to study the problem more accurately.

As has been pointed out, each material requires a separate chart for the overstrain region. It is however a fortunate fact that, provided the overstrain is not carried too far, and if the stresses in the chart are plotted as ratios of the shear yield stress, the lines on the charts are very nearly coincident. Fig. 2.13 shows the overstrain region of a chart on which the curves for four separate and very different materials have been drawn. These are the En 25 material used by

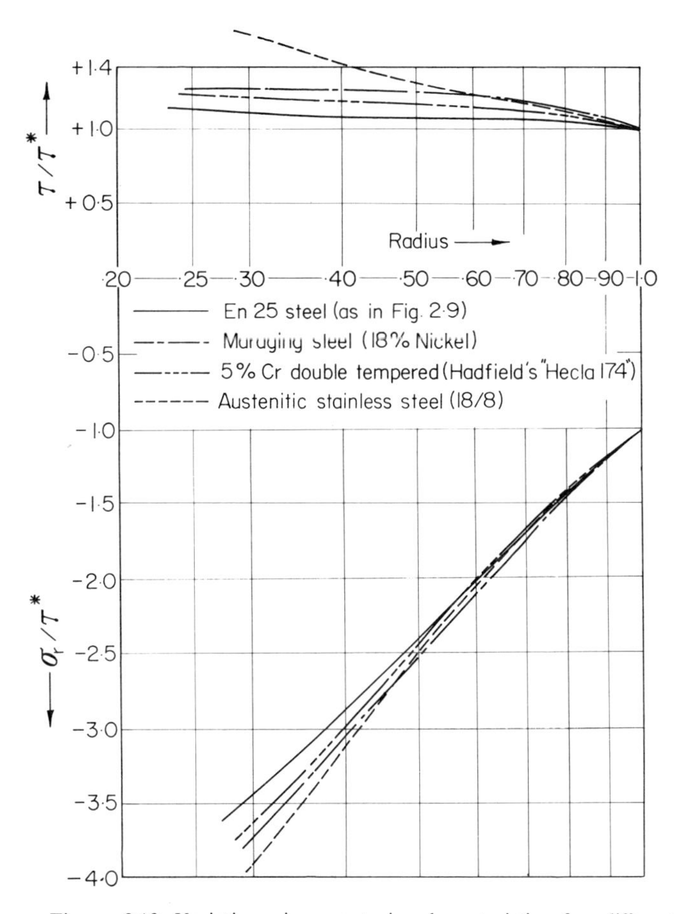

Figure 2.13 Variations in overstrain characteristics for different materials

Crossland and Bones[2.4] and by Franklin and Morrison[2.13]; a maraging steel containing 18% nickel and 8% cobalt; a special high tensile steel manufactured by Hadfields Ltd. under the trade name Hecla 174; and an austenitic stainless material. The approximate analyses and mechanical properties are given in Table 2.5, from which it will be seen how widely the materials vary in strength, and also in other characteristics.

For instance, as we have seen, the En 25 has a wide range of deformation at almost constant shear stress before beginning to strain-harden: by contrast, the maraging steel has a rather gradual yield in shear, after which it has a wide range of deformation at constant stress right up to fracture. The austenitic material is very soft initially, but it has an exceptionally large capacity for strain-hardening, and – as can be clearly seen in Fig. 2.13 – the point of inflexion in the σ_r/τ^* curve

Table 2.5

Material	*Approximate analysis*	*U.T.S. (lbf/in²)*	*U.T.S. (bars)*	*Shear yield stress (bars)*
En 25	0·3% C, 2·57% Ni, 0·58% Cr, 0·60% Mo	122,500	6,470	4,095
Maraging (Maraged at 480°C)	18% Ni, 8% Co, 0·6% Ti	265,000	18,300	9,100
Hadfield's "Hecla 174" double tempered	5% Cr, 1·5% Mo, 1% Va	260,000	17,800	7,750
Austenitic Stainless (softened)	18% Cr, 8% Ni	85,000	5,900	1,200

In cases where no torsion properties are given the yield in shear has been taken as the 0·2% Proof Stress in tension divided by $\sqrt{3}$.

has not been reached even at a radius of 0·3, which corresponds with a shear strain of more than 10%. The Hadfield's material has a high strength with a moderate strain-hardening capacity and it is included here because it has already been used in a small scale apparatus by Thomas, Turner and Wall[2.17] up to 25,000 bars.

For small scale apparatus, the autofrettage operation can very conveniently be effected by pushing oversized drifts through the bore. If these are highly polished and generously lubricated before each pass the result is to produce a very good finish on the bore surface. The usual practice is to work with each drift about 0·005 in/in of diameter greater than the bore, and—if the part being treated is a simple cylinder—the autofrettage can be controlled by noting the change in external diameter. This has also been used by Proskuryskov and Karasev[2.18] and has been referred to as "broach planishing".

Heat Treatment. A number of authorities, particularly Macrae[2.2], have recommended a low temperature heat treatment (usually in the range 250 to 300°C), and it appears that this was a regular procedure in arsenals where guns were being treated. More recently, however, doubts have arisen as to the value of this. Franklin and Morrison[2.13] tried it on some of their specimens and concluded that it did have a slightly beneficial effect, but that its virtues had been considerably exaggerated. It is suggested therefore that heat treatment should be carried out after each application of autofrettaging pressure (where this is done in more than one stage), but only if it can be done easily and cheaply; and it is, of course, essential that the temperature should be properly controlled and that the heating and cooling should be slow and uniform. It is also obvious that the temperature must never be high enough to draw the temper of the material or otherwise affect its microstructure.

2.10 Approximate Analytical Solution of Autofrettage Problems

This depends on two assumptions: first that, within the range of overstrain involved, the material behaves according to the criterion of true plasticity, i.e. of straining at constant shear stress, and secondly that the changes in dimensions can be neglected. The shear stress curve is then similar to that shown in Fig. 2.9b (without the strain-hardening rise at the right-hand end) and if r_c is the radius of the transition from plastic to elastic, we have—for the plastic zone—from eq. (2.7):

$$(\sigma_r)_c-(\sigma_r)_i = 2\int_{r_i}^{r_o} \tau^* d\ln r = \tau^* \ln k_c^2 \tag{2.49}$$

where $k_c = r_c/r_i$. Also, in the elastic zone, we know from eq. (2.29) that an applied pressure of $-(\sigma_r)_c$ will produce a shear stress of τ^*, or:

$$-(\sigma_r)_c = \tau^* \frac{r_o^2-r_c^2}{r_o^2} \tag{2.50}$$

Then on eliminating $(\sigma_r)_c$ and substituting K for r_o/r_i, k_c for r_c/r_i, and p_i for $-(\sigma_r)_i$ we get:

$$p_i = \tau^* \ \ln k_c^2 + \tau^* \frac{r_o^2-r_c^2}{r_o^2}$$

or

$$p_i = \tau^* \ (1+\ln k_c^2 - k_c^2/K^2) \tag{2.51}$$

This enables the complete solution to be obtained if we assume the constant cross-sectional area condition to hold in the plastic region. Applying this to the example given in Section 2.8, we have $p_i = 5{,}870$ bars, $\tau^* = 4{,}095$ bars, and $K = 2{\cdot}40$. Substituting into eq. (2.51) and solving for k_c gives it the value 1·52, as compared with 1·51 obtained from the chart, and 1·57 determined experimentally by releasing the strains by boring out. As has been remarked in the previous section, the experimental technique involved here calls for very accurate measurements and very skilled manipulation and, notwithstanding the excellence of the practice in the Bristol University laboratories and workshops, it would seem that a difference of 4% could easily occur, and the authors also mention variations in hardness of the bar from which this cylinder test piece was machined. One would expect the result of this approximate method of calculation to imply a greater amount of overstrain than would be obtained in practice or by a method of calculation (such as the chart) which takes strain-hardening into account. Thus the slightly greater overstraining implied by the approximate method is just what would be expected. The fact that the actual experiment indicated still greater overstrain suggests that the material may have been slightly softer than those from which the chart was prepared. Even so the agreement between all three results is good enough for most design projects.

The strains at the outside can be calculated by regarding the elastic portion (diameter ratio K/k_c or 1·58) as a separate cylinder loaded by an internal pressure of:

$$\tau^* \frac{K^2-k_c^2}{K^2}, \text{ or } 2{,}450 \text{ bars.}$$

Then the value of $(\varepsilon_\theta)_o$ will be given by:

$$E(\varepsilon_\theta)_o = \tau^* \frac{K^2-k_c^2}{K^2} \frac{k_c^2}{K^2-k_c^2}(2-\nu) \tag{2.52}$$

or, in this case, $(\varepsilon_\theta)_o = 1{\cdot}38\times10^{-3}$ in/in.
Thus, on an initial outside diameter of 1·80 in, this will amount to a swelling of $1{\cdot}80\times1{\cdot}38\times10^{-3}$ or 0·0025 in.

To obtain the bore strain we can calculate the value of ε_θ at the transition radius, i.e. at the maximum elastic radius by similar consideration as used above to find the external diameter changes, and then, by assuming that the cross-sectional area is unchanged in the overstrain region, it is clear that the tangential strain varies inversely as the square of the radius. Now, in an elastic cylinder of diameter ratio K, the value of ε_θ at the inner surface is given (see eq. (2.27)) by:

$$E(\varepsilon_\theta)_i = \frac{p_i}{K^2-1}\{(K^2+1)+\nu(K^2-2)\} \tag{2.53}$$

so in this case, by substitution:

$$(\varepsilon_\theta)_{i\ elastic} = \frac{\tau^*\{(K^2+k_c^2)+\nu(K^2-2k_c^2)\}}{EK^2}$$

Here where $k_c = 152$, we have:

$$(\varepsilon_\theta)_{i\ elastic} = 2{\cdot}93\times10^{-3}$$

Thus the value of ε_θ at the actual bore will be:

$$2{\cdot}93\times10^{-}3\times k_c^2 = 6{\cdot}77\times10^{-3} \text{ in/in}$$

The initial bore was 0·75 in and therefore the increase while the pressure is acting will be $6{\cdot}77\times0{\cdot}75\times10^{-3}$, 0·0051 in.

It is evident therefore that in this particular example the approximate treatment is entirely adequate so far as practical considerations are concerned. For thicker cylinders, however, the divergence may become more serious, especially if the material involved shows a greater degree of strain-hardening.

Appendix to Chapter 2

Large Strain and Ultimate Bursting Pressure of Cylinders

The treatment here is based on the three assumptions stated in Section 2.8, which can be combined into the following two conditions:

(i) The cross-sectional area of the wall remains constant.

(ii) There is a unique relationship between shear stresses and shear strains which can be found from a torsion test.

The assumptions of symmetry and geometrical perfection also apply in the overstrain region and under conditions where part of the section is overstrained and part still elastic, which means that cylinders made from materials such as low carbon steels will not usually conform to this assumption and cannot therefore be reliably analysed by the considerations which follow. On the other hand, even these usually return to a fair approximation to circular symmetry when fully overstrained and consequently allow their Ultimate Bursting Pressures to be calculated with reasonable accuracy.

When partially overstrained the elastic part is at the outside and its external pressure therefore is, under normal conditions, zero; consequently its axial strain cannot be zero and there must be some discontinuity at the transition between the elastic and overstrain zones. Axial strain in an elastic cylinder can only disappear if $p_i = K^2 p_o$ but the errors introduced by assuming it to be zero are evidently small.

Now the condition that the cross-sectional area stays constant if there is a radial shift u_i at the bore surface (radius r_i) will be:

$$\pi(r+u)^2 - \pi(r_i+u_i)^2 = \pi r^2 - \pi r_i^2$$

from which:

$$u^2 + 2ur = u_i^2 + 2u_i r_i \qquad (2A.1)$$

and, evidently, if r happens to be the outside radius r_o, then:

$$u_o^2 + 2u_o r_o = u^2 + 2ur = u_i^2 + 2u_i r_i \qquad (2A.2)$$

From this the shear strain at any radius can be calculated in terms of, say, the tangential strain at the outside surface, which is:

$$(\varepsilon_\theta)_o = \frac{u_o}{r_o} \qquad (2A.3)$$

Now the correlation of strains between the cylinder and the torsion test is an easy matter so long as they are small, but the bursting of a cylinder of ductile material by steady pressure may involve very large strains. Fortunately we are dealing with plane strain in each case and are assuming that there is no change in density. Also we are concerned with the progressive increase in the same kind of loading in each case, and there is therefore no need to use the incremental strain functions. Any unique function of *two* strains will suffice in such comparisons, so that—for instance—the relation for small strains, i.e.

$$_1\gamma_3 = \varepsilon_1 - \varepsilon_3 \tag{2A.4}$$

could be used throughout. In practice however the strain in the torsion test is most easily described by the tangent of the angle of twist, i.e.

$$\gamma = \frac{r\theta}{L} \tag{2A.5}$$

where r is the radius of the specimen, θ the angle turned through by one end of the gauge length relative to the other end and distant L from it.

The corresponding tangent formula for the heavily strained cylinder can be derived quite simply in the following manner. If we add r^2 to each part of eq. (2A.1) and take the square root, we have, for the left hand part:

$$r+u = \pm\sqrt{(r^2+\lambda^2)} \tag{2A.6}$$

where $\lambda^2 = u_o^2 + 2u_o r_o$.

Consequently

$$1+\frac{u}{r} = 1+\varepsilon_\theta = \pm\sqrt{\left(1+\frac{\lambda^2}{r^2}\right)} \tag{2A.7}$$

But the constancy of area condition demands that:

$$(1+\varepsilon_\theta)(1+\varepsilon_r) - 1 = 0 \tag{2A.8}$$

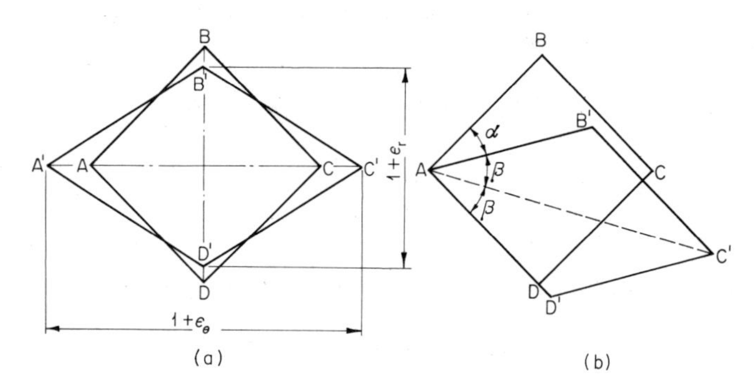

Figure 2A.1 Diagram illustrating relationship of large strains

from which:

$$1+\varepsilon_\theta = \frac{1}{1+\varepsilon_r} \tag{2A.9}$$

If we now consider a small square in the unstrained cross-section as represented by ABCD in Fig. 2A.1a, with diagonals of unit length, and let it be strained into the rhombus A′B′C′D′, such that the new diagonals A′C′ and B′D′ have lengths $1+\varepsilon_\theta$ and $1+\varepsilon_r$ respectively, the shear strain will be twice the angle through which the side AB has been rotated to reach the position A′B′. This can be seen more easily if we apply the rhombus to the original square in such a way that A′D′ lies along AD with A′ coinciding with A as in Fig. 2A.1b. The shear strain is evidently given by the angle BAB′, or—according to our present convention—by its tangent. If we denote BAB′ by α and B′AC′ by β it is evident that:

$$\alpha + 2\beta = \frac{\pi}{2}$$

whence

$$\tan\alpha = \tan\left(\frac{\pi}{2} - 2\beta\right) = \cot 2\beta$$

and therefore:

$$\gamma = \tan\alpha = \frac{1 - \tan^2\beta}{2\tan\beta} \qquad (2A.10)$$

But angle B′AC′ in Fig. 2A.1b is, from the construction used, identical with angle B′A′C′ in Fig. 2A.1a, from which we see that its tangent is:

$$\frac{\frac{1}{2}(1+\varepsilon_r)}{\frac{1}{2}(1+\varepsilon_\theta)}$$

Thus:

$$\tan\beta = \frac{1+\varepsilon_r}{1+\varepsilon_\theta} \qquad (2A.11)$$

and

$$\gamma = \frac{1}{2}\left\{\frac{(1+\varepsilon_\theta)^2 - (1+\varepsilon_r)^2}{(1+\varepsilon_\theta)(1+\varepsilon_r)}\right\} \qquad (2A.12)$$

If ε_θ and ε_r are both small compared with 1, eq. (2A.12) evidently reduces to:

$$\gamma = \varepsilon_\theta - \varepsilon_r$$

as in eq. (2A.4). But, again while ε_θ and ε_r are small, eq. (2A.9) reduces to:

$$\varepsilon_\theta + \varepsilon_r \simeq 0$$

and thus:

$$\gamma \simeq 2\varepsilon_\theta \simeq \frac{2u}{r} \qquad (2A.13)$$

In the more general form where larger strains are involved, substitution for $1+\varepsilon_\theta$ and $1+\varepsilon_r$ in eq. (2A.12) from (2A.7) and (2A.8) gives:

$$\gamma = \frac{1}{2}\left\{\left(\frac{r^2+\lambda^2}{r^2}\right) - \left(\frac{r^2}{r^2+\lambda^2}\right)\right\} \qquad (2A.14)$$

These shear strains in a cylinder wall are then, according to our assumptions, directly comparable with those given by eq. (2A.5) for the torsion test.

For virtually all practical cases of overstrained cylinders as pressure containers, i.e. in autofrettage construction, the strains are small enough to be calculated from the approximate relation of eq. (2A.13) which, by substitution in terms of u_o and r_o from eq. (2A.7), becomes:

$$\gamma \simeq \frac{2u_o r_o}{r^2} \tag{2A.15}$$

On the other hand, for determining the Ultimate Bursting Strength of cylinders, especially with materials with high work-hardening capacity such as austenitic stainless steels in an initially softened condition, it may be necessary to extend the range covered considerably and the more exact relation of eq. (2A.14) becomes necessary. The need for this normally arises when the difference between the results of eq. (2A.14) and (2A.15) rises to about $2\frac{1}{2}\%$ which occurs when u/r is of the order of 3% with most engineering materials.

The calculation of the true shear stress from the torque-twist record resulting from a torsion test on a solid round specimen is not a matter of simply dividing by the section modulus when the material is overstrained, as it is for elastic conditions: instead an analysis due to Nadai[2.19] must be used.

Thus if T be the torque in the torsion test when the twist over the gauge length L has reached θ, the corresponding maximum shear stress is given by:

$$\tau = \frac{4}{\pi D^2}\left(\theta\frac{dT}{d\theta} + 3T\right) \tag{2A.16}$$

This evidently involves plotting the torque twist record with sufficient accuracy to determine the gradients at a number of points, but this is not usually a very troublesome matter.†

For purposes of comparison it is advantageous to plot the shear stress as the ratio of its limiting elastic value τ^*. Then both the stress and the strain are in the form of dimensionless quantities. This has been done in Fig. 2A.2 which shows both the torque and the derived shear stress.

Having prepared such a record of shear stress against shear strain, it becomes a simple—and not unduly laborious—matter to work out the data for a chart for the study of autofrettage problems (such as that given in Fig. 2.10, p. 33) and also to extend it so that the Ultimate Bursting Pressure as well as the deformation immediately prior to rupture can be calculated.

The procedure is to assume some particular radial shift, either at the inside or outside surface, and then to derive the shear strains at successive radii outwards or inwards, as convenient. When dealing with an autofrettage chart, it is usually easiest to suppose that the shear strain is just at the limit of elasticity for unit

† It may be worth pointing out that a torsion test on a thin tube gives the true shear stress directly if one can assume that the stress does not vary across the thickness of the specimen. Unfortunately, however, tests of this kind are apt to be vitiated by the collapse of the specimen, or at any rate by serious departure from the circular section. Nevertheless, Franklin and Morrison[2.13] did succeed in continuing such tests well into the overstrain region, and their results amply confirm the Nadai method, which itself depends on certain assumptions the justification for which would otherwise be difficult to assess.

radius. Then for radii greater than unity, the system is elastic and the simple universal chart of Fig. 2.7, p. 28 obtains and can be drawn in easily.

For the region of overstrain the critical radius is given the value 1·0 and u the value which gives a shear strain equal to τ^*/G, where G is the Modulus of Rigidity. The strains are still very small in this region and eq. (2A.15) can be used.

In this case the values of r which are being considered get progressively smaller and in consequence they will reach values where the more precise formula of (2A.14) has to be used, usually when r is less than about 0·5.

Essentially the procedure is to derive the appropriate values for differences in σ_r by integrating the curve of shear stress, but this is done arithmetically. Thus, referring to eq. (2.6), in the deformed section:

$$(\sigma_r)_1 - (\sigma_r)_2 = 2\int_{(r+u)_1}^{(r+u)_2} \frac{\tau}{r+u}\, d(r+u) \tag{2A.17}$$

Then, if we divide the radii into fairly closely spaced rings, it can be assumed that the curve is near enough to being straight between $(r+u)_2$ and $(r+u)_1$, so that the integral is near enough to being the average of the two values of $\tau/(r+u)$ multiplied by the difference between the $(r+u)$ values. The rest of the procedure should be clear from the first few lines of Table 2A.1, which is based on the torsion results of Crossland and Bones[(2.4)] with an alloy steel (to which the Autofrettage Chart of Fig. 2.10 relates). In this the limiting elastic shear stress was found to be 4,095 bars (59,400 lbf/in^2) and the shear modulus $0{\cdot}795 \times 10^6$ bars. Thus the shear strain at this limit γ^* will be:

$$\frac{4{,}095}{795{,}000} = 5{\cdot}16 \times 10^{-3}$$

and if r_o be taken as unity this will also be the value of λ^2 to a sufficient degree of accuracy.

The method of deriving the table will now be clear for conditions of relatively small strain. We begin with the unstrained radius $r = 1{\cdot}0$, for which $r+u$ (the corresponding strained radius) is 1·0026. The appropriate values of u and γ are then inserted. The shear stress is next read from a curve such as the lower one in Fig. 2A.2, and as this is in terms of ratios of its limiting elastic value, evidently it must have unit value at this point. We then divide by $r+u$ and record in column VII.

This procedure is then repeated for a series of decreasing radii as shown. We can thus complete column III by putting in the numbers in brackets which are the differences between the successive values of $r+u$. Column VIII can also be evaluated, being twice the average of each pair of numbers in column VII. The reason for its being twice is that there is a factor of 2 in the integral of (2A.17), which simplifies the problem, because each entry in this column is merely the sum of each pair in column VII. The final column is obtained by multiplying each item in column VIII by the difference figure in column III. This is evidently twice the area of the thin strip of the curve of $\tau/\tau^*(r+u)$ cut off between the strained values of radii 1·0 and 0·95. These items are also progressively summed in column IX.

It will also be noted that the first entry in this column is 1·000. The explanation of this is that it represents the value of the radial stress σ_r† at that radius where the shear stress is just at its limiting value, the thickness of the cylinder being assumed to extend infinitely outwards. A glance at Fig. 2.10 should make this quite clear.

The various curves in Fig. 2.10 are given in Table 2A.1 as follows, in the area marked "Plastic Zone" in that diagram: tangential strain, column V values divided by 2; shear stress, column VI; radial stress, summation figures in column IX.

The continuation of this table down to values of r of 0·25 or less is necessary

Table 2A.1

I	II	III	IV	V	VI	VII	VIII	IX
r	$r^2+\lambda^2$	$r+u = \sqrt{(r^2+\lambda^2)}$ *and differences*	u	$\gamma (= 2\varepsilon_\theta)$ *for small values*	τ/τ^*	$\tau/\tau^*(r+u)$	*Twice average of col. VII*	*Product of col. VIII and differences*
1·0	1·00516	1·0026	0·0026	0·0051	1·000	0·997		1·000
		(0·0499)					2·048	0·102
0·95	0·90766	0·9527	0·0027	0·0057	1·0015	1·051		1·102
		(0·0498)					2·162	0·108
0·90	0·81516	0·9029	0·0028	0·0064	1·003	1·111		1·210
		(0·0499)					2·289	0·114
0·85	0·72766	0·8530	0·0030	0·0071	1·0045	1·178		1·324
		(0·0498)					2·430	0·121
0·80	0·64516	0·8032	0·0032	0·0080	1·006	1·252		1·445
		(0·0199)					2·538	0·051
0·78	0·61356	0·7833	0·0033	0·0085	1·008	1·286		1·496
		(0·0199)					2·607	0·052
0·76	0·58276	0·7634	0·0034	0·0089	1·009	1·321		1·548
		(0·0199)					2·679	0·053
0·74	0·55276	0·7435	0·0035	0·0094	1·010	1·358		1·601
		(0·0199)					2·756	0·054
0·72	0·52356	0·7236	0·0036	0·0099	1·011	1·398		1·655
		(0·0199)					2·836	0·056
0·70	0·49516	0·7037	0·0037	0·0105	1·012	1·438		1·711
		(0·0199)					2·920	0·058
0·68	0·46756	0·6838	0·0038	0·0111	1·013	1·482		1·769
		(0·0199)					3·009	0·060
0·66	0·44076	0·6639	0·0039	0·0118	1·014	1·527		1·829
		(0·0199)					3·103	0·061
0·64	0·41476	0·6440	0·0040	0·0126	1·015	1·576		1·890
		(0·0199)					3·204	0·063
0·62	0·38956	0·6241	0·0041	0·0133	1·016	1·628		1·953
		(0·0198)					3·312	0·065
0·60	0·36516	0·6043	0·0043	0·0143	1·0175	1·684		2·018
		(0·0199)					3·427	0·068
0·58	0·34156	0·5844	0·0044	0·0153	1·019	1·743		2·086

† This is a compressive and therefore negative stress and is shown thus in Fig. 2.10.

in order to calculate the Ultimate Bursting Pressure of cylinders whose initial diameter (or radius) ratio is 2 or more, but it presents no difficulties and is only moderately laborious. With the relatively low value of λ^2 in this particular case the use of the approximate formula for the shear strain (eq. 2A.15) can be continued at least down to $r = 0{\cdot}3$. It should be noted also that accuracy in the estimation of shear strain becomes less important when the curve relating stress and strain is relatively flat, as it is here (see Fig. 2A.2).

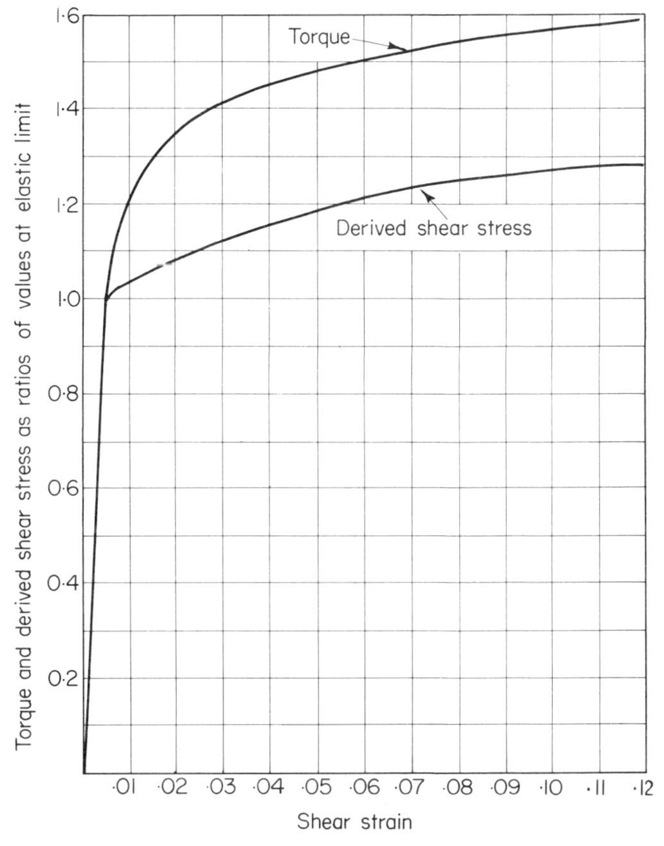

Figure 2A.2 Torque-twist and derived shear stress v shear strain curves of torsion test with B.S.En 25 steel *(Crossland & Bones)*

To calculate the value of the Ultimate Bursting Pressure for a cylinder of given initial diameter ratio, say 1·6 to 1, we proceed as follows. Take various radii whose ratio is 1·6; to start with, for instance, 1·0 and 0·625. The appropriate values of radial stress are 1·000 and 1·937 (the latter by interpolation in column IX of Table 2A.1). The corresponding values of u/r are 0·0026 at the outside and 0·0041 at the inside. The difference in radial stress, corresponding to the difference in pressure, is then $1{\cdot}937 - 1{\cdot}000 = 0{\cdot}937$ times τ^* (which is 4,095 bars). This is equivalent to a pressure difference between the outside and inside surfaces of the cylinder. We must now repeat the procedure for a series of different values of r and see which gives the biggest difference in radial stress; this has been done in Table 2A.2. As will be seen, the figures in column VII pass through a maximum when $r_i = 0{\cdot}225$ and $r_o = 0{\cdot}36$, the value being about 1·01, i.e. $1{\cdot}01 \times 4{,}095$, or say 4,140 bars, which is then regarded as the Ultimate Bursting Pressure. It is worth noting that Crossland and Bones burst a cylinder of this diameter ratio and measured the pressure as 4,200 bars; also that the last measurement of the outside diameter made just before the specimen burst showed a value for $(u/r)_o$ of 0·0365 which is certainly very close to the value 0·037 shown in Table 2A.1 for $r_o = 0{\cdot}36$.

Table 2A.2

I	II	III	IV	V	VI	VII
r_i	r_o	$(u/r)_i$	$(u/r)_o$	p_i/τ^*	p_o/τ^*	$\Delta p/\tau^*$
0·625	1·00	0·0041	0·0026	1·937	1·000	0·937
0·5	0·80	0·0103	0·0040	2·386	1·445	0·941
0·4	0·64	0·0160	0·0063	2·869	1·890	0·979
0·3	0·48	0·0283	0·0111	3·483	2·486	0·997
0·275	0·44	0·0350	0·0250	3·670	2·669	1·001
0·25	0·40	0·075	0·030	3·876	2·869	1·007
0·225	0·36	0·091	0·037	4·101	3·093	1·008
0·20	0·32	0·113	0·047	4·349	3·345	1·004
0·175	0·28	0·144	0·060	4·621	3·632	0·989

The middle curve of Fig. 2A.3 is a plot of columns VII and IV of Table 2A.2, together with measured points from Crossland and Bones' paper.[2.4] After an external strain of only 0·0026 (corresponding to a maximum shear strain at the bore of 0·008 which is only some 60% above the limiting elastic strain) the pressure is

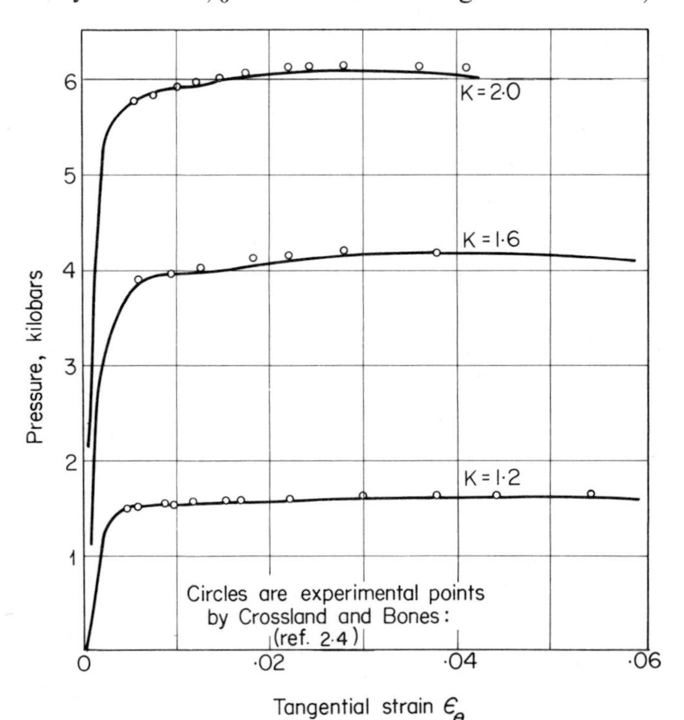

Figure 2A.3 Derived relation between applied pressure and resulting tangential strain

within 8% of the U.B.P., whereas the strain at the outside increases 14 times before the material actually ruptures. This rapid increase of strain with comparatively small increase in pressure has been termed "ballooning" or "collapse" by Crossland and Bones, and the pressure at which this process begins the "ballooning pressure". For this case the value of the latter is about 3,800 bars, and it is interesting to note that it corresponds quite closely to the value of:

$$2\tau^* \ln K$$

which represents the condition when overstrain has just penetrated to the outer surface without any strain-hardening and neglecting the effect of the deformation on K. In this case the natural logarithm of 1·6 is 0·470, giving a pressure of 3,850 bars. From the design point of view this may be a better criterion for autofrettaged cylinders, although it is not always so clearly marked with thicker walls, especially where the material has an appreciable capacity for strain-hardening.

Fig. 2A.3 also shows corresponding pressure versus tangential strain curves for cylinders of initial diameter ratios 1·2 and 2·0 in the same material. By plotting the maxima of these against their initial diameter ratios we obtain the design curve shown in Fig. 2.12, p. 37.

REFERENCES

2.1. LAMÉ, G. and CLAPEYRON, B. P. E., *Mém. présentés par Divers Savants,* **4,** Paris 1833.
2.2. MACRAE, A. E., *Overstrain of Metals,* H. M. Stationery Office, 1930.
2.3. COOK, G., *Proc. Inst. Mech. E.,* **126,** 407, 1934.
2.4. CROSSLAND, B. and BONES, J. A., *Proc. Inst. Mech. Eng.,* **172,** 777, 1958.
2.5. MAXWELL, J. CLERK, Letter to William Thomson dated Dec. 18, 1856, quoted by F. K. Th. van Iterson in *Plasticity,* Blackie & Son Ltd. 1947.
2.6. VON MISES, R., *Nachr. K. Ges. Wiss. Göttingen,* **4,** 582, 1913.
2.7. HAIGH, B. P., *Brit. Ass. Rep.,* 1919 and 1920.
2.8. RÔS, M. and EICHINGER, A., *Eidgenössiche Material-prüfungsanstalt an der E. T. H., Zürich,* No. 34.
2.9. COOK, G. and ROBERTSON, A., *Engineering,* **92,** 786, 1911.
2.10. BRIDGMAN, P. W., *Phil. Mag.,* **24,** 63, 1912.
2.11. JACOB, L., *Résistance et Construction des Bouches à Feu,* Doin et Fils, Paris, 2nd Ed. 1920, 157.
2.12. STEELE, M. C. and YOUNG, J., *Trans. Am. Soc. Mech. Eng.,* **74,** 355, 1952.
2.13. FRANKLIN, G. J. and MORRISON, J. L. M., *Proc. Inst. Mech. Eng.,* **174,** 947, 1960.
2.14. MANNING, W. R. D., *Engineering,* **169,** 479, 509 and 562, 1950.
2.15. FAUPEL, J. H. and FURBECK, A. R., *Trans. Am. Soc. Mech. Eng.,* **75,** 345, 1953.
2.16. WELLINGER, K. and UEBING, D., *Mitt. V. G. B.,* **66,** 134, 1960.
2.17. THOMAS, S. L. S., TURNER, H. S., and WALL, W. F., Conference on High Pressure Engineering, London, Sept. 1967, Paper No. 11, *Proc. Inst. Mech. Eng.,* **182,** Pt 3C, 271.
2.18. PROSKURYSKOV, YU. G. and KARASEV, N. A., *P.E.R.A. translations,* **33,** No. 4.
2.19. NADAI, A. and WAHL, A. M., *Plasticity,* McGraw-Hill, 1931, 126.
2.20 LANGENBURG, F. C., *J. Am. Soc. Steel Treating,* **8,** 447, 1925.

3 Multi-Component Cylinders

3.1 General

Attention has already been drawn, in the preceding sections, to the very uneven stress distribution across the wall of an elastic cylinder, which is clearly illustrated in Fig. 2.5a. This was appreciated amongst the ordnance designers and manufacturers as soon as Lamé's work had revealed how the stresses were actually distributed, and, by the middle of the nineteenth century, at least two proposals had been put forward for improving the situation. Both aimed at introducing a residual compressive stress in the material near the inner surface. One was to build up the cylinder from two or more concentric tubes, each initially too small to slip over the next smaller one, and then to heat the larger until it had expanded sufficiently to do so; afterwards, as it cooled, the larger would press against the outer surface of the smaller, thus putting it into compression. The other method was to wind wire under high tension round a core tube to achieve the same result.

Both methods were used extensively in making guns, and to a lesser extent with industrial pressure vessels, and both have been modified recently. The use of components which have themselves been subjected to autofrettage has also been suggested by Wilson and Skelton.[3.1]

The main difficulty in appraising these various possibilities is to decide whether the design should correctly be based on shear or direct stresses. As we have seen in Chapter 2, there is little doubt that with high tensile, low alloy, steels – the class of material normally specified for high pressure vessels – shear stress criteria are the most satisfactory, both within the elastic range and also for partially overstrained conditions. On the other hand, with various types of compound construction, materials of much greater hardness, e.g. hard drawn wire for winding, or tungsten carbide liners, may be involved, and the design criteria will need some reconsideration. This applies particularly to a recently devised system of construction for ultra-high pressure service (i.e. above 20 kilobars) in which a ring of sector-shaped bars† is inserted in the wall between a core tube and an outer wall; the sectors are usually made of tungsten carbide since they are – in theory, at any rate – subjected only to compressive stresses.

Another system which has great theoretical attractions is that of internal jackets. Thus, instead of developing residual compressive stresses by the metal-to-metal contact pressure of surfaces, a narrow annular void is left into which is pumped fluid at a controlled pressure. In theory a system with a number of such concentric voids can be visualised, each with a different supporting pressure, but in practice the complications of design become very severe if more than one is introduced. Even so, however, this has the great merit of enabling the supporting pressure to be reduced as the working pressure is lowered. As will be seen, one of the limitations of shrink compounding is that the residual stresses tend to become bigger than the working stresses, which introduces a very definite pressure ceiling for this type of construction.

Most of the forms of reinforcement to be considered here for ultra-high pressure work involve very thick walls – overall diameter ratios of 10 or more – and this raises the question whether the ultimate strength of such cylinders is in reality so much greater than that of a simple cylinder of the same size, made from the best modern materials. Unfortunately, there is no certain answer to this question since it could only result from tests to destruction, and these would obviously be expensive, as well as difficult and potentially dangerous to carry out. Even so one would think they might be worth while (at any rate for thick simple cylinders) from the point of view of steel makers and forge masters, who might thereby be able to extend their products into much higher ranges of pressure.

3.2 Compounding by Shrinkage

The theory of this is complicated at the outset by the fact that it is virtually impossible to mate a pair of tubular components in this way without setting up stresses in the axial direction; and these stresses must depend on many factors, some of which will always be beyond our control. The usual practice is to assume that, until the internal working pressure is applied, there is no axial stress, even if this is obviously an over-simplification. It might be possible to achieve it with a duplex cylinder if one could subject the inner (cool) component to an axial force which would extend it by the same amount as the temperature had extended the length of the outer (hot) component, and then gradually to reduce this force as the two components approached the same temperature. The mating of such tubes is however a difficult operation anyway and one can hardly imagine such a complication being introduced by a manufacturer without adding a substantial amount to his price. In practice therefore we must accept an inevitable axial stress of unpredictable magnitude, and the only reasonable assumption about it is that it will generally be intermediate algebraically between the tangential and

† This is frequently referred to as "segmented" construction although the geometrical shape of the components is certainly a sector, and not a segment. Poulter's original patent for this design[3.2] uses the term "sector" correctly.

radial stresses, so that the maximum shear stress will still be controlled by these two.

Crossland and Burns[3.3] carried out some investigations with small duplex cylinders and showed that this last condition is justified, although axial stresses certainly were set up and these varied from specimen to specimen, as might be expected. They also showed that these stresses did not interfere to any serious extent with the theoretical treatment, although their presence in unknown intensity meant that it was virtually impossible to use any criterion in design which involved three-dimensional stresses, as for instance with Maxwell's hypothesis; on the other hand the Tresca criterion, being concerned only with stresses in the transverse plane, could be used, so in what follows we shall for the most part use the latter, and it should be remembered that it leads to a more conservative basis for design.

It may be worth while therefore to examine the simplest case of multiple compounding, viz. when each component is made of the same material and when the maximum shear stress in each has the same value while the working pressure is acting. Let there be n components, and let the diameter ratios of each be k_1, k_2 etc. up to k_n; also let the contact pressure between components 1 and 2 be ${}_1R_2$, that between 2 and 3 ${}_2R_3$ and so on, Fig. 3.1. Then, if the working pressure be p_i, the

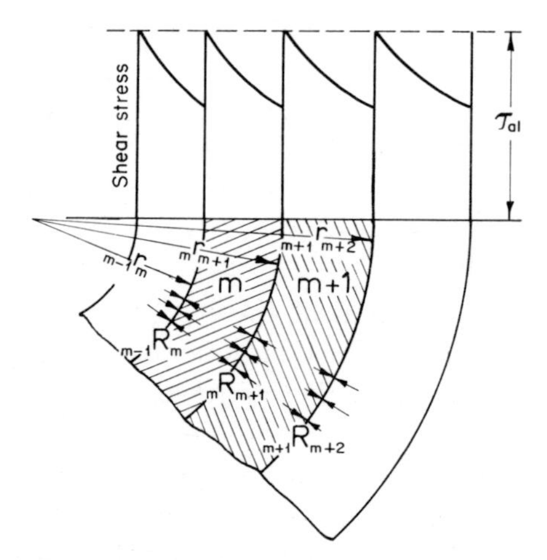

Figure 3.1 Diagram of stresses and contact pressures in multi-component shrunk construction cylinder

maximum shear stress τ, in the first will be given by:

$$\tau_1 = (p_i - {}_1R_2)\frac{k_1^2}{k_1^2-1} \tag{3.1}$$

and in the second:

$$\tau_2 = ({}_1R_2 - {}_2R_3)\frac{k_2^2}{k_2^2-1} \tag{3.2}$$

and for the last:

$$\tau_n = ({}_{n-1}R_n - 0)\frac{k_n^2}{k_n^2 - 1} \tag{3.3}$$

assuming there is no pressure on the outermost surface.

Now the basic assumption is that the shear stresses shall have the same maximum value in each component and that this shall equal the allowable working shear stress τ_{al}. Then:

$$\tau_1 = \tau_2 = \ldots = \tau_n = \tau_{al} \tag{3.4}$$

Now if we consider two adjacent components, say m and $m+1$, there will be three contact pressures involved, namely ${}_{m-1}R_m$, ${}_mR_{m+1}$, ${}_{m+1}R_{m+2}$, and two diameter ratios, namely k_m and k_{m+1}. If now this pair is considered separately, and if we write X for the product $k_m \times k_{m+1}$, then it becomes a simple matter to eliminate ${}_mR_{m+1}$ from the equation for the shear stress which by our initial condition must have the same maximum value in each. In fact we can use this value, which then is given by:

$$\tau_{al} = \frac{({}_{m-1}R_m - {}_{m+1}R_{m+2})X^2k_m^2}{(2X^2k_m^2 - X^2 - k_m^4)} \tag{3.5}$$

The condition that k_m is the optimum value for this pair (assuming that X, representing the combined diameter ratio of the two components, is fixed) can be obtained by differentiating eq. (3.5) with respect to k_m and equating to zero. This gives:

$$X = k_m^2 \tag{3.6}$$

which means that k_m and k_{m+1} must have the same value. Consequently the diameter ratio of each component must be the same, and if the overall diameter ratio of the assembly is K, then:

$$k_1 = k_2 = \ldots = k_n = K^{1/n} \tag{3.7}$$

or the contact radii increase in geometric progression. It is clear then, from equations (3.1) to (3.4), that if the diameter ratios of all the components are the same, and the maximum shear stress in each is the same, the difference between the contact pressures at the surfaces must be the same, i.e.

$$p_i - {}_1R_2 = {}_1R_2 - {}_2R_3 = \ldots = {}_{n-1}R_n - 0 \tag{3.8}$$

and the contact pressures while the working pressure is acting are therefore in arithmetic progression.

Combining these results leads to a simple result, the relation:

$$\tau = \frac{p_i}{n}\frac{K^{2/n}}{(K^{2/n} - 1)} \tag{3.9}$$

for the maximum shear stress in each component. If then we give this its maximum value, the maximum allowable pressure $(p_i)_{al}$ will be given by:

$$(p_i)_{al} = \tau_{al}\frac{n(K^{2/n}-1)}{K^{2/n}} \tag{3.10}$$

It is interesting to note that in the limit as the number of components becomes very large eq. (3.10) approaches the form:

$$(p_i)_{al} = \tau_{al} \ \ln K^2 \tag{3.11}$$

which is the relation for a cylinder with constant shear stress across its section (compare eq. (2.7b), p. 13).

In arriving at these relations we have ignored the residual stresses, although it is a fact that more than one attempt to obtain an extra strong cylinder by multiple shrinking has failed because of overstrain in the compounding opera-

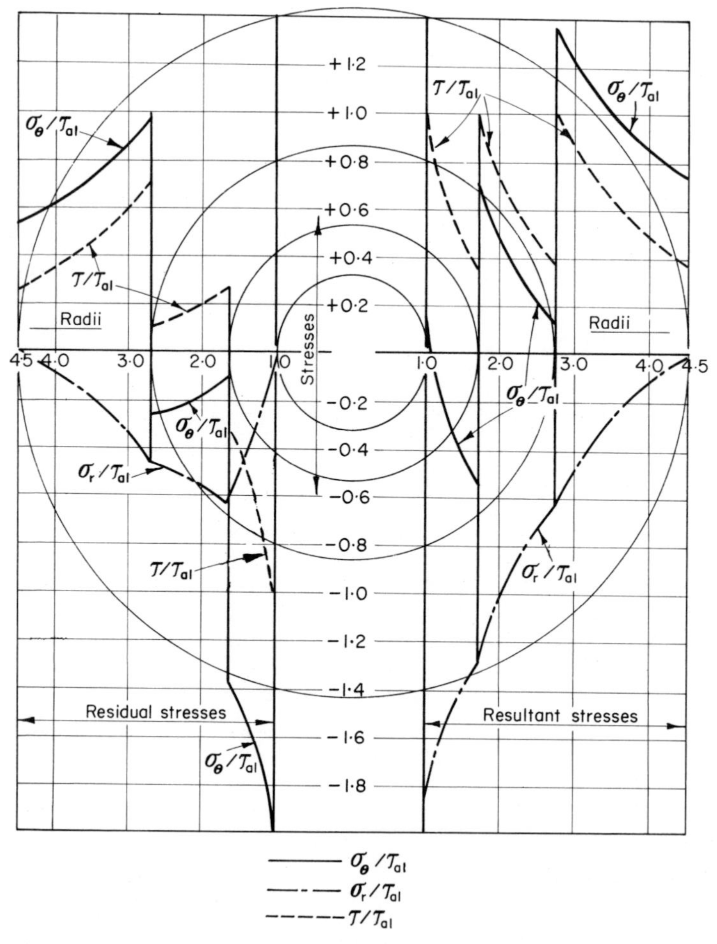

Figure 3.2 Stress distribution in ideal 3-component cylinder; under working conditions, and residual when working pressure is removed

tions. Fig. 3.2 shows the stress distribution in a 3-component, ideally mated, cylinder with an overall diameter ratio of 4·5, both with the internal pressure acting and when only the residual stresses are present. One sees at a glance that the most intense residual shear stress is that at the inside surface of the innermost component. In making these calculations we assume that the stresses are additive; thus, if the compound cylinder behaves elastically, applying or removing the working pressure will be the same as adding or subtracting the stresses which would be generated in a simple cylinder of the same overall diameter ratio, assuming that material of sufficient strength to carry these without overstrain could be found.

Thus the limiting case will be that in which the maximum shear stress is reversed as the internal pressure is applied or removed; and the shear stress in the simple cylinder will be:

$$\tau = p_i \frac{K^2}{K^2-1}$$

Thus the residual stress will be the difference between that and the value given by eq. (3.10). In the ideal limiting case this difference will be the same as the stress given in eq. (3.9), viz.

$$p_i \frac{K^2}{K^2-1} - p_i \frac{K^{2/n}}{n(K^{2/n}-1)} = p_i \frac{K^{2/n}}{n(K^{2/n}-1)}$$

or

$$\frac{n(K^{2/n}-1)}{K^{2/n}} \leqslant \frac{2(K^2-1)}{K^2} \tag{3.12}$$

which is the condition that the residual shear stress shall nowhere exceed its maximum value when the allowable working pressure is acting.

In Fig. 3.3 are plotted the family of curves obtained by putting various values of n into eq. (3.10), and the curve representing the right-hand side of eq. (3.12) is also drawn in. The area on the diagram above and to the left of this latter curve is shaded to indicate that anywhere in that region represents a position where unacceptable stresses will occur during the compounding operation.

It will be seen that this condition cannot arise with two-component systems, and that when there are three components the limiting condition almost exactly occurs when $K = 4{\cdot}5$; the ratio $(p_i)_{al}/\tau_{al}$ then has the value 1·90, which is the situation dealt with in Fig. 3.2. With a fourth component the overall diameter ratio at the optimum condition is reduced to approximately 3·4, and $(p_i)_{al}/\tau_{al}$ to 1·83.

The cost of these shrinking operations is considerable and it is unlikely that, in the restrictive assumption of this analysis (i.e. all components made from the same material), the use of more than four components could be economically justified.

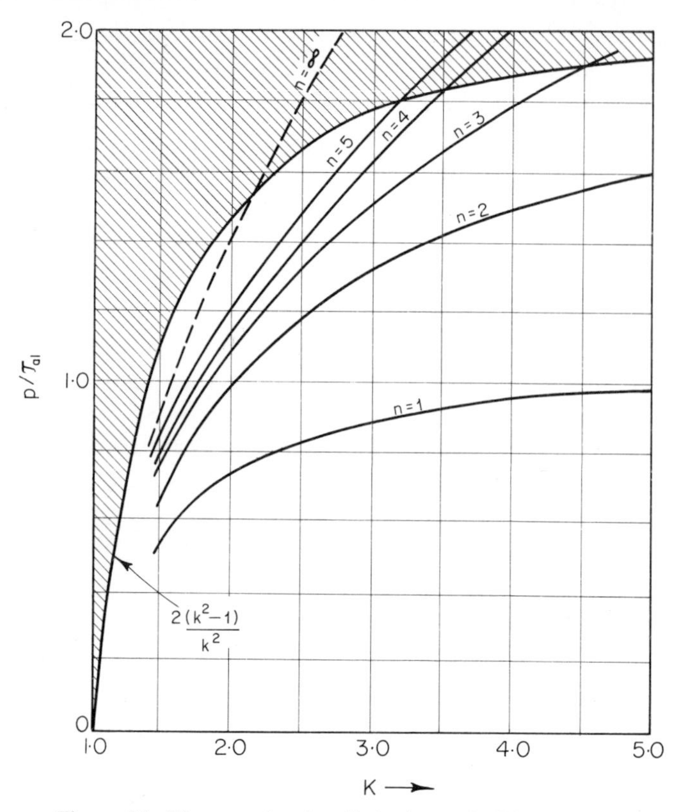

Figure 3.3 Diagram showing limitations of shrink construction; when the conditions lie in the shaded area, the residual stresses will be higher than the working stresses

There remains to derive an expression for the necessary shrinkages, which turns out to be remarkably simple when axial stresses are neglected. We know the contact pressures which operate when the working pressure is acting (according to the assumptions made above), and we can therefore evaluate the residual contact pressures by subtracting from them the radial stresses that would occur in the equivalent simple elastic cylinder at the corresponding radii if the same internal pressure were acting. Thus if ${}_1(\sigma_r)_2$ be the radial stress in the equivalent simple cylinder at the radius corresponding to the outside of the innermost component (let it be ${}_1r_2$) we have:

$$ {}_1(\sigma_r)_2 = -p_i \frac{K^2-k_1^2}{k_1^2(K^2-1)} \tag{3.13} $$

and consequently the residual contact pressure at this surface will be given by:

$$ {}_1(R)_{2(resid)} = {}_1R_2 - p_i \frac{K^2-k_1^2}{k_1^2(K^2-1)} \tag{3.14} $$

Now the amount of interference we must allow will be the amount by which the external diameter of the innermost tube is reduced by this external contact pressure plus the amount by which the second component expands as the result of the residual contact pressure between it and the innermost. This can be derived from a relation similar to that of eq. (2.27), except that here we are neglecting the axial stress, which has the effect of changing the factor applied to the first term from $(1-2v)$ to $(1-v)$. The actual algebra is given by one of the authors (W.R.D.M.)[3.4, 3.5] and will not be repeated here, but the result is surprisingly simple. Each pair of adjacent components requires the same degree of interference per unit of diameter; thus:

$$\frac{w}{r} = \frac{2p_i}{nE} \tag{3.15}$$

where w is the shrinkage between the components in contact at radius r.

Example. Consider a three-component cylinder of this kind with a bore of 6 in diameter made throughout from the steel used by Franklin and Morrison[2.13] in which the allowable shear stress is to be $\frac{2}{3}$ of the limiting value as determined in a torsion test, i.e. $4{,}095 \times \frac{2}{3}$ or 2,730 bars. The value of E for this is $2{\cdot}03 \times 10^6$ bars. Thus the maximum allowable pressure in the ideal three-component assembly with an overall diameter ratio of 4·5 is 1·90 times the allowable shear stress, or 5,190 bars. The shrinkage allowance w/r will then be given by:

$$\frac{2 \times 5{,}190}{3 \times 2{\cdot}03 \times 10^6} = 0{\cdot}00170 \text{ in/in}$$

Thus if the required bore diameter is 6 in, the diameter ratio of each component will be $\sqrt[3]{4{\cdot}5}$, or 1·65; and the outside diameters of the three components will be 9·900 in, 16·335 in, and 27·00 in, and if the inner tube is finished to an outside diameter of exactly 9·900 in the inside of the middle component should be bored to:

$$9{\cdot}900 - 9{\cdot}900 \times 0{\cdot}00170, \text{ or to } 9{\cdot}883 \text{ in}$$

By the same argument, if the outside of the middle component is finished to 16·335 in diameter, the bore of the outer component should be finished to:

$$16{\cdot}335 - 16{\cdot}335 \times 0{\cdot}00170, \text{ or to } 16{\cdot}307 \text{ in}$$

In the assembly, the procedure will normally be to mate the inner pair first, and the theoretical minimum temperature difference required, assuming a coefficient of linear expansion for this steel of $1{\cdot}20 \times 10^{-5}$ in/in deg C will only be about 150°C. In practice however it is advisable to make the temperature difference as large as one conveniently can to give the largest possible clearance. In this case the steel tempering temperature is around 600°C, so there would be no risk in heating it to say 500°C, at which temperature the clearance would be (assuming that the inner member remains at 20°C):

$$9{\cdot}883(1+480 \times 1{\cdot}2 \times 10^{-5}) - 9{\cdot}900 = 0{\cdot}040 \text{ in}$$

Thus it will be seen that shrinking is a decidedly delicate operation with only 40 thousandths of an inch clearance on about 10 inches.

The second shrinkage becomes even more tricky because the mating of the first two will have caused some increase in the outside diameter over which the next shrinkage operation has to be clear, and it may be of some use by way of example to calculate what we can expect the final diameter of the middle component to be, after it has been shrunk on to the innermost.

We assume that no axial stresses are produced by these shrinkage operations although we know that this is only approximately true; however, Crossland and Burns[3.3] have shown that such errors as this assumption may introduce are not likely to be serious. It is easiest to consider the problem in terms of the radial shift of the various surfaces. Thus, if ${}_1u_o$ be the radial shift of the outside of the innermost component, and ${}_2u_i$ that of the inside of the middle one, and if ${}_1r_2$ is the radius of contact (i.e. the outside of the innermost) then:

$$\frac{{}_1w_2}{{}_1r_2} = \frac{{}_2u_i - {}_1u_o}{{}_1r_2} \tag{3.16}$$

Now let ${}_1R'_2$ be the residual contact pressure caused only by the mating of these two. Then:

$$E\frac{{}_2u_i}{{}_1r_2} = \frac{{}_1R'_2}{k_2^2-1}(1-\nu)+\frac{{}_1R'_2k_2^2}{k_2^2-1}(1+\nu)$$

or

$$\frac{{}_2u_i}{{}_1r_2} = \frac{{}_1R'_2}{E(k_2^2-1)}\{(1-\nu)+k_2^2(1+\nu)\} \tag{3.17}$$

and

$$E\frac{{}_1u_o}{{}_1r_2} = -\frac{{}_1R'_2k_1^2}{k_1^2-1}(1-\nu)-\frac{{}_1R'_2}{k_1^2-1}(1+\nu)$$

or

$$\frac{{}_1u_o}{{}_1r_2} = -\frac{{}_1R'_2}{E(k_1^2-1)}\{k_1^2(1-\nu)+(1+\nu)\} \tag{3.18}$$

and since $k_1 = k_2 = k$ these equations can be combined thus:

$$\frac{{}_1w_2}{{}_1r_2} = \frac{{}_1R'_2}{E(k^2-1)}\{(1-\nu)+k^2(1+\nu)+k^2(1-\nu)+(1+\nu)\}$$

or

$$\frac{{}_1w_2}{{}_1r_2} = \frac{2\,{}_1R'_2(k^2+1)}{E(k^2-1)}$$

Whence:

$${}_1R'_2 = \frac{{}_1w_2 E(k^2-1)}{2\,{}_1r_2(k^2+1)} \tag{3.19}$$

but we can substitute for w/r from eq. (3.15) and:

$${}_1R'_2 = \frac{p_i}{n}\frac{k^2-1}{k^2+1} \tag{3.20}$$

Then the value of the radial shift at the outside of the middle component, i.e. at $r = {}_2r_3$, will be given by:

$$E\frac{{}_2u_o}{{}_2r_3} = \frac{2\,{}_1R'_2}{k^2-1} \tag{3.21}$$

or

$$\frac{{}_2u_o}{{}_2r_3} = \frac{2p_i}{nE(k^2+1)} \tag{3.22}$$

In this case $p_i = 5{,}170$ bars and $E = 2{\cdot}03 \times 10^6$ bars, $k = 1{\cdot}65$ and $n = 3$, whence:

$$\frac{{}_2u_o}{{}_2r_3} = \frac{2 \times 5{\cdot}17 \times 10^{-3}}{3 \times 2{\cdot}03 \times 3{\cdot}72} = 4{\cdot}56 \times 10^{-4} \text{ in/in}$$

and ${}_2r_3$ is initially $\frac{1}{2} \times 16{\cdot}335$ in, so that the diameter swelling will be:

$$4{\cdot}56 \times 16{\cdot}335 \times 10^{-4} = 0{\cdot}0074 \text{ in}$$

This represents the equivalent of only about an extra 40°C temperature difference.

3.3 Internal Liners

The use of liners of special corrosion or wear resisting material in cylinders is of considerable importance in practice. Usually the object is to reduce the cost by using only a relatively thin liner of what may be very expensive material, leaving the backing cylinder or mantle, which may itself be of compound construction, to resist the pressure stresses.

One particular example of this, which brings with it some rather special problems of its own, is the use of tungsten carbide liners in the cylinders of pumps and compressors. The main object of this is to provide a very hard and wear-resistant surface against which moving seals such as piston rings can rub for long periods without needing renewal; and it is of course much cheaper to renew a liner than the whole cylinder when this does eventually become necessary. Tungsten carbide is in some ways a very well adapted material for this purpose because, in addition to its extreme hardness, it has a low coefficient of thermal expansion and a very high elastic modulus. The first enables the liner to be withdrawn merely by heating up the assembly, and the second enables the range of pressure endured by the mantle to be reduced, with consequent increase in fatigue resistance; it also decreases the range of strain in the bore during each stroke of the machine, thereby increasing the life of the piston rings. Against this however must be set the disadvantage that the material is utterly unreliable in tension and it is necessary to ensure that the stresses in the liner are always compressive. In spite of this the liner can absorb a considerable pressure difference, as can be seen from the following considerations. The position in which tensile stresses are most likely to arise is, of course, the inner surface in the tangential direction, and at the time when the internal pressure is at a maximum. If this is denoted by p_i as before, and if R_w is the contact pressure between the liner and the mantle while the internal pressure is acting and k_L is the diameter ratio of the liner, then according to eq. (2.23):

$$\sigma_\theta = \frac{-R_w k_L^2 + p_i}{k_L^2 - 1} - \frac{(R_w - p_i)k_L^2}{k_L^2 - 1} \tag{3.23}$$

and this must never be positive—or in the limiting case the right-hand side can be zero. Thus:

$$R_w \geqslant p_i \frac{k_L^2 + 1}{2k_L^2} \tag{3.24}$$

Liners of this sort are seldom less than $\frac{1}{2}$ in thick, since otherwise their manufacture is difficult; a possible size that has been used successfully is approximately 4 in bore by 5 in outside diameter, working up to a maximum of 3,000 bars. Then $k_L = 1{\cdot}25$, and R_w is about 2,500 bars, or a reduction of 16% in the pressure to be resisted by the mantle.

The calculation of interferences, etc. becomes considerably complicated by the differences in the moduli and Poisson's Ratio values, but in view of its importance in design the theory is included here.

We have the condition that, when the working pressure is removed, the change of radial shift at the outside of the liner must be the same as that at the inside of mantle. So let the radial shifts of the liner be u_w and u_R respectively when there is an internal working pressure acting and when there is only the residual contact pressure; and let v_w and v_R be the corresponding radial shifts in the mantle. We can then write down the relations for these changes as follows, bearing in mind that we are *again neglecting axial stresses*; thus if E_L and E_M, and ν_L and ν_M are the values of Young's Modulus and Poisson's Ratio for the liner and mantle materials respectively, and if k_M is the diameter ratio of the mantle, we have:

$$E_L \frac{u_w}{r_c} = \frac{-R_w k_L^2 + p_i}{k_L^2 - 1}(1-\nu_L) - \frac{(R_w - p_i)}{k_L^2 - 1}(1+\nu_L) \tag{3.25}$$

$$E_L \frac{u_R}{r_c} = \frac{-R_R k_L^2}{k_L^2 - 1}(1-\nu_L) - \frac{R_R}{k_L^2 - 1}(1+\nu_L) \tag{3.26}$$

$$E_M \frac{v_w}{r_c} = \frac{R_w}{k_M^2 - 1}(1-\nu_M) + \frac{R_w k_M^2}{k_M^2 - 1}(1+\nu_M) \tag{3.27}$$

$$E_M \frac{v_R}{r_c} = \frac{R_R}{k_M^2 - 1}(1-\nu_M) + \frac{R_R k_M^2}{k_M^2 - 1}(1+\nu_M) \tag{3.28}$$

where r_c is the radius of contact, and R_R is the contact pressure due only to residual shrinkage stresses when no working pressure is acting. Now the condition for continuity of strains is:

$$u_w - u_R = v_w - v_R \tag{3.29}$$

and therefore

$$\frac{r_c}{E_L}\left\{\frac{p_i - (R_w - R_R)k_L^2}{k_L^2 - 1}(1-\nu_L) + \frac{p_i - (R_w - R_R)}{k_L^2 - 1}(1+\nu_L)\right\} =$$

$$\frac{r_c}{E_M}\left\{\frac{(R_w - R_R)}{k_M^2 - 1}(1-\nu_M) + \frac{(R_w - R_R)k_M^2}{k_M^2 - 1}(1+\nu_M)\right\}$$

or

$$\frac{2p_i}{E_L(k_L^2 - 1)} = (R_w - R_R)\left\{\frac{k_L^2(1-\nu_L) + (1+\nu_L)}{E_L(k_L^2 - 1)} + \frac{k_M^2(1+\nu_M) + (1-\nu_M)}{E_M(k_M^2 - 1)}\right\} \tag{3.30}$$

Now if we write:

$$C_1 = k_L^2(1-\nu_L)+(1+\nu_L) \;:\; C_3 = k_M^2(1+\nu_M)+(1-\nu_M)$$
$$C_2 = 1/E_L(k_L^2-1) \;:\; C_4 = 1/E_M(k_M^2-1)$$

eq. (3.30) simplifies to:

$$R_w - R_R = p_i \frac{2C_2}{C_1C_2+C_3C_4} \tag{3.31}$$

As an example we may consider a type of compressor offered some years ago by the American firm of Clark Bros. of Olean, N.Y., in which gas was compressed in one stage from about 250 to 2,400 bars, an unusually large ratio by comparison with more recent practice. The cylinders were of nickel-chrome-vanadium-molybdenum steel, treated by autofrettage, and subsequently fitted with liners of tungsten carbide. The steel cylinders were 12 in outside diameter by 3 in bore and the liners were of $2\frac{1}{8}$ in bore, i.e. a diameter ratio of 1·41. The contact pressure would have therefore to be not less than $0{\cdot}75p_i$ to prevent tensile stresses appearing in the tungsten carbide—see eq. (3.24).

For tungsten carbide Young's Modulus is about $6{\cdot}0 \times 10^6$ bars and Poisson's Ratio 0·25; the corresponding figures for the steel are likely to be, say $2{\cdot}0 \times 10^6$ and 0·29. Substituting in these figures, together with 1·41 for k_L and 4·0 for k_M, we get:

$$C_1 = 2{\cdot}75$$
$$C_2 = 16{\cdot}67 \times 10^{-8}$$
$$C_3 = 21{\cdot}35$$
$$C_4 = 3{\cdot}33 \times 10^{-8}$$

Thus

$$R_w - R_R = p_i \frac{2 \times 16{\cdot}67 \times 10^{-8}}{2{\cdot}75 \times 16{\cdot}67 \times 10^{-8} + 21{\cdot}35 \times 3{\cdot}33 \times 10^{-8}}$$

$$= 0{\cdot}284\; p_i = 682 \text{ bars}$$

and, since $R_w = 0{\cdot}75\; p_i = 1{,}800$ bars, R_R must be 1,118 bars. We can neglect the inlet pressure in this case and regard the pressure range per stroke as 2,400 bars. The range of pressure through which the bore of the steel cylinder is loaded per stroke is however only $R_w - R_R$ or 682 bars, which is less than $\frac{1}{3}$ the range a similar cylinder without tungsten carbide liners would be required to undergo. As will be seen in Chapter 6, this will have a very important effect in enhancing the resistance to fatigue.

The shrinkage required is:

$$\frac{u_R - v_R}{r_c}$$

which we see from equations (3.26) and (3.27), after substituting the constants to simplify the arithmetic, gives:

$$\frac{u_R - v_R}{r_c} = R_R(C_1C_2 + C_3C_4) \tag{3.32}$$

This gives a value of approximately $1{\cdot}31 \times 10^{-3}$ in/in, which on a diameter of 3 in means an allowance of about 0·004 in. In practice the shrinkage is probably made greater to ensure that there are no tensile stresses in the liner. A fortunate circumstance with tungsten carbide is that its coefficient of expansion is much less than that of steel (about 7×10^{-6} per °C compared with 12×10^{-6}) and there is therefore a differential expansion of 5×10^{-6} in/in deg C. This means that, provided the shrinkage requirement can be kept below about 2×10^{-3} in/in, a worn or damaged liner can usually be removed from its cylinder merely by heating the assembly through about 450°C, when it will drop out. In the case considered above it would seem reasonable to use a shrinkage of say 1·5 to $1{\cdot}6 \times 10^{-3}$ in/in for this operation.

3.4 Corrosion-Resistant Liners

These are usually made from expensive alloys and for that reason they are thin and not expected to carry any appreciable share of the resistance to pressure. The shrinking is merely to hold the liner in position and can often be applied by cooling by, say, solid CO_2. This is often made easier by the fact that the corrosion-resistant materials generally have considerably greater coefficients of expansion than the carbon or high tensile steels normally used for pressure containers. Only very slight interference is necessary since the pressure will in due course force the liner out at any place where it is not fully supported.

The difference in expansion coefficients between the materials of the liner and the container can sometimes lead to trouble in an assembly that operates over a considerable range of temperature, since heating will cause the contact pressure to be increased; and if the heating up is done before the working pressure is applied, there may be a risk of collapsing the liner owing to instabilities produced by the contact pressure on its outer surface.

The behaviour of thin cylinders subjected to external pressure is difficult to predict accurately, but Southwell[3.6] has suggested the following formula which seems to give reasonably satisfactory results. If the wall thickness be t and the mean radius r, and if L be the length between any stiffening devices such as flanges or internal support rings which hold the section circular, then:

$$p_o^* = (Et/r)\left\{\frac{n^2-1}{12(1-\nu^2)}\left(\frac{t}{r}\right)^2 + \frac{\pi^4}{n^4(n^2-1)}\left(\frac{r}{L}\right)^4\right\} \tag{3.33}$$

where p_o^* is the collapse pressure and n is a positive integer greater than one.

Collapse under these circumstances takes the form of inward bulges or nodes, and n is the number of these in the section. Some question may arise in the case of a liner as to whether the embracing container, by preventing any deformation outward, may not also prevent any inward bulging, but experience suggests that collapse inwards does sometimes occur and it will in general be wisest to take this into account by means of the Southwell formula or some similar relation.

As these liners are always likely to be thin (those that are not will be in no danger of this mode of failure) the stress in the tangential direction can be regarded as constant, so that the simple relation:

$$\sigma_\theta = \frac{p_o r_o}{t} = \frac{p_o r_o}{r_o - r_i} \simeq \frac{p_o r_i}{r_o - r_i} \tag{3.34}$$

will be sufficiently approximate. Then if D_o be the outside diameter and ΔD_o the change in it due to this external pressure:

$$\frac{\Delta D_o}{D_o} = \varepsilon_\theta = \frac{\sigma_\theta}{E_L}$$

and this will also be the change in diameter brought about by the differential expansion, i.e.

$$\varepsilon_\theta = (\alpha_L - \alpha_M)\Theta$$

where α_L and α_M are the coefficients of expansion of the liner and mantle materials respectively and Θ is the range of temperature. Then in the case where the resulting pressure is approaching the collapse pressure we shall have:

$$p_o^* = E_L \frac{r_o - r_i}{r_i}(\alpha_L - \alpha_M)\Theta \tag{3.35}$$

Referring back now to eq. (3.33), the second term in the brackets can usually be disregarded in most of the examples we are here concerned with, and then we need only consider the smallest value of n, i.e. 2; eq. (3.33) is thus reduced to:

$$p_o^* = \frac{E_L}{4(1-\nu_L^2)}\left(\frac{r_o - r_i}{r_i}\right)^3 \tag{3.36}$$

which can be combined with eq. (3.35) to show that the maximum safe temperature difference (on the basis of these considerations) is given by:

$$\Theta = \frac{\left(\frac{r_o - r_i}{r_i}\right)^2}{4(1-\nu_L^2)(\alpha_L - \alpha_M)} \tag{3.37}$$

Thus, if one had a stainless steel liner $\frac{1}{8}$ in thick to fit into a 12 in diameter steel vessel, and if the liner's material had a coefficient of expansion of 18×10^{-6} and a Poisson's Ratio value of 0·30, compared with a coefficient of 12×10^{-6} for the steel of the vessel, the critical temperature difference would work out at only 20°C. There is probably a considerably greater margin of safety than this in practice, but even so it would be safer to use rather thicker material for the liner, and attention should always be given to the possibility of trouble with thin liners rucking in service due to this cause.

3.5 "Multi-Layer" Construction

A method of building up the thick walls of pressure vessels was devised by the A.O. Smith Corporation of Milwaukee, U.S.A.,† during the 1930's. It consisted in wrapping a series of sheets of relatively thin metal tightly round one another over a core tube and holding each with a longitudinal weld as shown in Fig. 3.4. The length which could thus be made was limited to the capacity of the wrapping machine, but longer vessels were obtained by welding together two or more such lengths, with circumferential welds going right through the whole thickness of the wall thus built up. Heavy forged ends could also be attached by similar methods.

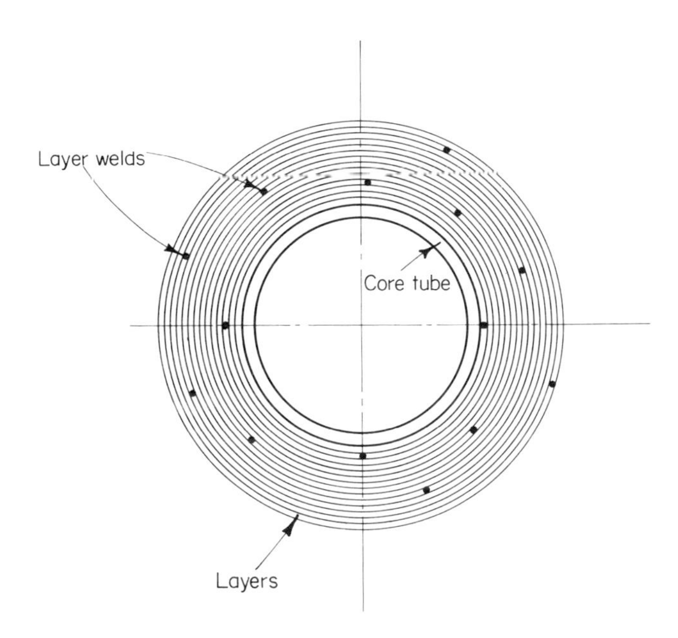

Figure 3.4 Multi-layer construction; section through cylinder

A description of the process and the results of some tests are given by Jasper and Scudder.[3.7] It would appear from this that the A.O. Smith Corporation's original object was to produce a thick wall from much thinner material, thereby easing the task of ensuring sound material thoughout; it was only later, thanks mainly to Jasper's work, that these vessels were found to have greater elastic strength than was expected from a similar vessel made in one piece. Later, Manning[3.4, 3.5] showed that cylinders made in this way did approach quite closely to what would be expected from a compound cylinder made up from a large number of concentric tubes in accordance with eq. (3.10), the shrinkages being obtained by the contraction of the longitudinal welded seams. Table 3.1 gives the details of the two cylinders which were tested to destruction, and it will

† The U.S. manufacture by this process is now operated by the Struthers Wells Corpn. of Titusville, Pa.

be noted that in each case the elastic strength of the "Multi-Layer" vessel is appreciably greater than that of a similar solid-walled cylinder.

The difference in the case of Cylinder A was over 30%. Cylinder B on the other hand was only about $13\frac{1}{2}$% better, the difference evidently being that the annealing at 620°C had reduced the internal supporting pressures between the layers, and it is difficult to see what value this operation could possibly have.

Table 3.1

(Experimental figures from Jasper and Scudder)[3.7]

Vessel designation	*A*	*B*
Diameter of bore Diameter of outside Core tube thickness No. of wrappings Thickness of wrappings	19 in 26 in $\frac{1}{2}$ in 12 $\frac{1}{4}$ in	19 in 26 in $\frac{1}{2}$ in 12 $\frac{1}{4}$ in
Stress-relieving temperature	not treated	620°C
0·2% proof stress of material (tensile)	2,440 bars (39,770 lbf/in^2)	2,600 bars (37,670 lbf/in^2)
Actual yield pressure	925 bars	793 bars
Calculated yield pressure for monobloc cylinder	704 bars	698 bars
Calculated yield pressure for actual cylinder assuming perfect mating (eq. (3.10))	976 bars	928 bars
Actual Ultimate Bursting Pressure	1,295 bars	1,200 bars
Calculated Ultimate Bursting Pressure	1,310 bars	1,310 bars

It will also be noticed that the Ultimate Bursting Pressure in each case is not greatly different from that to be expected with a monobloc cylinder of the same overall dimensions. Once again however the annealing or stress-relieving treatment is shown to have had a deleterious effect.

The agreement between the actual and theoretical bursting pressures in the case of Cylinder A is remarkable, the difference being only 1% compared with a one-piece cylinder; and even the annealing operation only raises the difference to about 8%, a comparison which takes no account of the reduction in strength which this treatment would cause in the material of the one-piece cylinder.

The fact that the compounding operation appears to have virtually no effect on the Ultimate Bursting Pressure is at first sight surprising although Crossland and Burns[3.3] demonstrated it in their experiments. The explanation is no doubt to be found in the very large strains which occur in a test to destruction of a cylinder made from really ductile material, and which are so much larger than any set up by the compounding procedures. Thus we may conclude that compounding, even in a multi-layer cylinder, does indeed raise the elastic range to a marked extent, but it has virtually no effect on the bursting strength.

In the earlier examples of this type of construction some failures were reported in the big circumferential welds used to join the cylindrical pieces into longer units and to secure the forged ends. More recently some of the German manufacturers who have used modifications of the multi-layer or laminar form of construction have introduced stepped rings as shown diagrammatically in Fig. 3.5, in order to separate the junctions of the various layers, see Werzner.[3.20]

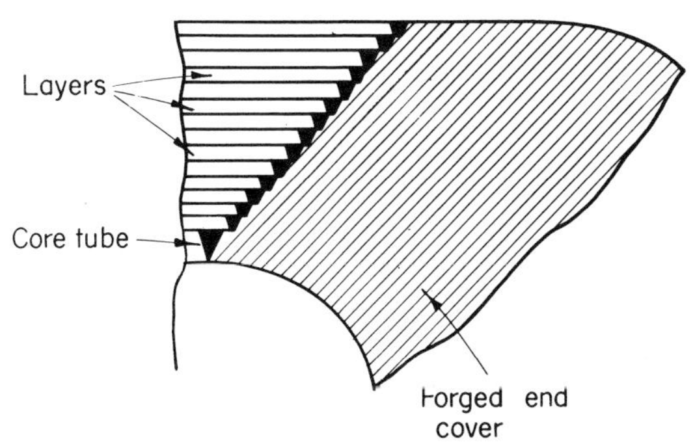

Figure 3.5 End connection for multi-layer vessel
(Fried. Krupp GmbH, Essen, Germany)

In this way they evidently avoid the main difficulty although at the expense of extra machining. Developments in welding thick metal, the electro-slag process for instance, may however favour a return to the simpler procedure, more or less as originally practised.

3.6 Effect of Errors in Shrinkage

We have seen that, when the number of components is fairly large – as in the vessels to which Table 3.1 refers – even considerable deviations from the ideal requirements stated in Section 3.2 make comparatively little difference. In those cases the shrinkages were quite impossible to determine from the data given by Jasper and Scudder[3.7], and there was considerable departure from the condition of geometric similarity (i.e. diameter ratios the same); all the shrunk-on layers were of the same thickness and the core tube had a ratio more than twice that of the adjoining layer. Even so, however, with the cylinder which was not annealed after fabrication, the difference between the ideal and the actual elastic strength was only a little over 5 %.

The ideally required shrinkage depends in direct ratio on the number of components – see eq. (3.15) – and thus one might expect the matter to be more critical with fewer components. In Ref. 3.4 the case is considered of a three-component cylinder with a bore of 6 in diameter designed to carry a working pressure of 25 tonf/in^2 (3,550 bars) with a maximum shear stress not exceeding $14\frac{1}{4}$ tonf/in^2 (2,200 bars). The required overall diameter ratio was given as 3·72 and the external diameters of the components were as follows:

Innermost	9·30 in
Middle	14·42 in
Outermost	22·33 in

It was then assumed that the workshop in which the machining operations were to be carried out could not guarantee an accuracy better than $\pm 0{\cdot}001$ in on any of the diameters, external or internal, a not unreasonable tolerance requirement.

Table 3.2

Combination	*Shrinkage between inner and middle*	*Shrinkage between middle and outer*
A	High	High
B	High	Low
C	Low	High
D	Low	Low

Thus there could be four extreme conditions as shown in Table 3.2 and the contact pressures and resulting shear stresses under the operating pressure were worked out for each case. It was found that the worst combination occurred with B when the stress in the middle tube rose nearly 12% above the stipulated maximum. This shows the importance of accuracy in these machining operations and of the gauging and inspection of the finished surfaces.

Another matter of great importance in the preparation of component cylinders for shrinking together is the constancy of diameter. Even a very small degree of taper may cause the relative movement of the components in service, especially if this involves pressure cycling. It is probably in part due to the longitudinal stresses caused by the original shrinkage, but—as has been stated above—the elimination of these is almost impossible.

It should also be emphasised that the operation of assembling by shrinkage is difficult and can be dangerous in inexperienced hands. In particular, oil or grease on the cool surface can be compressed and ignited by the hot surface, causing a decidedly unpleasant explosion.

In some cases, if the interference is large or the material such that its temperature could not be raised sufficiently to provide an adequate clearance, a slight taper may be provided deliberately and the contact pressures generated by forcing one component into the other. Again there is the risk of movement of one component relative to the other and cases have been reported where the inner cylinder has been violently ejected. This is particularly liable to occur if the mating is done cold using, say, the ram of a hydraulic press to provide the force, and if heavy grease is used as a lubricant. Bett[3.8] reports that he has found carbon tetrachloride very effective for this purpose. It is a fairly good lubricant under high contact pressure, but it also appears to have some mildly corrosive action on the surface of most steels which, after a few minutes, is sufficient to increase the friction very greatly and so fix the two components together.

Another procedure which can be used with small assemblies is to force one part into the other without making use of any temperature difference. In this it is advisable to draw the inner component in from the front end rather than push it in from the back, so that the driving force tends to reduce the shrinkage rather than to increase it. The effect is not large, but the situation is generally of the kind where every little helps. A short taper at one end of each cylinder is usually necessary to give the inner a lead into the outer, and if this procedure is adopted for more than two components it is advisable to begin with the outermost pair and draw in the smaller ones in sequence. But with this method of assembly the accuracy of the diameters is even more important and it is recommended that the surfaces be finished by honing.

Where the components are made from materials with markedly different coefficients of expansion, it is important that the final measurement of the diameters be carried out when the parts are at substantially the same temperature. For instance, most of the austenitic stainless steels have a coefficient of about $1{\cdot}8 \times 10^{-5}$ per degree C, so that if one is working to limits of 0·0001 in a mere 5° or 6° change in temperature would produce a significant effect on component dimensions.

3.7 Strip and Wire Winding

The idea of reinforcing a tube for carrying high pressure by winding on wire under tension had been discussed by ordnance designers from about 1850 onwards. In Britain it was enthusiastically championed by J. A. Longridge who wrote an extensive treatise on the subject, published in 1884[3.9], and big guns incorporating the idea became more or less standard in many of the world's arsenals from about 1900 onwards – at least until the end of World War I.

The application to pressure vessels was described by Newitt in 1936.[3.19] The main trouble in the application to guns had been the weakness in the longitudinal direction, which became serious as the muzzle velocities and barrel lengths increased, causing a marked deterioration in the accuracy of their shooting, but with static vessels the end forces could be taken by external means. Various designs for doing this were suggested and tried during and after World War II – see, for instance, a method involving portal frames used by Birchall and Lake[3.10] but probably the neatest was a method devised by J. Schierenbeck in Germany[3.11] and first successfully used for a large pressure vessel in 1938. This consisted in using an interlocking strip, as shown in Fig. 3.6, instead of wire

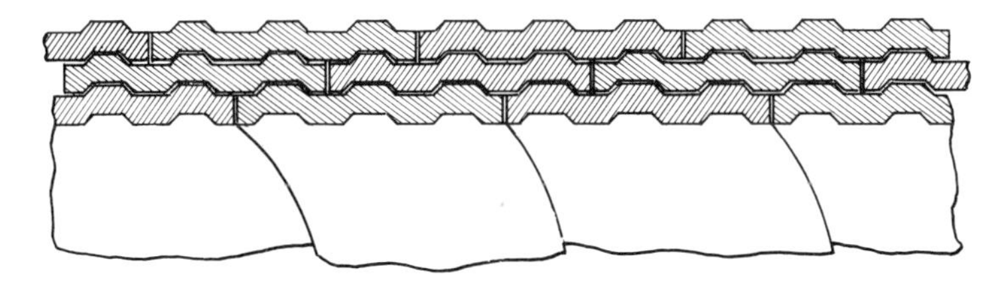

Figure 3.6 Interlocking strip winding *(Schierenbeck)*

or flat ribbon. The method has been applied to very large vessels (with cubic capacities of more than 20 m^3) and for pressures up to 700 bars, and even higher pressure units have been tested on a pilot-plant scale. The method was certainly used extensively by the Germans during the second world war primarily for hydrogenation plants, and also in plastics and synthetic ammonia production, and a number of vessels have since been made by a similar process in the Newcastle upon Tyne (Elswick) Works of Messrs. Vickers Ltd. Further developments have been introduced both here, in Germany, and elsewhere, as described in various articles (see especially Ref. 3.12).

This method retained the basic advantages of strip or wire winding, namely the use of material of small section whose mechanical properties could be treated to give exceptional strength as well as allowing ease and certainty of inspection and quality control, and also of eliminating the difficulty of avoiding errors in shrinkage, while largely overcoming the inherent shortcoming inevitable with plain strip or wire, namely the longitudinal weakness. Also, in the years following World War II, when large guns were no longer so much in demand, this provided useful work for very large gun lathes and boring equipment. Vickers, in fact, made most of their strip-wound vessels on a converted gun lathe, see Fig. 3.8.

There are several limitations to the use of this method of reinforcement, but it is potentially very valuable, especially in cases where the overall diameter ratio must not exceed say 2·5. A brief consideration of the theoretically ideal solution for any particular set of actual conditions may therefore be worth while.

The main advantage is that the strip can be of hard-drawn material, allowing a considerably higher tensile working stress than would be acceptable in, for instance, a large forging. By using suitably chosen winding-on tensions for each layer it is theoretically possible to construct an assembly in which the windings all carry the same tangential stress when the internal pressure is acting, and this stress can be the highest allowable (here denoted σ_{al}).

Essentially the process must be regarded as a means of reinforcing a core tube and this must be capable of taking a considerable proportion of the working pressure, as well as resisting the high external pressure of the windings as residual stresses when the pressure is removed from the bore. It must therefore be designed on a shear stress basis, as described in Chapter 2, with the maximum value denoted τ_{al}.

Fig. 3.7a shows the tangential, radial, and shear stresses in the transverse plane in an ideally constructed cylinder of overall diameter ratio 2, when it is carrying an internal pressure of 6,000 bars, and when the shear stress in the core is allowed to reach 4,000 bars and the tension in the windings 8,500 bars. The diameter ratio of the core tube is then 1·33, and that of the windings 1·5.

It has been assumed throughout these considerations that the stress in the axial direction σ_z is intermediate in value between the tangential stress σ_θ and the radial stress σ_r, so that the maximum shear stress will occur in the transverse plane. Its value will however usually be indeterminate because of uncertainties as to the value of σ_z, and this means that the Maxwell criterion cannot be used: on the other hand there will be enough information available for the maximum shear stress, or Tresca, criterion and what follows is therefore based on this.

Considering now the stresses in the core tube when the working pressure p_i is acting, we have:

$$(p_i - R_w)\frac{k_c^2}{k_c^2 - 1} = \tau_{al} \tag{3.38}$$

where R_w is the contact pressure of the windings on the outside of the core tube whose outside and inside radii are respectively r_2 and r_1, and where $r_2/r_1 = k_c$. Similarly, when the working pressure is removed, the shear stress in the core tube will be exactly reversed in an ideal arrangement, so that:

$$(0-R_R)\frac{k_c^2}{k_c^2-1} = -\tau_{al} \tag{3.39}$$

where R_R is the residual contact pressure of the windings.

Now if the assembly can be considered as wholly elastic, the difference between R_w and R_R will be equal to the radial stress at $r = r_2$ in an equivalent elastic simple cylinder of the same overall dimensions, when carrying the same internal pressure, i.e.:

$$R_w - R_R = p_i\frac{(r_3^2-r_2^2)r_1^2}{(r_3^2-r_1^2)r_2^2} \tag{3.40}$$

where r_3 is the external radius both of the strip wound assembly and of the equivalent elastic cylinder. It should be noted that the sign has been changed to allow for the fact that p_i, R_w and R_R are all pressures and not stresses.

In the windings, while the internal pressure is acting, the tangential stress is, by our ideal condition, constant and equal to σ_{al}, and thus:

$$R_w r_2 = \sigma_{al}(r_3-r_2)$$

or

$$R_w = \sigma_{al}(k_s-1) \tag{3.41}$$

where $k_s = r_3/r_2$, the diameter ratio of the windings.

It is also worth noting that the shear stress in the strip zone τ_s will be given by:

$$\tau_s = \tfrac{1}{2}(\sigma_\theta-\sigma_r)$$

and that this will have its maximum value at $r = r_2$, when $\sigma_\theta = \sigma_{al}$ and $\sigma_r = -R_w$. On substituting from eq. (3.41) we see that:

$$\tau_{s\,max} = \tfrac{1}{2}k_s\sigma_{al} \tag{3.42}$$

so that if the diameter ratio of the strip windings exceeds 2·0, the shear stress in the inner layers will be greater than the maximum allowable tensile stress, and this state of affairs could hardly be accepted.

We can now deduce certain relations from the conditions represented by the above equations. Thus by adding equations (3.38) and (3.39), we get:

$$p_i = R_w + R_R \tag{3.43}$$

Then, if we write K for the overall diameter ratio, so that:

$$K = k_s \times k_c$$

by substitution into eq. (3.40) we get:

$$R_w - R_R = p_i \frac{k_s^2 - 1}{K^2 - 1} \tag{3.44}$$

which, together with eq. (3.43) above, gives:

$$R_R = \frac{p_i}{2} \frac{K^2 - k_s^2}{K^2 - 1}$$

Then eq. (3.39) can be written in terms of K and k_s thus:

$$R_R = \tau_{al}(K^2 - k_s^2)/K^2 \tag{3.45}$$

from which we see that:

$$K^2 = \frac{2\tau_{al}}{2\tau_{al} - p_i} \tag{3.46}$$

thus showing that the overall diameter ratio is controlled only by the required working pressure and by the allowable shear stress in the core tube.

Then, substituting for R_w from eq. (3.41), for R_R from eq. (3.45), and for K^2 from eq. (3.46) into eq. (3.43) gives a quadratic equation for k_s, the solution of which is:

$$k_s = \frac{\sigma_{al} \pm \sqrt{\{\sigma_{al}^2 - 2(2\tau_{al} - p_i)(\sigma_{al} + p_i - \tau_{al})\}}}{(2\tau_{al} - p_i)} \tag{3.47}$$

This will evidently be satisfied only by a limited range of values of p_i, τ_{al}, and σ_{al}, since k_s must obviously be greater than unity and less than K; also it must be less than 2 if we are to avoid having a shear stress greater than the biggest direct stress. The following example shows how this theory can be used.

Example. It is required to contain a pressure of 6,000 bars in a cylinder made from material for which the maximum allowable shear stress is 4,000 bars, reinforced with windings of strip which can carry a maximum tensile stress of 8,500 bars. What will be the required section if the bore diameter is to be 12 in?

Substituting 6,000 for p_i and 4,000 for τ_{al} into eq. (3.46) gives

$$K^2 = \frac{2 \times 4{,}000}{2 \times 4{,}000 - 6{,}000} = 4{\cdot}0$$

and consequently

$$K = 2{\cdot}0$$

Similarly, from eq. (3.47):

$$k_s = \frac{8{,}500 \pm \sqrt{\{(8{,}500)^2 - 2(8{,}000-6{,}000)(8{,}500+6{,}000-4{,}000)\}}}{(8{,}000-6{,}000)}$$

and

$$k_s = 1{\cdot}5$$

consequently

$$k_c = 1{\cdot}33$$

Thus the core tube must have an external diameter of 16 in and the windings an external diameter of 24 in, and the cross-sectional area of the parts will be 251 and 88 in^2 for the windings and core tube respectively. Since the weights will be proportional to the cross-sectional areas, the finished cylinder will contain about 75% of its weight in windings and the remaining 25% in the core. It is worth noting that, if a strip capable of working at a stress of 10,000 bars could have been used, k_s would have come down to 1·4, but the overall diameter ratio would remain the same.

The stresses in the walls while the internal pressure is acting, in the case of the windings stressed to 8,500 bars, are shown in Fig. 3.7a, and those remaining after the pressure has been blown down in Fig. 3.7b. As will be seen, the windings cause

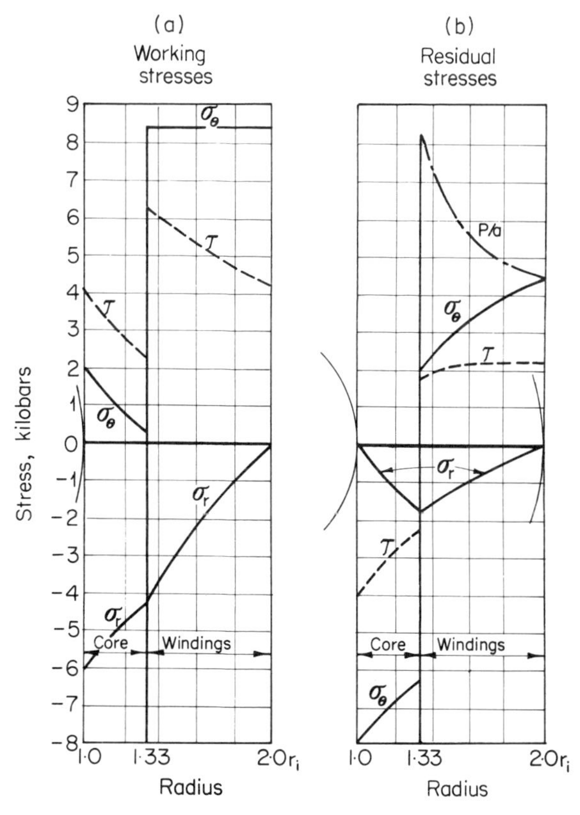

Figure 3.7 Stresses in ideal strip or wire wound vessel; (a) pressure acting (b) residual

high compression stresses in the core tube, and the resulting shear stress is equal in intensity (and opposite in sign) to that produced by the working pressure.

It may be of interest to compare this example with what would be required for similar conditions if the reinforcement was to be achieved by means of multiple shrink compounding. The ratio of pressure to allowable shear stress is 1·5, and we can see at a glance, using Fig. 3.3, that in a two-component assembly the necessary diameter ratio would be 4·0, and even with four components it would still have to be about 2·8. There is therefore a very considerable saving of weight when the strip reinforcement is used.

There remains to determine the exact winding tension required for each layer, remembering that the subsequent layers will reduce it considerably. Moreover, since the winding operation cannot be carried out while the core tube is already carrying the working pressure, the pattern of stress we have to set up is that of the residual stresses shown in Fig. 3.7b. The easiest way to look at the problem is to imagine that it has already been solved and that under the working pressure the pattern is as we require it, i.e. as shown in Fig. 3.7a. Then the necessary residual stress system is found by subtracting the working stresses at each radius, both in the core and in the windings, from that which would be set up in the equivalent simple elastic cylinder; this has been done to obtain Fig. 3.7b. If now we imagine each successive layer being unwound, the tension of all those remaining will increase, and that in each exposed layer will be that required when it was put on. It must be remembered that the assembly will at all stages behave elastically, so that the effect of removing a layer will be the same as removing the corresponding external pressure, and if r is the radius of the layer just exposed and if σ_{rR} is the radial stress it was carrying, then the tangential stress in the exposed layer will have been increased thereby to the extent of:

$$\sigma_{rR}\frac{r^2+r_1^2}{r^2-r_1^2}$$

noting that r_1 here is the inside radius of the composite tube, which is of course that of the core tube. The values of σ_{rR} can either be calculated or taken from the curve shown in Fig. 3.7b. Then if P be the pull necessary in winding on a particular layer at radius r, and if a be the cross-sectional area of the strip, then:

$$\frac{P}{a}=\sigma_{rR}\frac{r^2+r_1^2}{r^2-r_1^2}+\sigma_{\theta R} \tag{3.48}$$

$\sigma_{\theta R}$ being the residual tangential stress at that radius, which again can be read off from Fig. 3.7b, which also shows as a chain-dotted line the values of P/a.

In practice, especially with strip of interlocking section, the thickness of each layer may be too great for the stress system to be represented by the smooth curve obtained by the above means; instead it will proceed in a series of steps. However, unless the thickness per layer is 5% or more of the total wall thickness, this is not a serious consideration, and even then the bending stresses set up in the winding operation tend to even out these discontinuities. The more practical consideration is the extent to which the ideal conditions specified above can be realised. This depends very largely on the skill and experience of the manufacturer, but at best there should be no need to allow more than say 10% margin, assuming that the allowable stress levels have been estimated with due regard for the working conditions, see Siebel and Schwaigerer[3.13].

The interlocking strip method has other advantages in that each layer can have its strip end fastened by welding, and the whole assembly is then held together without risk of the ends slacking off and reducing the residual stresses. It is also possible to thicken up the ends by adding extra windings, and – if properly applied – the composite wall is so well interlocked that it can be drilled and tapped to take studs for securing the end covers. Messrs. Vickers Ltd. have recently devised an alternative method in which forged flanges are screwed on over the windings. The latter are screwed internally with a thread of the same pitch and of a profile which can be formed by trueing up the top layer of the strip. The flanges are then heated before screwing on, so as to produce a slight shrinkage over the end windings. Fig. 3.8 shows a vessel of this kind which was designed and built at Elswick by Messrs. Vickers Ltd. It has an internal diameter of 1·52 m, weighs 176 tonnes and was tested to 490 bars.

Some of the earlier German vessels were designed to carry the end loads of the covers by longitudinal stresses in the core tube, but this involved very high stresses with consequent deformation which tended to decrease the residual

Figure 3.8 Large vessel using interlocking strip construction *(made by Vickers Ltd, Elswick, Newcastle-upon-Tyne)*

stresses in the windings. However, if for any reason the strip (or wire) reinforcement has no interlocking action, then it is evidently necessary to support the end loads by the core tube, or alternatively by some external structure. This latter can take the form of an outer mantle, lightly shrunk over the windings. This serves the purpose of keeping the windings in position, thereby preserving their respective tensions, and also of resisting bending or other longitudinal forces. It is not normally considered from the point of view of radial strength although it usually has some small beneficial effect.

It may be appropriate to end this Section with a warning about the potential dangers of using materials of different expansion coefficients if the vessel is to be operated at different temperatures. Only a very small change is then sufficient to upset completely the elaborately contrived system of internal stresses, winding tensions, etc.

3.8 Pressure Jacketing (Cascading)

Several authors, notably Berman[3.14] and Wilson and Skelton[3.1], have suggested using a series of concentric cylinders with annular jackets between each, into which controlled intermediate pressures could be fed. This is essentially a device for enabling the external pressures to be varied and thus avoiding the condition where the residual stresses exceed those occurring when the main internal pressure is acting. For conditions of elasticity throughout, it is ideally equivalent to permitting entry into the shaded area of Fig. 3.3; in addition it avoids the effects of errors in shrinking which – as we have seen (Section 3.6) – can easily lead to stresses at least 10% higher than those allowed for in the design calculations.

Unfortunately, however, the practical complications of applying this system are such as to make it only economically worth while in very special cases; in particular for the control of strains rather than stresses. It has been used with great success by Johnson and Newhall[3.15] to give a controlled clearance between the piston and cylinder in a dead-weight piston pressure gauge (see Chapter 14) for pressures up to 15 kilobars. For general pressure containers – apart from the shrinkage and residual stress limitations – it is evident from Fig. 3.3 that the chief advantages would be gained by the use of at least four components, which means three separate pressurising systems, a complication which could hardly be considered for any industrial process. Thus, there seems little purpose in considering the matter further here, although this could be a fertile field for the exercise of design ingenuity.

3.9 Sectored Vessels

In this arrangement the cylindrical section is again divided into a number of concentric rings, one of which is made up of sector-shaped pieces. This was first suggested by Poulter who took out a patent for it in 1951.[3.2] The object is to prevent tensile stresses from developing therein, and, since the circumferential area of the surface varies with the radius, it is evident that the stress will be in inverse ratio. Also, if the claim that only compressive stresses are present is substantiated, there is no reason why ultra-hard materials such as tungsten carbide should not be used for the sectors, in which case it should be possible to hold pressures at least up to the crushing strength of that material, or say 60 kilobars. In fact their tapered section fulfils much of Bridgman's "massive support" requirement[3.16] and therefore might be expected to resist even higher pressures.

Fig. 3.9 shows a typical section for a design of this kind, with a thin-walled core tube (whose function is principally to prevent seepage up the crevices between adjacent sectors), the ring of sectors, and the outer mantle, here shown as a duplex-

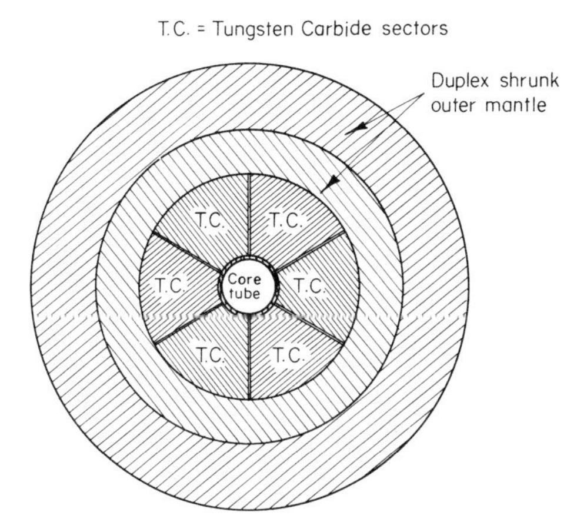

Figure 3.9 Schematic section through sectored cylinder

shrunk ring of 2 to 1 diameter ratio. The sector ring has a diameter ratio of 4 to 1 and the core tube of 1·4. If we neglect the pressure difference across the core, the pressure at the inside of the sectors will be four times that at the outside, and the maximum pressure that the duplex outer cylinder would withstand, assuming an acceptable working shear stress of 4,000 bars, is (according to eq. (3.10)—when $n = 2$) numerically the same, viz. 4,000 bars; and the pressure difference across the ring of sectors will be in proportion to their diameters, viz. 4 to 1, so the working pressure could be 16,000 bars (or 16 kilobars).

The design problem here is principally concerned with the exact shape of the sectors, whose radial flanks must be slightly curved so that when the assembly is subjected to the full pressure they are flat straight planes everywhere in contact with their neighbours. To achieve this with a material like tungsten carbide, which is very liable to be unhomogeneous, requires both skill and good fortune, but it would appear that the problems involved have been solved, at any rate by some of those who are using this principle, e.g. Barogenics Inc. of New York City, U.S.A.[3.17] and the A.S.E.A. organisation in Sweden.[3.18]

REFERENCES

3.1. WILSON, W. R. D. and SKELTON, W. J., Conference on High Pressure Engineering, London, Sept. 1967, Paper No. 5, *Proc. Inst. Mech. Eng.*, **182,** Pt 3C, 1.
3.2. POULTER, T. C., U.S. Patent No. 2,554,499; application date Sept. 8, 1947, granted May 29, 1951.
3.3. CROSSLAND, B. and BURNS, D. J., *Proc. Inst. Mech. Eng.*, **175,** 1083, 1961.
3.4. MANNING, W. R. D., *Engineering,* **163,** 349, 1947.
3.5. MANNING, W. R. D., *Engineering,* **170,** 464, 1950.
3.6. SOUTHWELL, R. V., *Phil. Trans. Roy. Soc.* **A213,** 187, 1913.
3.7. JASPER, T. MCL. and SCUDDER, C. M., *Trans. Am. Inst. Chem. Eng.*, **37,** 885, 1941.
3.8. BETT, K. E., *private communication,* 1968.
3.9. LONGRIDGE, J. A., *A Treatise on the Application of Wire to the Construction of Guns,* E. and F. N. Spon 1884.
3.10. BIRCHALL, H. and LAKE, G. F., *Proc. Inst. Mech. Eng.*, **146,** 340, 1947.
3.11. SCHIERENBECK, J., *Brennstoff-Chemie,* **31,** 375, 1950.
3.12. Unsigned article in *Engineering,* **187,** 155, 1949.
3.13. SIEBEL, E. and SCHWAIGERER, S., *Chem.-Ing.-Tech.*, **24,** 199, 1952.
3.14. BERMAN, L., *Trans. Am. Soc. Mech. Eng.*, **88,** 500, 1966.
3.15. JOHNSON, D. P. and NEWHALL, D. H., *Trans. Am. Soc. Mech. Eng.*, **75,** 301, 1953.
3.16. BRIDGMAN, P. W., *Proc. Roy. Soc.*, **A203,** 1, 1950.
3.17. Barogenics Co. Inc. of New York City, U.S.A. (formerly the Engineering Supervision Co. Inc.), British Patent No. 899,524; application date Jan 7, 1960, from U.S. application dated Jan. 8, 1959; patent granted June 27, 1962.
3.18. Allmänna Svenska Elektriska A/B, French patent No. 1,346,567, granted Dec. 26, 1962.
3.19. NEWITT, D. M., *Trans. Inst. Chem. Eng.*, **14,** 100, 1936.
3.20. WERZNER, K., *Mitt Krupp Werksberichte,* **20,** 273, 1962.

4 The Sphere

4.1 General

The sphere is the only other shape which is susceptible to simple analytical treatment, both in the elastic and plastic states. By its very nature it is obviously difficult to reinforce by shrinking or any external support other than a pressure-filled outer jacket, but fortunately it is very much stronger for a similar diameter ratio than the corresponding cylinder. Moreover, this greater strength is present in both the elastic and plastic states. We shall therefore confine our attention here to elastic and partially plastic conditions and ignore any possibilities of external reinforcement.

The sphere's most common use is in the form of hemispherical ends to cylindrical vessels where this greater strength is easily demonstrated; if the wall thickness is the same throughout, overstrain certainly begins well away from the ends, which may in fact act as local reinforcement. Unfortunately there is no published experimental work to confirm or dispute the theoretical treatment of the stresses and strains in spheres, but it is sufficiently similar to that applied by Lamé to cylinders to allay most fears that it might dangerously overestimate their strength.

4.2 The Elastic Sphere

The static equilibrium of a sphere about a diametral plane is evidently given by:

$$p_o \pi r_o^2 = 2\pi \int_{r_i}^{r_o} \sigma_\theta r dr + p_i \pi r_i^2 \tag{4.1}$$

Then, treating this in a manner similar to that applied to cylinders in Section 2.2, we have:

$$2\int_{r_i}^{r_o} \sigma_\theta r dr = (\sigma_r)_o r_o^2 - (\sigma_r)_i r_i^2 = \Big[\sigma_r r^2\Big]_{r=r_i}^{r=r_o} \tag{4.2}$$

which, on differentiation with respect to r, gives:

$$2\sigma_\theta r = \frac{d}{dr}(\sigma_r r^2) \tag{4.3}$$

or

$$\sigma_\theta - \sigma_r = \frac{r}{2}\frac{d\sigma_r}{dr} \tag{4.4}$$

But, as before:

$${}_\theta\tau_r = \tfrac{1}{2}(\sigma_\theta - \sigma_r)$$

and thus, on integration and substitution for σ_r at the surfaces, we get:

$$p_i - p_o = 4\int_{r_i}^{r_o} {}_\theta\tau_r d\ln r \tag{4.5}$$

which should be compared with eq. (2.7).

This treatment is quite general and makes no conditions as to the physical state (e.g. elastic, plastic, etc.) of the material. When we introduce the assumption of true elasticity we can at once write down the stress-strain relations in the principal directions, and the situation is simplified by the fact that the symmetry of the system makes the tangential relation the same in all directions and thus we only have two stress-strain equations:

$$E\varepsilon_\theta = \sigma_\theta - \nu(\sigma_r + \sigma_\theta)$$
$$E\varepsilon_r = \sigma_r - 2\nu\sigma_\theta$$

We can again express these strains in terms of a single parameter, the radial shift u, as in equations (2.12) and (2.13), so that:

$$E\frac{du}{dr} = \sigma_r - 2\nu\sigma_\theta \tag{4.6}$$

and

$$E\frac{u}{r} = \sigma_\theta - \nu(\sigma_r + \sigma_\theta) \tag{4.7}$$

which, together with eq. (4.4), give three equations containing u, σ_θ and σ_r as the only unknowns.

Subtracting eq. (4.7) from eq. (4.6) gives:

$$E\left(\frac{du}{dr} - \frac{u}{r}\right) = Er\frac{d}{dr}\left(\frac{u}{r}\right) = -(1+\nu)\ (\sigma_\theta - \sigma_r)$$

and, by substitution from eq. (4.4), we get:

$$E\frac{d}{dr}\left(\frac{u}{r}\right) = -\frac{1+\nu}{2}\frac{d\sigma_r}{dr} \tag{4.8}$$

Then, differentiating eq. (4.7) with respect to r gives:

$$E\frac{d}{dr}\left(\frac{u}{r}\right) = \frac{d\sigma_\theta}{dr} - \nu\left(\frac{d\sigma_r}{dr} - \frac{d\sigma_\theta}{dr}\right) \tag{4.9}$$

and subtracting this from eq. (4.8) eliminates u, so that:

$$\frac{1-\nu}{2}\frac{d\sigma_r}{dr}+(1-\nu)\frac{d\sigma_\theta}{dr}=0$$

or

$$2\frac{d\sigma_\theta}{dr}+\frac{d\sigma_r}{dr}=0 \tag{4.10}$$

Integrating, and writing $3A$ for the constant of integration gives:

$$2\sigma_\theta+\sigma_r=3A \tag{4.11}$$

Combining this with eq. (4.3):

$$\frac{1}{r^2}\frac{d}{dr}(r^3\sigma_r)=3A$$

which on integrating and adding B as a second constant of integration gives:

$$\sigma_r=A+B/r^3 \tag{4.12}$$

which, with eq. (4.11), gives:

$$\sigma_\theta=A-B/2r^3 \tag{4.13}$$

Once again we have to evaluate A and B from the boundary conditions; if $\sigma_r=-p_i$ at $r=r_i$ and $-p_o$ at $r=r_o$, we get:

$$A=\frac{-p_oK^3+p_i}{K^3-1} \tag{4.14}$$

and

$$B=\frac{(p_o-p_i)K^3}{K^3-1}r_i^3 \tag{4.15}$$

where K is again the ratio r_o/r_i.

Thus the principal stresses are given by:

$$\sigma_\theta=\frac{-p_oK^3+p_i}{K^3-1}-\frac{(p_o-p_i)K^3}{2(K^3-1)}\left(\frac{r_i}{r}\right)^3 \tag{4.16}$$

and

$$\sigma_r=\frac{-p_oK^3+p_i}{K^3-1}+\frac{(p_o-p_i)K^3}{K^3-1}\left(\frac{r_i}{r}\right)^3 \tag{4.17}$$

Also:

$${}_\theta\tau_r=\frac{-3(p_o-p_i)K^3}{4(K^3-1)}\left(\frac{r_i}{r}\right)^3 \tag{4.18}$$

and

$$\varepsilon_\theta = \frac{-p_o K^3 + p_i}{E(K^3-1)}(1-2\nu) - \frac{(p_o - p_i)K^3}{2(K^3-1)}\left(\frac{r_i}{r}\right)^3 (1+3\nu) \tag{4.19}$$

4.3 Stress Distribution in the Elastic Sphere

Compared with the cylinder this is not quite so simple. There is a similar hydrostatic stress, as represented by A, but the other term has values of B/r^2, $-2B/r^2$ and $-2B/r^2$ for the three principal directions. As a result the two main shear criteria, i.e. those of Tresca and Maxwell, lead to the same result, viz. $3B/4r^2$, which at its maximum value when $r = r_i$ is:

$$\frac{3(p_o - p_i)K^3}{4(K^3-1)}$$

Thus overstrain is likely to be reached when this function has the value $\sigma_T^*/2$ (where σ_T^* is the limiting elastic stress in a tensile specimen), and consequently:

$$p_o - p_i = \frac{2\sigma_T^*(K^3-1)}{3K^3} \tag{4.20}$$

The two other comments in Section 2.4 apply equally to the sphere, i.e. the most intense stresses occur at the inner surface whether $p_o - p_i$ is positive or negative, and their values depend (according to this theory) only on the ratio of the diameters and not on their absolute magnitudes.

Comparing the two shapes and the values of the differences between the applied pressures (i.e. $p_o - p_i$) at which overstrain is likely to begin, these are most likely to be given by eq. (4.20) above for the spheres, and by eq. (2.34) for the cylinders. Thus the ratio of the limiting elastic stress in the sphere to that in the cylinder of the same diameter (or radius) ratio is:

$$\frac{2(K^2+K+1)}{\sqrt{(3)}K(K+1)}$$

from which it will be seen that this varies between the limits of 1·732 as K approaches unity, i.e. as the wall becomes very thin, and 1·155 as K becomes large. We see therefore that if, for instance, we fit a cylindrical vessel with a hemispherical end, keeping the wall thickness the same, then the stresses in the end will always be less than those in the cylindrical barrel; and a consequence of this is that the strains will also be less so that the ends will have a supporting effect on the parallel portion.

There appears to have been very little experimental work on the stress-strain relations in thick-walled spheres, and no doubt the reason for this is the relative unimportance of vessels of this shape. Spherical containers such as "Horton spheres" have been studied in more detail, but they are only for low pressure storage and hardly come within the scope of this book. On the other hand, it would seem that thick-walled spherical specimens, made in two equal halves and lapped together, would be comparatively easy to make and test by subjecting to external pressure under conditions which would make observation and measurement also comparatively easy. In this way one could go a long way towards verifying this theoretical treatment.

4.4 Approximate Solution of Partially Overstrained Sphere

The assumption here is that the shear stress remains constant over the region in which overstrain has taken place, and that its value is then the same as its limiting elastic value, denoted here by τ^*.

Then if r_c be the radius of the transition surface between the elastic and plastic regions (which is assumed to be perfectly symmetrical and concentric with the surfaces of the sphere), we have from eq. (4.5) for the plastic zone:

$$(\sigma_r)_c - (\sigma_r)_i = \tau^* \ln(r_c/r_i)^4 \tag{4.21}$$

and, writing k_c for r_c/r_i and substituting for $(\sigma_r)_i$, we get:

$$(\sigma_r)_c + p_i = \tau^* \ln k_c^4 \tag{4.22}$$

In the elastic zone we know that a pressure of $-(\sigma_r)_c$ acting inside a sphere of external and internal radii r_o and r_c respectively will produce a maximum shear stress equal to τ^*; whence:

$$(\sigma_r)_c = \frac{4\tau^*(r_o^3 - r_c^3)}{3r_o^3}$$

which becomes, if K is the overall diameter ratio,

$$(\sigma_r)_c = \frac{4\tau^*(K^3 - k_c^3)}{3K^3} \tag{4.23}$$

Combining this with eq. (4.22) gives:

$$p_i = \tau^* \left\{ \ln k_c^4 + \frac{4}{3}\left(1 - \frac{k_c^3}{K^3}\right) \right\} \tag{4.24}$$

In many problems the principal requirement is to find the extent of the overstrain (i.e. to find k_c) for any particular condition, and it is then easier to transpose eq. (4.24) thus:

$$\ln k_c - \tfrac{1}{3}\left(\frac{k_c}{K}\right)^3 + \tfrac{1}{3} = \frac{p_i}{4\tau^*} \tag{4.25}$$

Applying this to a sphere of similar diameter ratio and of the same material as the cylinder considered in the example in Section 2.10 we have $p_i = 5{,}870$ bars, $\tau^* = 4{,}095$ bars, and $K = 2{\cdot}40$. Substituting into eq. (4.25) gives a value of 1·055 for k_c, which compares with 1·52 for the cylinder, giving a good illustration of the much greater strength of the sphere.

If it had been overstrained to a value for k_c of 1·52, then—from eq. (4.24)—it is seen that the necessary pressure would be about 10,900 bars, or an increase of 85 %, compared with the cylinder.

In regard to the Ultimate Bursting Pressure of spheres, no experimental results have been published (so far as the authors can discover) so this can only be guessed very approximately. The shear strain increases much more rapidly as the plastic zone penetrates outwards from the inside surface than it does in the corresponding cylinder, and it may well be that a test to destruction with the strains in equilibrium throughout – i.e. with the stress-strain relations continuing to conform to the theory – would never be possible. The computation would also be considerably more involved since the strains could only be found by assuming that the volume of the metal remained constant. In view of these difficulties and uncertainties it is probably as good as anything to assume that failure will occur when the overstrain has penetrated right through to the outside, i.e. when $k_c = K$. As eq. (4.24) shows, we then get:

$$p_i = \tau^* \ln K^4 \tag{4.26}$$

and, in the example considered here, this means that the greatest value for p_i is 4,095 × 4 × ln2·4, or say 14,300 bars. This could be regarded as an absolute maximum and compares with 7,900 bars for the cylinder, again showing that the sphere is theoretically much the stronger; and experience certainly shows that in thick-walled vessels with hemispherical ends on parallel cylindrical barrels the former are definitely the stronger parts and tend to reinforce the latter.

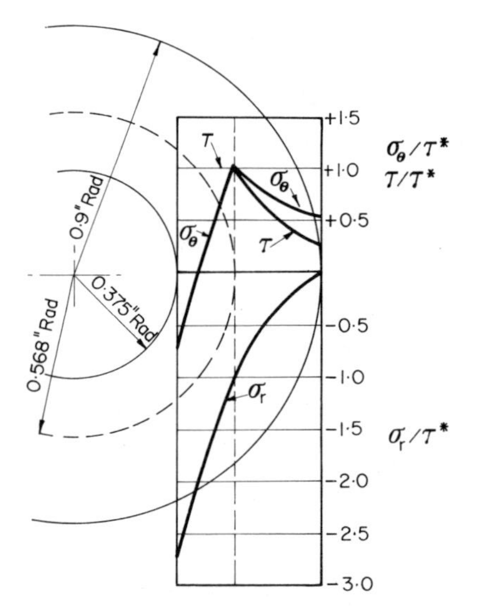

Figure 4.1 Stresses in partially overstrained sphere

Fig. 4.1 shows the stresses which would be set up, according to this approximate theory, in a sphere of diameter ratio 2·4 if overstrained to a surface of radius 1·52 times that at the bore – or very approximately to the geometric mean radius. If we compare this with Fig. 2.11, the difference between the two shapes is clearly seen, although it must be remembered that the latter diagram is supported by experiment and the former depends upon a good deal of guesswork. It should be noted also that the near equality of the shear and tangential stresses at the critical radius is fortuitous in this example.

5 The Effects of Temperature

5.1 General

There are three effects which may need consideration in the design and operation of high pressure equipment. The first of these concerns the stresses which are set up in the walls of a container when different parts of it are at different temperatures; the second is high temperature creep and the many complex problems this may involve. The third, which is less generally recognised, concerns the reduction in strength, both elastic and ultimate, which occurs in metals due to temperatures well below the levels at which creep can be detected.

The first effect is capable of a reasonably simple solution for the particular case of steady and symmetrical heat flow through a cylindrical wall, as—for instance—with a high pressure heat exchanger. Unfortunately, many instances arise in practice in which these conditions do not apply and their solution is then usually a matter of assumption and approximation; moreover, the steady flow condition is one of *minimum* stress and consequently, during starting up or shutting down, some parts of the equipment will be subjected to appreciably more severe stressing. These difficulties will be discussed in Section 5.2.2, although no simple solution is possible.

The whole problem of creep under conditions of complex (three-dimensional) stress is at present only capable of solution in special cases and there is still no adequate experimental check on the various theoretical treatments which have been offered. In the case of thick cylinders however, Skelton and Crossland[5.1] have evolved an experimental technique which enables creep in thick-walled cylinders to be studied over long periods and with an accuracy amply good enough to check many of the theories which have been put forward. At the time of writing the amount of published data from these workers is small, but they are known to be continuing with this work (at the Queen's University in Belfast) and it is likely that further results will be available by the time this book is published.

The weakening effects of temperature at levels where creep can be disregarded is well known although not much has been published concerning its influence on the strength of pressure containers; once again therefore it is necessary to work mostly by analogy with other types of loading.

5.2 Temperature Stresses in Cylinders

5.2.1 *Steady and Symmetrical Heat Flow*

We begin here with the following assumptions:

(i) The temperature distribution is symmetrical about the axis of the cylinder and does not vary in the axial direction (i.e. all points at the same radius are at the same temperature).

(ii) The temperature at any and every point in the wall does not vary with time.

(iii) The axial strain across the section is constant.

(iv) The values of Young's Modulus (E), Poisson's Ratio (ν) and the coefficient of linear expansion (α) are constant over the temperature range with which we are concerned.

The general condition of symmetrical heat flow in a cylindrical wall is given by:

$$r\frac{\partial T}{\partial t} = \frac{K_h}{c\rho}\frac{\partial}{\partial r}\left(r\frac{\partial T}{\partial r}\right) \tag{5.1}$$

where T is the temperature at any radius r, and K_h, c and ρ are respectively the thermal conductivity, the specific heat, and the density of the material. Then, as soon as we apply assumption (ii) above, the left-hand side of the equation becomes zero and:

$$\frac{\partial}{\partial r}\left(r\frac{\partial T}{\partial r}\right) = 0 \tag{5.2}$$

Now the solution of this is:

$$T = A + B\ln r \tag{5.3}$$

where A and B are constants of integration, to be evaluated from the boundary conditions, namely $T = T_o$ at $r = r_o$, and $T = T_i$ at $r = r_i$. When A and B are thus eliminated:

$$T = \frac{T_i \ln r_o/r + T_o \ln r/r_i}{\ln r_o/r_i} \tag{5.4}$$

The condition of static equilibrium remains the same as that derived in Section 2.2, namely:

$$\sigma_\theta - \sigma_r = r\frac{d\sigma_r}{dr} \tag{5.5}$$

but the strains have to be modified by the expansion effect so that, for instance, the tangential strain due to the stress alone is $\varepsilon_\theta - \alpha T$ and in consequence the relations of equations (2.8), (2.9) and (2.10) become:

$$E(\varepsilon_\theta - \alpha T) = \sigma_\theta - \nu(\sigma_z + \sigma_r) \tag{5.6}$$

$$E(\varepsilon_z - \alpha T) = \sigma_z - \nu(\sigma_r + \sigma_\theta) \tag{5.7}$$

$$E(\varepsilon_r - \alpha T) = \sigma_r - \nu(\sigma_\theta + \sigma_z) \tag{5.8}$$

Then, if there are no end constraints:

$$2\pi \int_{r_i}^{r_o} r \,.\, \sigma_z \,.\, dr = 0 \tag{5.9}$$

We can again write ε_θ and ε_r in terms of the radial shift u and the stresses can then be evaluated. The resulting relations, writing Θ for $T_o - T_i$, giving it a positive sign when the outer wall is hotter than the inner, i.e. when heat is flowing inwards, and K for r_o/r_i as usual, are:

$$\sigma_\theta = \frac{E\alpha\Theta}{2(1-\nu)}\left[\frac{K^2}{K^2-1}\left(1+\frac{r_i^2}{r^2}\right) - \frac{1-\ln r_i/r}{\ln K}\right] \tag{5.10}$$

$$\sigma_z = \frac{E\alpha\Theta}{2(1-\nu)}\left[\frac{2K^2}{K^2-1} - \frac{1-2\ln r_i/r}{\ln K}\right] \tag{5.11}$$

$$\sigma_r = \frac{E\alpha\Theta}{2(1-\nu)}\left[\frac{K^2}{K^2-1}\left(1-\frac{r_i^2}{r^2}\right) + \frac{\ln r_i/r}{\ln K}\right] \tag{5.12}$$

From these equations the stresses can be evaluated in terms of the coefficient $E\alpha\Theta/2(1-\nu)$, and of the symbols contained therein E, α and ν are specific to a particular material. Table 5.1 lists the values of $E\alpha/2(1-\nu)$ for a few materials, in bars per degree C and also in lbf/in^2 per degree C.

Table 5.1

Material	$E\alpha/2(1-\nu)$	
	Bars/degC	*lbf/in^2/degC*
Cast iron	7·55	110
Pure iron	14·4	210
0·15% C steel	17·2	250
En25 steel[1]	18·5	268
13% Cr steel[2]	18·3	265
Austenitic stainless	20·8	302
18% Ni Maraging[3]	13·4	195

[1] 0·3%C, 2·57% Ni, 0·58% Cr, 0·60% Mo, 124,000 lbf/in^2. Ultimate Tensile Strength.
[2] 0·14% C, 12·8% Cr, 0·4% Ni, 90,000 lbf/in^2. Ultimate Tensile Strength.
[3] 0·027%C, 18·08% Ni, 4·80% Mo, 7·10% Co, 0·40% Ti, 0·13% Al. Aged at 480°C. 264,000 lbf/in^2. Ultimate Tensile Strength.

The resulting stress distributions across the wall of a cylinder of K = 3·0 are shown in Fig. 5.1, while the variations of the maxima with K are shown in Fig. 5.2. The ordinates in each case are factors which, when multiplied by the appropriate value of $E\alpha/2(1-\nu)$ and by the temperature difference between the metal at the two surfaces, will give the actual stress. It should be noted that, in most cases where heat is flowing from one fluid to another through the wall, the actual temperature difference in the metal (Θ) will be appreciably less than that between the two fluids, owing to the temperature drop across the surface films; and this will mean that the stresses will be overestimated when Θ is taken as the difference between the temperatures of the fluids. A slight margin of safety may be introduced thereby, although this would be small when the fluids are in motion.

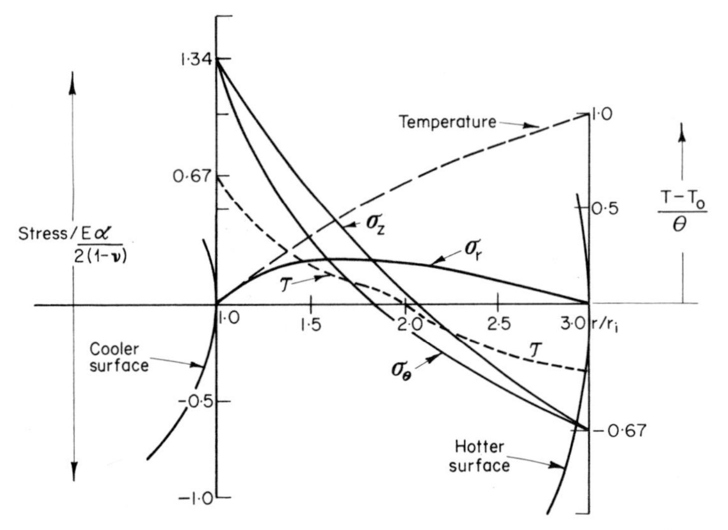

Figure 5.1 Distribution of temperature stresses across wall of 3 to 1 diameter ratio cylinder

In every case, as Fig. 5.1 shows, the stresses at the bore are tensile when the heat flow is inwards, i.e. when the outer wall is the hotter. This means that they will normally be acting in the same sense as those due to the internal pressure, and they could therefore be the cause of overstrain in a cylinder which would otherwise have withstood the applied pressure elastically. The obvious conclusion from this is that, wherever possible, vessels should be heated internally, since the tendency will then be for the thermal stresses to counteract those due to the pressure.

The problem could be considerably more complicated where very thick and partially overstrained walls are concerned, since the stress distribution will then have been greatly altered from that existing under elastic conditions; and in vessels built up by shrink-compounding there will be increased resistance to heat flow at the contact surfaces between the components, and this will complicate the stress pattern still further.

5.2.2 *Varying Surface Temperatures with Symmetrical Heating*

The problem becomes much more troublesome when any other condition than that of steady heat flow is encountered. If, for instance, we consider a steam-jacketed high pressure vessel which on starting up suddenly receives a supply of steam into the jacket which was previously empty, there will be an immediate rise in temperature of the outside wall. Heat will then flow rapidly inwards, heating up the metal as it does so, and if the contents of the vessel can absorb the heat at a rate comparable with that required in steady running it is clear that the heat flow past the outer surface will be appreciably greater until the temperature conditions represented by eq. (5.4) are reached; consequently there will be greater temperature gradients, with resulting greater stresses, than in the steady running state of affairs.

A graphical solution of the problem has been devised by R. A. Strub[5.2], but the work becomes rather troublesome if heat is also flowing out of the wall through the other surface. An approximation can, however, be obtained by assuming a simplified temperature distribution. For instance, shortly after the steam had been admitted to the jacket we might assume that the outer surface of the vessel was within a few degrees of the steam temperature, while inside the wall and, say, 10% of its thickness from the outside the metal was only a degree or two higher than ambient temperature (if the vessel had been standing for some time). The maximum stresses due to this temperature difference in 10% of the radial thickness of the wall can then be read off from Fig. 5.2; and, by

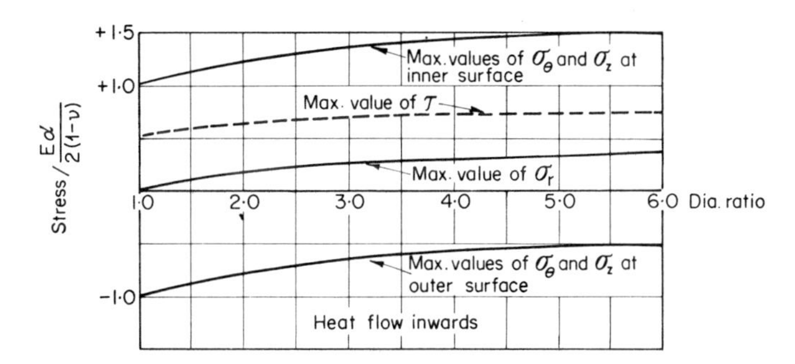

Figure 5.2 Variation of maximum temperature stresses with diameter ratio

reference to eq. (5.8), the radial strain can be obtained, and from that the stresses in the remainder of the wall necessary to produce a similar radial strain. In this way we can get a fair idea of what these stresses would be, although even this sort of approximation could be quite laborious, and unless the conditions are realistically known it is doubtful whether the practical results would justify the effort.

The problem becomes somewhat simpler if it is assumed that heat is flowing in at one surface, but that no heat is either entering or leaving at the other. A mathematical solution of this was given by Jaeger[5.3], and some applications are to be found in the work of Carslair and Jaeger[5.4], while curves to simplify the solution for particular cases were given in 1959.[5.5]

Unfortunately, however, the stresses are greatly influenced by variations in the initial temperature distribution within the walls, and this can hardly ever be known accurately. Also it is seldom possible to maintain temperature symmetry, and the result is further uncertainty in any stress calculations. Thus, from the practical point of view, it seems evident that we can hardly hope to know the stress intensities which are introduced by transient variations of temperature, and the first and most obvious lesson to be learnt is that heating up (and cooling down) should be carried out as slowly and as carefully as possible. For instance, referring again to the steam-jacketed vessel mentioned above, if the working temperature of the contents is well above ambient, then the inner surface should also be warmed, at least up to about the temperature at which the contents will be introduced, and time should be allowed for the walls to come within a few degrees of that temperature. In this way one can avoid the rapid heating of the internal metal with consequent large and unknown temperature gradients.

Another point to consider in the detailed design of such a vessel is the arrangement of steam inlets. These should if possible be fitted with baffles so that a blast of steam cannot be projected against one or two localised regions of the vessel wall. Otherwise there is always the risk of unskilled operation of the valves leading to the sudden generation of hot spots on the wall with consequent unsymmetrical temperature gradients of unknown magnitudes, a situation which can lead to overstrain and even crack formation in the worst conditions.

Having thus warned the reader of the difficulties and dangers that can arise under the more extreme conditions we must also point out that there are other factors which act in such a way as to reduce the seriousness of these effects. Let us consider for instance the starting up of the steam-jacketed vessel, assuming it to have been previously warmed so that most of the metal is at about the same temperature as the contents. Then, as the flow starts (or whatever heat-absorbing action is involved) and steam is admitted, the initial effect will be greatly to increase the temperature gradient just inside the outer wall. Referring now to the curve of temperatures for steady heat flow (shown as a broken line in Fig. 5.1) it is clear that, at the right-hand end, there will be a much steeper slope of the curve, but equally at the left-hand end the slope will be reduced. Thus, near the inside of the vessel the stresses will be reduced compared with those present under steady heat flow, and in this instance the stress increases are at the outside where the pressure stresses are much less severe. We see therefore that the starting-up stresses introduced by a procedure of this kind may not be such a serious matter.

The situation would, however, be very different if, for some reason, the temperature of the inside wall fell, as could happen if the contents of the vessel were warmer than their operating temperature and then suddenly fell to that level. In that case the changes in temperature gradients would be reversed from the situation suggested in the preceding paragraph, and this would result in an increase in stress above that for steady heat flow and in the region where the pressure stresses are highest.

Instances where there is a wide temperature variation in the walls at different angular positions are also not uncommon in practice. Some types of steam superheater for instance are subjected to intense radiant heat on one side and are more or less shielded on the other. With conditions of this sort, it is almost essential to allow the resulting expansions and distortions to proceed without resistance, since otherwise the stresses set up will be indeterminate and may easily be excessive.

5.3 Creep

5.3.1 *General Considerations*

If conditions are such that creep is to be expected, the designer is forced to accept a limited life for his equipment. Consequently he is more concerned with strain rates than with stresses. This is, of course, common to any stressed part in such conditions, but there is the further complication, when we come to deal with pressure containers, that their walls are certainly subjected to stresses in the three directions, whereas virtually all the available data is from uni-axial tensile tests. Various attempts have been made to obtain a satisfactory correlation for cylinders, but until recently these were mostly theoretical, and it was difficult to judge between them for lack of experimental verification.

It may help the reader to follow the discussion of the subject if we briefly recall the main features of the creep process and the usual tensile test procedure. Thus, the specimen is brought up to the required temperature and the load then applied, taking care to do this without shock. This results in an immediate extension similar to that caused by loading at room temperature; also it is reversible if the load is removed quickly. If left on, however, a slow "creep" or irreversible extension follows, but its rate steadily falls until a more or less constant value is reached, which may persist for very considerable periods of time; however, eventually the creep rate accelerates again and this is usually the prelude to rupture within a few hours. Fig. 5.3 shows a few typical creep records

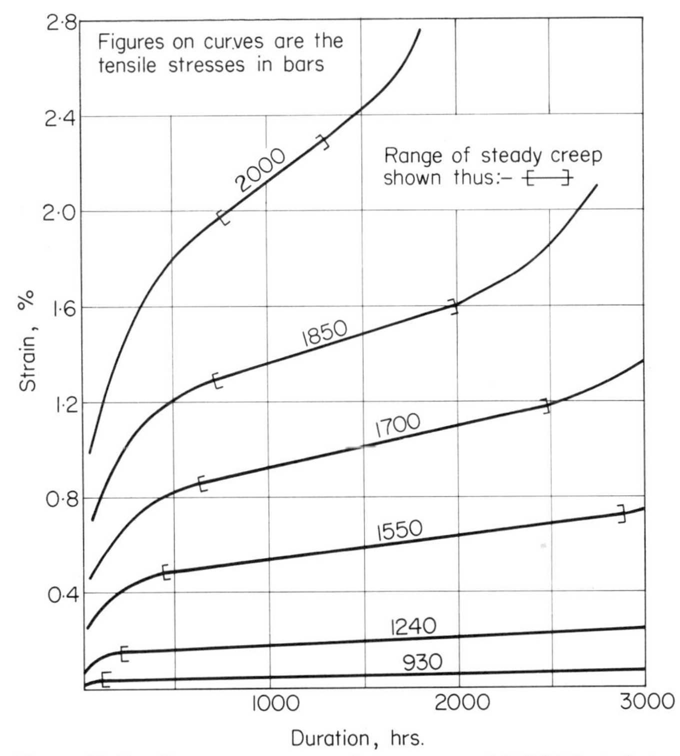

Figure 5.3 Tensile creep rates at various stresses of 0·19% C steel at 450°C

in which strain is plotted against time for constant stress and constant temperature conditions. It will be seen that, in curve *a*, the accelerating creep follows almost immediately after the initial slowing up, whereas in curve *b* there is an appreciable range of constant rate creep, and in curve *c* this range is much longer.

The first part of such a creep curve, in which the rate is falling, is known as the primary (or transient) stage, the constant rate part as secondary (or steady-state) creep, and that where the rate again increases as tertiary creep. From the point of view of virtually all design problems it is necessary to select a material which has a long range of secondary creep, since otherwise its life will be altogether too short. One may mention in passing that in some actual tests the secondary creep has persisted for 10^5 hours or more (over 11 years), but in others—as in the topmost curve of Fig. 5.3—it appears to be little more than a point of inflection between the primary and tertiary stages.

There seems little doubt now that, for most of the commoner engineering materials at any rate, it is the shear stresses which determine the creep behaviour, and it is also reasonably well established that creep is a constant density process; or, in other words, that the volume of any object which is exhibiting the phenomenon of creep will not be changed thereby. A number of investigators have studied this process theoretically, notably Johnson, Henderson and Khan[5.6] and there seems general agreement that the stress distribution in a cylinder wall changes its shape completely, from the elastic type at the outset to something similar to that of pure plasticity. Fig. 5.4 shows roughly how the stresses in the wall of a 0·19% carbon steel cylinder of diameter ratio 2, carrying 1,500 bars at 450°C, will change as creep proceeds. It should, however, be noted that this material would hardly last long enough for practical purposes under such conditions and the range of secondary creep is short, but it is a case dealt with theoretically by Johnson and his colleagues (loc. cit.) and it was actually studied experimentally by Skelton and Crossland[5.7] and their basic agreement indicates the general truth of this idea of the changing shape of these stress curves. This is also confirmed by the work of Mackie and his colleagues at Glasgow University as the result of creep experiments with cylinders of various non-ferrous metals, see especially King and Mackie[5.15]

One can also argue, on a common-sense basis, that something of the sort must happen. Referring to the elastic distribution, rapid creep must surely begin at the inner surface owing to the peak of shear stress there. The result of this will be for the material that is creeping to try to push the material outside it out of the way; this intermediate material is thus put under greater stress while at the same time relieving the most highly stressed inner layers. This process would go on until steady creep conditions are reached although, in the example considered here, it is doubtful whether such a cylinder would last long enough to reach this stage. (A similar specimen subjected to 1,400 bars lasted less than 2,000 hours at 450°C.) The reason for this is that creep brings about changes in dimensions which increase the internal diameter more rapidly than the external, thus reducing the ratio and increasing the bore, and both these effects increase the stress if the pressure is kept constant.

The steady-state stress distribution is consistent with an equilibrium condition

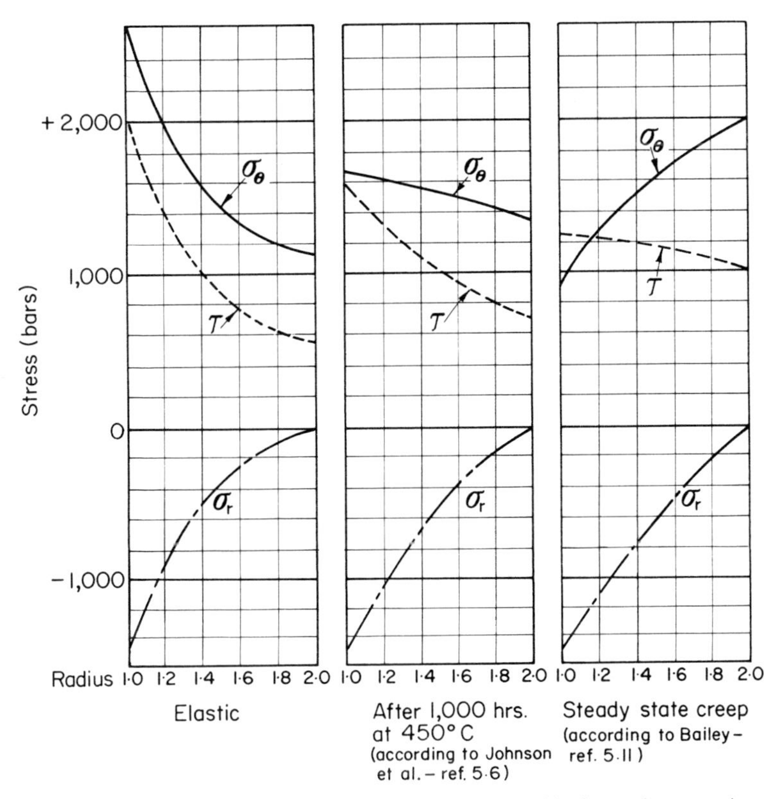

Figure 5.4 Changes in stress distribution in cylinder with time, when creeping owing to internal pressure: (a) at application of pressure, (b) after 1,000 hours, (c) steady state conditions

and it seems reasonable to accept it as the most likely condition for cylinders which have a long life under creep. Another point which shows clearly in Fig. 5.4 is that the tangential stress curves representing the distribution after various periods of time all pass through—or very close to—the same point at a radius of about 1·45 times the bore, i.e. near the mean radius, and at a stress of 1,500 bars. Thus the product of the stress at this point by the thickness of the wall will remain constant and the "mean diameter formula"

$$\sigma_\theta = \frac{(K+1)}{2(K-1)}\, p_i$$

can be used. This has been done by Soderberg[5.8] in a proposed design formula, in which he assumes that this point of constant tangential stress occurs at the arithmetic mean radius.

5.3.2 *Conventional Basis of Design*

In the past, designs have almost always been based on one of three criteria derived from tensile creep tests. These are:

(a) Stress to cause a specified amount of creep in a specified time, usually 1%

in 10^5 hours (or 10^{-7} inch per inch per hour) (σ_c),

(b) Stress to cause the onset of the tertiary stage of creep in a specified time (again usually 10^5 hours),

(c) Stress to cause rupture in a specified time (also usually 10^5 hours) (σ_{Ru}),

all, of course, suitably reduced by a safety factor. Of these (b) is generally considered not to be sufficiently distinct, while (a) has also led to disagreement and lack of correlation; (c) has therefore been preferred. A further difficulty at once arises, however, because much of the tensile data comes from experiments which have been going for no more than 10^4 hours, and resort has sometimes had to be made to extrapolation to a period 10 times as long as the longest test. Fortunately, this procedure, when combined with a safety factor seldom less than 1·5, has so far proved reasonably successful, and in some cases the creep tests have now reached as much as 30,000 hours (nearly $3\frac{1}{2}$ years) without any suggestion that the previous extrapolation had led to an incorrect forecast. On the other hand it is beginning to look as if some of the safety factors may have been excessive.

One of the most extensive studies of the application of tensile data to cylinder design under creep conditions is that carried out by the Electrical Research Association at their laboratories in Leatherhead, some of the results of which were published in 1963 by Chitty and Duval.[5.9] This work consisted in carrying out simultaneous creep-rupture tests in tension and in cylinders of the same material at the same temperature, and from this they concluded that the pressure (p_{Ru}) which would burst a cylinder in a certain time could be related to the tensile stress (σ_{Ru}) which would break a tensile specimen in the same time by the "mean diameter formula":

$$p_{Ru} = \frac{2(K-1)}{K+1}\sigma_{Ru} \tag{5.13}$$

where K is the diameter ratio†. This apparently holds for values of K as high as 1·7, but in most cases the failures occurred in 5,000 hours or less.

Soderberg[5.8] suggested a rather similar formula although his was related to strain rates. Thus the pressure which would cause the mean radius of a cylinder to creep at a given rate—let it be denoted $p'_{c/mean\ rad.}$—should be related to the stress required to cause a tensile specimen to creep at the same rate—let it be denoted $(\sigma_T)_{c/mean\ rad.}$—by:

$$p'_{c/mean\ rad.} = \frac{4(K-1)}{\sqrt{3}(K+1)}(\sigma_T)_{c/mean\ rad.} \tag{5.14}$$

† The recommendation in B.S. 1515 is to use this formula allowing as the value of σ_{Ru} 66% of the tensile stress required to cause rupture in 10^5 hours.

His theory then predicts that the steady creep rate in the tangential direction in the cylinder—let it be denoted $(\dot{\varepsilon}_\theta)_{mean\ rad.}$†—will be related to the creep rate in the tensile specimen—let it be denoted $(\dot{\varepsilon}_T)$—by:

$$(\dot{\varepsilon}_\theta)_{mean\ rad.} = \frac{\sqrt{3}}{2}(\dot{\varepsilon}_T) \tag{5.15}$$

On the assumption that there is no change in density, the tangential creep rate at the outside surface, $(\dot{\varepsilon}_\theta)_o$, is approximately given by:

$$(\dot{\varepsilon}_\theta)_o = \left(\frac{2K}{K+1}\right)^2 (\dot{\varepsilon}_\theta)_{mean\ rad.} \tag{5.16}$$

This provides a comparatively simple means of testing the correlation, and this has been done by Norton and Soderberg[5.10], and also by Skelton and Crossland[5.7], and both find it reasonably satisfactory. It may also be worth noting that, if the creep rates are the same at the same stages of different creep tests, and if these lead to rupture in the same time, then Soderberg's theory, when related to rupture, gives:

$$p_{Ru} = \frac{4(K-1)}{\sqrt{3}(K+1)}\sigma_{Ru} \tag{5.17}$$

or a value about 15% greater than that suggested by Chitty and Duval, i.e. it is that much less conservative than the latter.

It will be appreciated that in both these cases the resulting design basis must inevitably be empirical, although Soderberg's has some theoretical basis, if we can accept that—in all cylinders undergoing creep—the stress and creep rate at the mean radius remain unchanged while the tangential stress distribution turns round from being highest at the bore to being highest at the outside. Unfortunately, Chitty and Duval were not equipped to measure the strains at any stage of their rests, and we cannot therefore compare their work with that of Norton and Soderberg. Incidentally, at least one of Norton's tests was extended to more than 12,000 hours.

5.3.3 *The Work of R. W. Bailey*

One of the earliest attempts at a solution of the problem of creep in thick cylinders was made by the late Dr. R. W. Bailey, who was then in charge of creep work at the Trafford Park, Manchester, laboratories of what was then the Metropolitan Vickers Co. (now part of A.E.I. Ltd.). The main features of this were fully described in a form suitable for practical application in *Engineering* during 1930.[5.11] The

† The Newtonian notation is used here; the dot above the symbol represents the rate, i.e.

$$\dot{\varepsilon} = \frac{d\varepsilon}{dt}$$

authors feel that this work and the ideas that lie behind it have hardly received the attention they deserve, and the reason for this is perhaps that the supporting experiments were carried out with cylinders of lead at atmospheric temperature. It is also a fact that Bailey afterwards devoted himself, so far as his interests in creep were concerned, mainly to attempts at an analytical solution of the problem (see for instance Ref. 5.12) and this work in some respects anticipated the subsequent work of Johnson, Henderson and Khan.[5.6]

The theory with which we are here concerned depended on two main assumptions:

(i) That there is no axial creep in a cylinder which is creeping in the transverse plane as a result of an applied internal pressure.

(ii) That there is a unique relationship between the shear stress and the secondary shear creep rate in any material at the same temperature.

The first of these is certainly a reasonable approximation. Such axial creep as has been found in experiments of this kind has always been much smaller than in the other principal directions. The second raises an interesting philosophical idea, because it implies that, if we can somehow or other impose a particular shear creep rate on a piece of material, then we must automatically generate in it the corresponding shear stress. There seems no obvious reason why this should not be true, any more than, for example, the unique relationship of stress and strain in the elastic region, but it has to be remembered that the creep we are concerned with is the steady creep in shear and the amount of direct experimental study of shear creep (e.g. with torsion tests) is not large. However, if the steady, or secondary, phase of tensile creep is a reality, a similar phase must surely exist in shear, although the resulting theoretical treatment can only be applied to cases where secondary creep can be relied upon to extend for at least as long as the working "life" of the part under consideration.

It must also be remembered that not every tensile creep test lasts long enough to reach the tertiary stage, and there is some evidence that this is in part a consequence of the geometry in a tensile specimen, as it may be also in a cylinder. However, most of the materials normally considered for service in pressure vessels, at temperatures where creep can be expected, certainly do have appreciable ranges of steady creep.

The essence of the Bailey solution is as follows. From his first assumption, coupled with the generally accepted view that creep is a constant density process, creep in the cylinder must occur without change in a cross-sectional area. Thus if we apply a particular linear creep rate to, say, the bore surface, we can find the shear creep rate at any radius; and, if we have a connection between shear stress and shear creep rate for the material at the appropriate temperature, we can find the shear stress at any radius, and so – from the basic equation (2.7) – we can by integration determine the internal pressure which was causing this particular creep rate. Then, by repeating this for several different bore creep rates we can plot a curve of internal pressure against maximum creep rate (since the creep in the bore will be the greatest in the system), and – from the designer's point of view – this is essentially the information required.

There is, however, one considerable difficulty, namely the obtaining of a shear

stress versus shear creep rate connection. This could most easily be deduced from a torsion creep record, but unfortunately very few reliable creep tests in torsion have been reported, and one usually has to make do with tensile records. Bailey reported that the stress normal to the transverse plane had no influence on the shear creep rate, and he therefore used the Tresca (maximum shear stress) correlation. It appears, however, that this conclusion was based mainly on his experiments with lead, and with ferrous metals the Maxwell (von Mises) correlation seems to be preferable. Fortunately the conversion of tension into shear data with either of these criteria is simple. In the case of Tresca it is a matter of halving the stress and multiplying the strain rate by 1·5, and in the case of Maxwell, as Shepherd[5.13] has shown, the tensile stress has to be divided by $\sqrt{3}$ and the strain rate multiplied by the same figure. Thus:
for the Tresca correlation:

$$\left.\begin{aligned} \tau &= \sigma_T/2 \\ \text{and} \qquad \dot{\gamma} &= 1{\cdot}5\dot{\varepsilon}_T \end{aligned}\right\} \tag{5.18}$$

and for the Maxwell correlation:

$$\left.\begin{aligned} \tau &= \sigma_T/\sqrt{3} \\ \text{and} \qquad \dot{\gamma} &= \sqrt{3}\,.\,\dot{\varepsilon}_T \end{aligned}\right\} \tag{5.19}$$

where σ_T is the stress and $\dot{\varepsilon}_T$ the creep rate in the tensile test.

Bailey showed that an exact analytical solution was possible for two cases, namely those where the tensile stress versus strain rate was of the forms:

$$\sigma_T = A_1 + C_1 \ln\dot{\varepsilon} \tag{5.20}$$

and

$$\sigma_T = A_2(\dot{\varepsilon})^{C_2} \tag{5.21}$$

were A_1, C_1 and A_2, C_2 are constants. In a number of actual records it appears that a relation of the kind given in eq. (5.20) fits the resulting curve very well, and in others that of eq. (5.21) may be preferred. On the other hand, there is no need to limit these considerations to cases and conditions where either of these simple relations approximately covers the experimental data, since a graphical method can be employed. In fact the situation is very similar to that used in Section 2.8 for partially overstrained cylinders, and explained in detail in the Appendix to Chapter 2; in this case we see that we only need the shear stress versus secondary shear strain rate curve.

The use of charts, prepared in accordance with the method generally described in that Appendix, is inevitably limited to cases where the cylinder is wholly at the same temperature. Bailey however also indicated[5.11] how a similar procedure could be applied where heat flow through the walls is taking place, and where in consequence the temperature varies along a radius. The procedure then requires data on creep rates for a number of intermediate temperatures between those at the two surfaces, and the integration cannot be condensed into a simple chart as it is with constant temperature conditions.

5.3.4 *Preparation of Creep Charts*

We assume here that the tensile data available do not fit closely enough to relations of the types given by either eq. (5.20) or eq. (5.21), so that graphical methods are required; the object of the chart is to simplify this procedure.

The first step is to derive from the tensile data the corresponding shear data according to eq. (5.18) for the Tresca correlation, or to eq. (5.19) for Maxwell's. In order to illustrate the method, we describe here the preparation of a chart from the data of Skelton and Crossland[5.7] for a 0·19% carbon steel at 450°C. This has the advantage that these authors also determined the relationship between pressure and diametral creep rate, and the usefulness of the method and its likely accuracy can therefore be judged by comparison with experiment. On the other hand it must be appreciated that one would not normally use a steel of this kind for such a high temperature. Also the fact that the stress versus log creep rate data happens to approximate very closely to a straight line law in this example must not be taken as implying that this is always or even commonly so. Since the cross-sectional area is assumed to remain constant, we have:

$$(1+\varepsilon_\theta)(1+\varepsilon_r)-1 = 0 \tag{5.22}$$

whence

$$\varepsilon_\theta = -\varepsilon_r$$

and

$$\gamma = \varepsilon_\theta-\varepsilon_r = 2\varepsilon_\theta = \frac{2u}{r} \tag{5.23}$$

where u is the radial shift at radius r.

But if u_o and u_i are the corresponding radial shifts at the outside and inside radii respectively, i.e. r_o and r_i, then the condition of constant area also requires that:

$$u_o^2+2u_or_o = u^2+2ur = u_i^2+2u_ir_i \tag{5.24}$$

and, as we are only concerned here with small strains, we can write as a sufficiently close approximation:

$$\gamma = \frac{2u_ir_i}{r^2} = \frac{2u_or_o}{r^2} \tag{5.25}$$

Then to obtain the rates of creep, we can differentiate with respect to time t and:

$$\frac{\partial\gamma}{\partial t} = \frac{2r_i}{r^2}\frac{\partial u_i}{\partial t} = \frac{2r_o}{r^2}\frac{\partial u_o}{\partial t}$$

or in the Newtonian notation:

$$\dot{\gamma} = \frac{2r_i}{r^2}\dot{u}_i = \frac{2r_o}{r^2}\dot{u}_o \tag{5.26}$$

From this we can write down the values of $\dot{\gamma}$ for any particular value we choose for either $\dot{u}_i$ or $\dot{u}_o$. The former will, of course, be the biggest creep rate in the section, but the latter will be the only one that is readily measurable. As we shall see, however, the chart allows us to obtain the values of both.

In Fig. 5.5 are shown the plot of tensile stress against direct strain rate for this material as obtained by Skelton and Crossland[5.7], the specimens being cut from the same bar as the cylinder creep specimens. The two lower curves are the derived shear stress versus shear creep rate curves, the upper one according to the Maxwell correlation and the lower to Tresca's. Thus the point shown with an

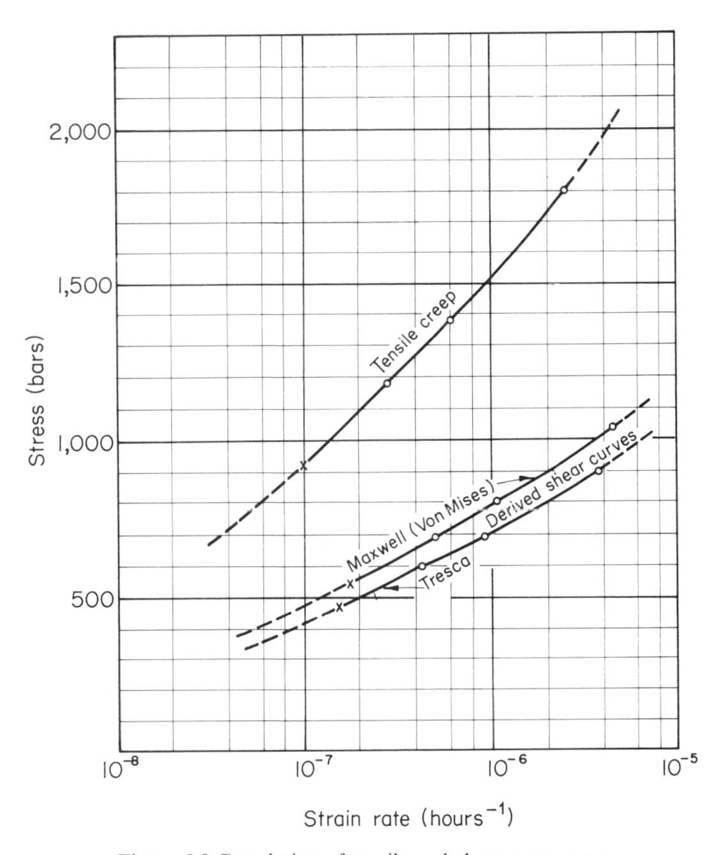

Figure 5.5 Correlation of tensile and shear creep rates

X on the tensile curve at about 6 tonf/in^2 (925 bars) and a strain rate of 10^{-7} in/in per hour corresponds (according to Maxwell) to a shear stress of $6/\sqrt{3}$, or 3·57 tonf/in^2 (535 bars), and the shear creep rate is $1{\cdot}732 \times 10^{-7}$; for the Tresca correlation the shear stress is 3 tonf/in^2 and the shear creep rate $1{\cdot}5 \times 10^7$. These points are also plotted as crosses, as will be seen.

In Fig. 5.6, the left-hand ordinate 1·0 represents a tangential strain rate of 10^{-5}, which corresponds to a shear strain rate of $2{\cdot}0 \times 10^{-5}$. From Fig. 5.5 we then see that the shear stress corresponding to such a shear strain rate is about 1,300 bars. Equation (5.26) then allows appropriate strain rates and stresses to be plotted for the corresponding radii. A further curve connecting τ and $\log r$ can then be drawn and the area under it between any two ordinates will represent one half the difference between the radial stresses at the two radii these ordinates represent. This has been done in Fig. 5.6, which shows at the top the shear stress and,

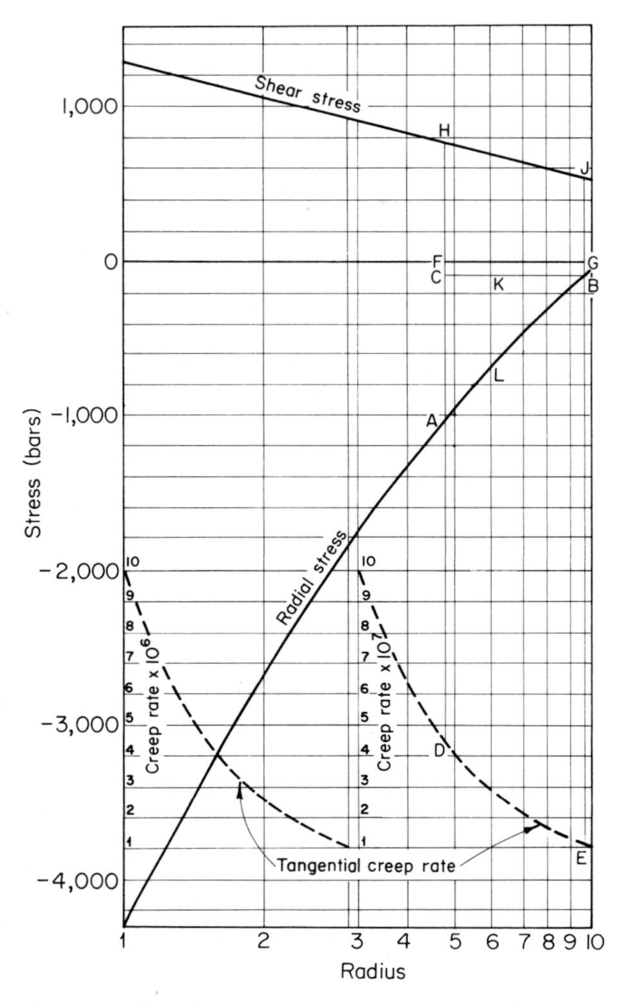

Figure 5.6 Chart for solving creep problems in cylinders of 0·19% C steel at 450°C

below the zero stress line *OFG*, the radial stress and the tangential creep rates to separate scales.

We can now illustrate the use of this chart by considering one of Skelton and Crossland's specimens; it had an outside diameter of 2 in and a bore of 1 in (i.e. diameter ratio 2·0) and it was tested at 450°C with an internal pressure of 6 tonf/in^2 (925 bars). The procedure is to take a rectangular card and mark off from the top left-hand corner a distance equal to the diameter ratio (2·0 in this example) to the scale of the chart; and then to mark off downwards from that corner a distance equal to the pressure (here 6·0 tonf/in^2). We then move the card over the chart, with its axes parallel to the chart's, until both the marks on the card fall on the radial stress curve, as at *A* and *B*. By projecting upwards from *A* to cut the zero stress line in *F* and the shear stress curve in *H*, and by drawing a horizontal through *B* to cut the first line in *C* we can read off the radial stress at any point by the intercept between the curve *AB* and the line *CB* at the appropriate radius. The corresponding shear stress curve is *HJ* relative to the line *FG*, and the tangential strain rates are found by projecting *HCA* and *JGB* to meet the appropriate curve in *D* and *E* respectively. From this we see that the tangential creep rate at the outside surface (represented by the point *E*) is about $1{\cdot}1 \times 10^{-7}$ and at the inner surface (point *D*) about $4{\cdot}4 \times 10^{-7}$. Skelton and Crossland determined the external rate in their experiment as $1{\cdot}135 \times 10^{-7}$, so the agreement is very satisfactory, although it is only fair to say that the corresponding comparisons for the higher pressure experiments were a good deal less so, the divergence being greater the higher the pressure. A possible explanation of this is that the secondary phase of creep is then more difficult to determine accurately because of its limited duration. On the other hand, the good agreement of the 6 tonf/in^2 experiment suggests that this procedure is likely to be adequate when dealing with conditions where the creep never exceeds 10^{-7} per hour, which is only a little under 0·1% in a year—and in 10 years this would mean a creep of nearly 1% which could probably be tolerated only in very special circumstances. It should also be noted that, if the correlation had been made according to the Tresca hypothesis the tangential creep rate at the outer diameter, as read from the corresponding chart, is about $1{\cdot}9 \times 10^{-7}$, thus showing that it is greatly overestimated thereby. This means, of course, that for a comparable creep rate the pressure would have to be limited to about 4·2 tonf/in^2 or say 650 bars, a decrease of 30%. Thus the Tresca correlation is very considerably more conservative than that due to Maxwell (von Mises) in this application, and the close agreement of creep rates derived from the latter with the results of careful experiment suggests that there is probably no need to use the more conservative method in actual design, at any rate for this material.

To complete the above example we can obtain the radial stress by measuring the vertical intercepts between the radial stress curve and the line *CB*, noting that the inner radius falls on the chart at $r = 4{\cdot}75$. Thus the intercept *KL*, lying on the radius $r = 6$, corresponds to a radius of approximately 0·63 in in the specimen.

The shear stress curve is obtained from the intercepts between the shear stress curve and the line *FG*, and the tangential creep rate—as we have seen—from the intercepts between DE and FG. The tangential stress, however, cannot be read direct from the chart, but must be calculated by means of the relation:

$$\sigma_\theta = 2\tau + \sigma_r$$

remembering that σ_r always has a negative value.

Fig. 5.7b shows the stress and creep rate distribution across the section of the experimental specimen when under steady creep at 450°C and under 950 bars internal pressure. The radii have been replotted on a linear scale.

5.3.5 *Creep in Cylinders Where the Temperature Varies Across the Wall*

This is the situation in a high pressure heat exchanger or any other vessel which is gaining or losing heat through its walls. In general the resulting creep is unlikely to cause serious effects unless it persists more or less steadily for a considerable time.

If we can assume symmetry in the distribution of temperature, the latter can be found from eq. (5.4). This would represent a stable state, and one can envisage situations where it might be maintained for periods of hundreds if not thousands of hours, long enough in fact to enable steady creep conditions to be reached and held until their effect became significant. It is then possible, if we have sufficient data, to evaluate the stresses and creep rates with some confidence.

Once again we can assume continuity of strain in the section so that by applying a tangential strain rate at one surface we can calculate the shear strain rates at any and every radius from eq. (5.25). Normally, however, we would not know the corresponding shear stress since this can only be obtained directly if we have stress versus strain rate (creep) curves for the temperatures at all points in the wall. Generally speaking, however, this can be dealt with accurately enough if we have stress versus creep curves for say 5 or 6 temperatures covering the range involved. With these the corresponding shear stress versus shear creep rate curves must be derived and plotted, using preferably the Maxwell correlation.

The temperature and the shear creep rate can now be found for any point within the wall, but it is no longer possible to read off the corresponding shear stress. Instead we must plot, from the family of shear creep curves, a connection between shear stress and temperature at constant shear creep rate, and from this the appropriate stress is obtained. By repeating this process at a number of points across the wall, we can obtain a curve connecting shear stress and radius, which on integration with respect to the natural logarithm of r will give half the difference in radial stress between the limiting radii.

In practical design cases the problem usually is to find the diameter ratio necessary to contain a given pressure and a given temperature difference for a given maximum creep rate, and there is no way of arriving at this directly by the procedure explained above. On the other hand, when once one has got a set of shear-stress versus shear-creep rate curves it is not unduly laborious to work out, say, three cases in which the tangential creep rate at the inner surface and the temperature difference across the wall are fixed. We then have three points on the curve relating internal pressure and diameter ratio, and if this extends on each side of the required internal pressure the required diameter ratio can easily be interpolated with sufficient accuracy for most practical needs, since the curvature will not be sharp.

Bailey[5.11] gives a graphical construction to aid a calculation of this sort, but he assumes a stress-creep relation of the type represented by eq. (5.20), which is a considerable limitation in practice. Also he is apparently more concerned to evaluate the stresses than the strain rates, and it is quite laborious to correlate these when the stresses are evaluated in this way.

5.3.6 *Effects of Intermittent Operation*

This is evidently a most important consideration; in some instances high pressure plant may for process reasons be taken off the line regularly after periods of less than 2000 hours (about 12 weeks) when they are usually blown down and sometimes cooled as well. As we have seen, if the time at temperature and pressure has been long enough for secondary creep to be established, a considerable re-distribution of stress will have taken place, even if the steady state condition (Fig. 5.7b) has not been fully reached. What happens then to the stresses, when the pressure is released?

Unfortunately we cannot answer this question from any experimental results; in fact it would be difficult to extract this information even if we had a specimen which had been under creep conditions long enough to produce these effects.

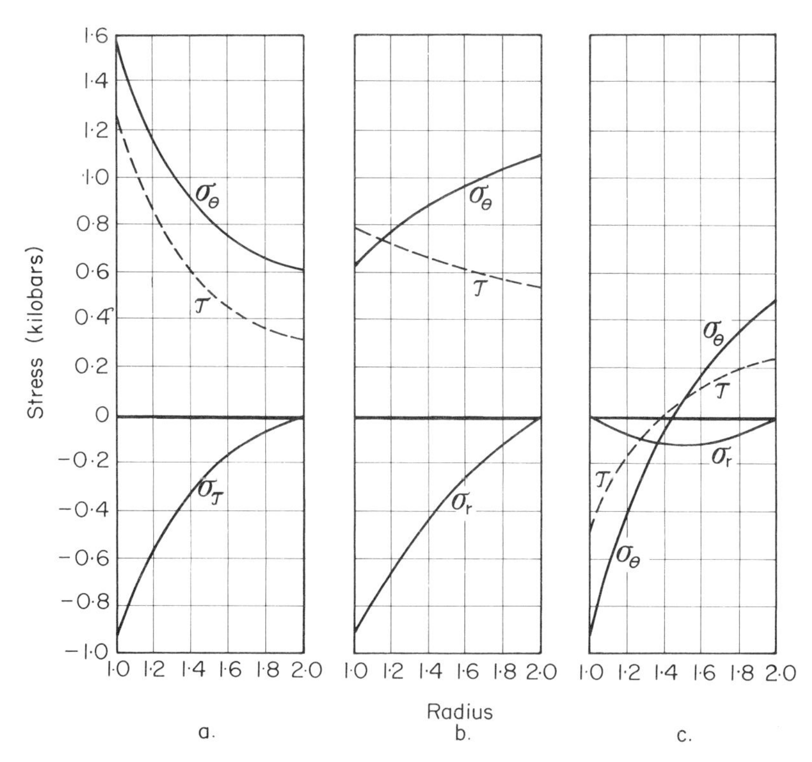

Figure 5.7 Stresses in cylinder of diameter ratio 2·0 (0·19% C steel) (a) before commencement of creep, (b) under steady state creep conditions at 450°C (p_i = 925 bars), (c) residual stresses after removal of internal pressure

Such answers as can be given must therefore be based largely on conjecture, helped perhaps by our knowledge of the somewhat analogous problems of autofrettage.

The removal of pressure, whether or not it is accompanied by a simultaneous removal of temperature, will be a very quick operation by comparison with the many hours which the creep required to change the shape of the stress distribution. Therefore it seems not unreasonable to expect it to occur as if everything was elastic, and – on that assumption – the effect will be to subtract from the stress system (developed as a result of the creep) that which would be produced by the same internal pressure in an elastic cylinder; in fact, the same stress system as we assume to be produced at the start of the operation, before even the primary phase of creep has had time to produce any effect.

Fig. 5.7 shows at (a) the elastic distribution referred to, and also at (c) the residual stresses which would be left after releasing the pressure. It will be seen that all these residual stresses are less than those produced by the original pressure application, or by the steady state condition after long creep. On the other hand, both the tangential and the shear stresses undergo a change of sign and the stress range is quite considerable in each case (2,400 bars in the former and 1,700 in the latter).

If we accept this picture of the stress behaviour during the shut-down operation, we must presumably accept also that, on restarting, the stress distribution will return more or less exactly to what it was before the shut-down; and there seems no reason why this should not happen without much change even after a considerable number of such operations. But it also seems likely that, if left standing at full temperature, but with no internal pressure, the residual stresses would be sufficient to start creep in the opposite direction, and we do not know what effect this would have on the life of the cylinder. It could perhaps lead eventually to fatigue damage if frequently repeated, but there is very little practical experience to go on, and this problem seems to be virtually untouched by experimenters.

Finally, one should not leave this subject without remembering that the above discussion has been concerned only with cylinders, a simple shape convenient for analytical consideration; but in any actual equipment there will certainly be irregularities of various kinds, e.g. screw threads, side connections, thermocouple pockets, etc., all of which will give rise to localised stress concentrations. There are good reasons for thinking that, as in the pressure testing of mainly elastic material, the resulting overstrain will tend to "spread" the stress and reduce the peaks of these concentrations; on the other hand, there have been failures for instance where nozzles have been welded into boiler drums, and this must remain a further doubt.

5.4 The Weakening Effect of Temperature

It is well known that rising temperature reduces both the yield stress and also the ultimate tensile strength of a material, and it must be presumed therefore that similar effects will occur in the walls of cylinders and spheres. Again we have to judge mainly by the results of tensile tests, but the few actual experiments seem to justify this philosophy.

The onset of overstrain in a cylinder is likely to be affected by temperature more or less pro rata with its effect on the elastic limit and yield; and correspondingly the ultimate tensile strength is the best guide to the effect on the ultimate bursting pressure. Crossland has found the latter to be true with high tensile alloy steels up to about 400°C.[5.14] For instance, the autofrettage example given in Section 2.8 would have its U.B.P. (and therefore its factor of safety) reduced by about 8% at 300°C and by 12% at 400°C.

Overstrain would commence at a pressure lowered by about the same amount, but the course of the shear-stress versus shear-strain curve in the overstrain region might be considerably altered. The actual limit of the elastic region would probably be reduced by about the same amount (i.e. 8% at 300°C), but the degree of strain-hardening would probably be less. In any case the problem here involves metallurgical considerations, which are outside the scope of this book, but the factor of safety should always be increased when elevated temperature working is envisaged.

So far as metallurgical effects are concerned, one must always use material which is unlikely to suffer major changes at the operating temperature, even when these are maintained for long periods. With hardened and tempered low alloy steels, the tempering temperature should always be at least 50°C higher than the maximum working temperature; and with precipitation hardening materials, e.g. maraging steels and high strength stainless steels, prolonged operation at temperatures near those recommended for the "soaking" treatment may actually result in increased strength, though usually at the expense of ductility and shock resistance.

References

5.1. SKELTON, W. J. and CROSSLAND, B., Conference on High Pressure Engineering, London, Sept. 1967, Paper No. 6, *Proc. Inst. Mech. Eng.*, **182,** Pt 3C, 151.

5.2. STRUB, R. A., *Trans. Am. Soc. Mech. Eng.*, **75,** 73, 1953.

5.3. JAEGER, J. C., *Phil. Mag.*, Series 7, **36,** 418, 1945.

5.4. CARSLAW, H. S. and JAEGER, J. C., *Conduction of Heat in Solids,* Clarendon Press, Oxford 1947.

5.5. Unsigned article in *The Engineer,* **207,** 56, 1959.

5.6. JOHNSON, A. E., HENDERSON, J. and KHAN, B., *Proc. Inst. Mech. Eng.*, **175,** 1043, 1961.

5.7. SKELTON, W. J. and CROSSLAND, B., Conference on High Pressure Engineering, London, Sept. 1967, Papers Nos. 3 and 7, *Proc. Inst. Mech. Eng.*, **182,** Pt 3C, 151 and 159.

5.8. SODERBERG, C. R., *Trans. Am. Soc. Mech. Eng.*, **61,** 737, 1941.

5.9. CHITTY, A. and DUVAL, D., Joint International Conference on Creep, 1963–4, Paper 2, *Proc. Inst. Mech. Eng.*, **178,** Pt 3A.

5.10. NORTON, F. H. and SODERBERG, C. R., *Trans. Am. Soc. Mech. Eng.*, **64,** 769, 1942.

5.11. BAILEY, R. W., *Engineering,* **129,** 772, 785 and 818, 1930.

5.12. BAILEY, R. W., *Proc. Inst. Mech. Eng.* **164,** 324, 1951.

5.13. SHEPHERD, W. M., *Proc. Inst. Mech. Eng.*, **159,** 95, 1948.

5.14. CROSSLAND, B., *High Temp. High Press.*, **1,** 133, 1969.

5.15. KING, R. H. and MACKIE, W. W., *J. Basic Eng., Trans. Am. Soc. Mech. Eng.*, **89D,** 877, 1967.

6 The Effects of Fatigue

6.1 Historical

The I.C.I. process for polymerising ethylene was discovered in 1933 in the Company's Northwich laboratories of what is now called their Mond Division. After considerable early difficulties a commercial process was established and production began in September 1939. The working pressure was around 1,500 bars, and this was much higher than that of any other industrial process operating at the time.

The product was required in large quantities for various war purposes, and time did not permit the complete testing and development of each plant unit. In particular, compressing the ethylene to well over a kilobar involved many uncertainties; no compressor manufacturer then had experience of building and running a machine under such conditions, and I.C.I. had to devise a mechanised version of the mercury displacement system which had been used extensively in the laboratory. The essential feature of this was that the pressure was generated in oil and transferred to the gas by means of a U-tube containing a lute of mercury.

Although there had been some misgivings about this type of machine at the outset, its performance proved very successful and virtually all I.C.I.'s wartime output of well over 4,000 tons of this material was made from gas compressed in machines of this kind. The first design, however, developed some unexpected fatigue troubles in the U-tubes after several months of satisfactory running, and it was these that led I.C.I., after the war, to approach Professor J. L. M. Morrison of the Mechanical Engineering Department at Bristol University with a view to carrying out a fundamental study of fatigue in containers subjected to repetitions of pressure.

The possibility of fatigue had certainly been appreciated when these machines were designed, and it was thought that an ample allowance had been made. One of the difficulties, however, was that most of the available fatigue data had been obtained from experiments with rotating beam machines of the Wöhler type, and this had to be correlated with the much more complex stress systems existing in the compressors. Actually, these failures all occurred in straight cylindrical parts and not – as might have been expected – in the neighbourhood of the valves where the stresses could not be analysed with any confidence. They were probably accelerated by poor machining in the bore and – in one instance, almost certainly – by non-metallic inclusions in the steel.

The particular feature which distinguished these fractures from any others that had been encountered in the course of this work was their appearance. The actual cleavage was smooth and straight in a radial plane containing the axis, and there was no sign of the permanent deformation which is always present when a tube of this material is statically tested to destruction. Fig. 6.1 is a photograph of one of the specimens tested at Bristol; for comparison Fig. 6.2 shows a cylinder of similar material burst by static pressure. The actual fatigue crack is the fan-shaped patch in the middle of the upper wall, and it is evident that it started at the centre on the inner surface and spread by radiating outwards in all directions in that plane. This went on until it just penetrated the outer surface, whereupon leakage occurred and the machine was automatically stopped. In the compressors, on the other hand, the cracks sometimes went unnoticed until

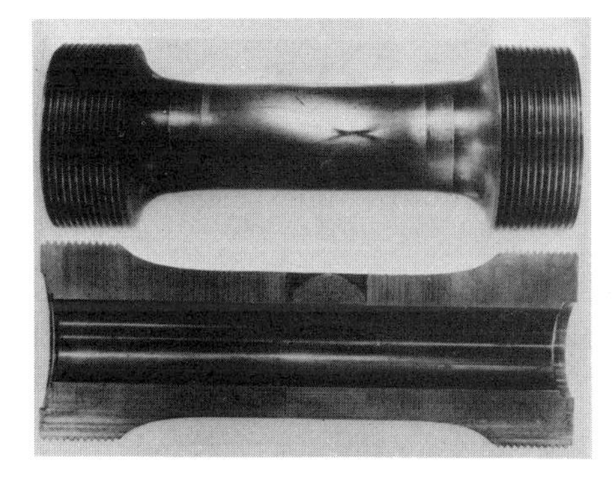

Figure 6.1 Fatigue failures in cylinders subjected to repetitions of pressure *(Bristol University Mechanical Engineering Department)*

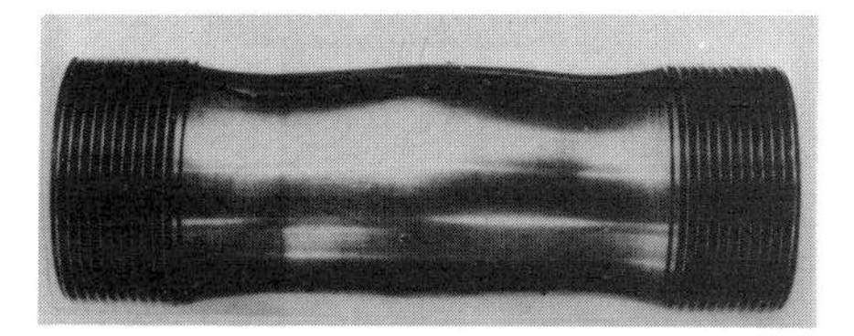

Figure 6.2 Cylinder burst by steadily increased internal pressure

they had spread for a considerable distance along the outer surface, the material being so strong and the crack so fine that there was no overstrain in any part of the section, and the crack became almost invisible when the internal pressure was removed. The slow development and gradual spreading of these cracks shows that the material had exceptional fracture toughness; by contrast, a series of specimens of an aluminium alloy (D.T.D. 364) tested at Bristol failed by what must have been almost instantaneous crack propagation, running right through the whole thickness of the wall.

The failures in the wartime compressors suggested that there was something badly wrong with the design procedure adopted. The fatigue strength of the steel, as measured by reversed bending, was known to be about $\pm 4{,}500$ bars, direct stress, while the maximum tensile stress in the walls was no more than 2,800 bars; thus the range would be from 0 to 2,800, whereas the material was apparently able to stand a total range of 9,000 bars, and this should have been more than enough to take care of any small stress-raisers that might have been present. However, as a result of the Bristol work, we now know that it is better to use a shear stress basis for design, and that—for repeated applications of pressure—the shear stress with this material should be kept below about 2,800 bars. In fact, the shear stress range in these compressors was probably about 2,300 bars, so there must have been some stress-raising effect, although this would hardly have caused trouble if the design basis had been justified.

Thus, the work at Bristol has revealed some very unexpected fatigue behaviour, which is still by no means fully understood.

6.2 The Work at Bristol University

This has now developed into a comprehensive study of the phenomena associated with the fatigue of pressure containers and therefore justifies a fairly complete summary here.

The first contacts were made in 1949, but the next two years were mostly spent in the design and construction of the machine, and in overcoming various minor difficulties with it. This does not concern us here, but the story is admirably told in the first of the papers published by Professor Morrison and his colleagues Professor B. Crossland and Dr. J. S. C. Parry[6.1], which also contains some of the earlier results. The machine, with only minor modifications, is now functioning almost perfectly, and many similar machines are in use in various parts of the world. The maximum of the pressure cycle was initially about 3,000 bars, but this has since been extended to more than 5,000.

The material first studied was the En 25T steel used in the compressors which failed, and which is similar to that used for the various static tests described in Chapter 2, except that it was specially selected by the manufacturers and therefore likely to be rather cleaner and more uniform than their ordinary product. The ultimate tensile strength of this material as tested was about 8,500 bars, but a similar composition was later heat-treated to 10,300 bars and similarly tested. A number of other materials were tested, as summarised in Table 6.1, and in each case conventional static and fatigue tests were carried out to give a valid comparison.

The strengthening effects of various surface treatments were dealt with as well as the weakening effect of radial holes and other irregularities.

6.2.1 *Preparation of Specimens*

The general shape of the specimens used is shown in Fig. 6.3. For the first series of tests, up to a diameter ratio of 2·0, the bore of each specimen was 1 in, but for the 3 to 1 ratio it was necessary to reduce this to 0·6 in since the section of the bar was not big enough. A few tests were also carried out with specimens $\frac{3}{4}$ in o.d. by

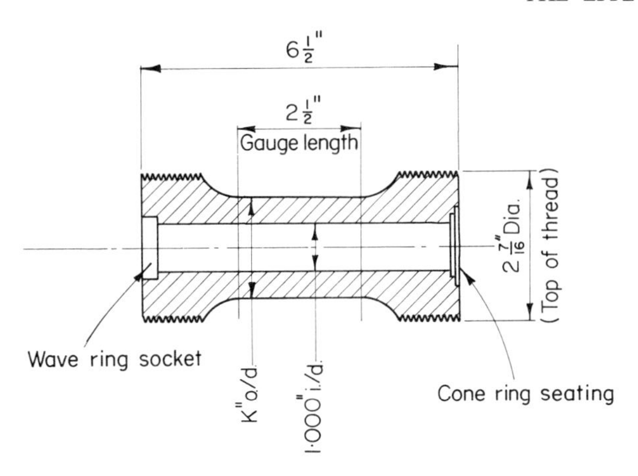

Figure 6.3 Specimen used by Morrison et al. (Ref 6.2)

$\frac{3}{8}$ in bore. All these are of course small, but they extend over a fair range of sizes and this shows no effect on the results.

Each specimen after machining was finished by honing with a diamond lap to a smoothness of 0·1μ or better, before stress relieving in vacuo, usually at a temperature not exceeding the tempering temperature less 50°C. It should be noted that this special treatment differs very considerably from ordinary machine shop practice, and it has been criticised on those grounds. Morrison however has continued to use it for the very good reasons that it gives more consistent results, and that these are *lower* than those obtained normally. Incidentally, tests using X-ray techniques on specimens honed, but not stress relieved, showed that the surface was carrying a compressive stress of the order of 3,000 bars, which would account for their greater endurance in these tests.

6.2.2 *Typical Results*

Fig. 6.4 is reproduced from the second paper published by Morrison and his colleagues[6.2] and shows a few typical results for the En 25 steel. In all cases the mean pressure was half the maximum.

There is a considerable scatter for each diameter ratio, but this is apt to occur with any kind of fatigue test. What is clear, however, is that there is a very definite "knee" in each curve and relatively few points representing fatigue failure to the right of the knee.

On the same diagram are plotted the results of four similar tests on commercially pure titanium. These are connected by a broken line and show quite different behaviour; the slope is much less steep and there is no sign of a break as at the so-called "knees" which are so clearly marked with the steel. This appears to be characteristic of most non-ferrous metals, implying that there is no stress so low that it will not cause fatigue if the repetitions go on long enough. The conception of an indefinite fatigue limit, i.e. a stress value below which no fatigue damage will result, is fairly generally accepted with steels for all kinds of cyclic loading, and thus the tests here described appear to conform—at least qualitatively—with our normal ideas on the subject.

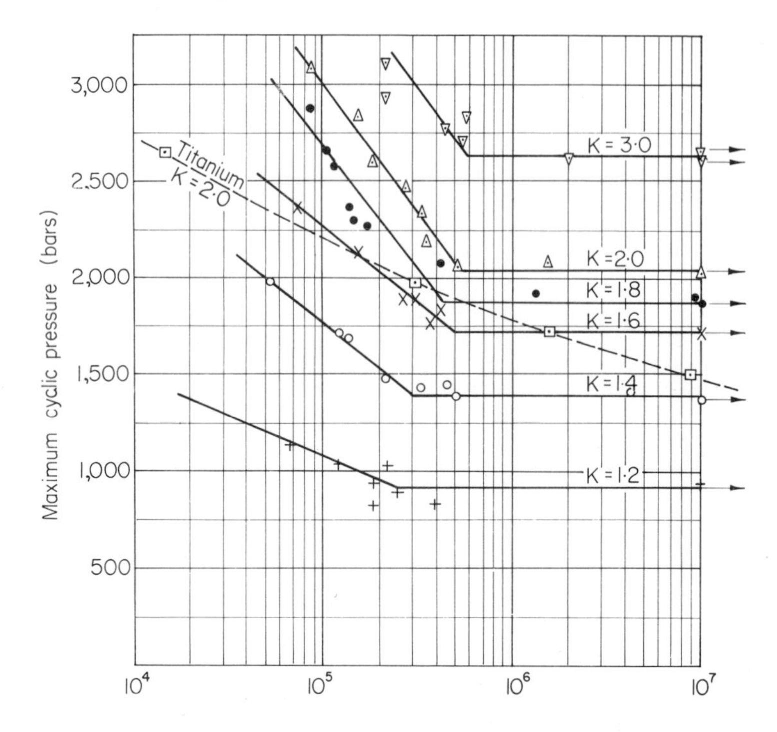

Figure 6.4 Correlation of internal pressure and log. (number of repetitions to failure) for various diameter ratios in B.S.En 25 steel – also in broken line for titanium metal of diameter ratio 2·0

Table 6.1(a)

Analyses and heat treatments

Material and code	*Analysis %*								*Heat treatment*	*Ultimate tensile strength bars*
	C	*Si*	*Mn*	*Ni*	*Cr*	*Co*	*Mo*	*Ti*		
Mild steel	0·15	0·047	0·66	—	—	—	—	—	Normalised	3,880
En 25T	0·30	0·15	0·66	2·55	0·58	—	0·59	—	Oil-hardened and tempered	8,600
En 25V	0·32	0·21	0·60	2·68	0·68	—	0·52	—	,,	10,300
En 40S	0·25	0·19	0·45	0·22	3·27	—	0·51	—	,,	8,300
En 56C	0·23	0·38	0·18	0·32	12·78	—	—	—	,,	7,100
Stainless	0·06	0·70	1·56	8·60	18·2	—	—	0·34	Softened[(1)]	5,900
En 31	1·05	0·32	0·36	—	1·35	—	—	—	Special[(2)]	17,200 to 19,300
Maraging	0·015	0·10	0·03	18·1	0·03	7·3	4·4	0·52	Special[(3)]	17,900

(1) From 1100°C water quenched.

(2) Austenitized for 30 min at 840–850°C in neutral salts, followed by quenching to 220°C and holding for 15 min, and air-cooling before tempering at 220°C for 9 hours.

(3) Annealed at 820°C after rolling, then age-hardening for 3 hours at 480°C.

Table 6.1(b)

Results of Static and Fatigue Tests

Material and code	Ultimate tensile strength	Limit of proportionality in torsion	Fatigue Limits				Ratios		
			Direct stresses		Shear stresses		Shear stress limit in cylinder repetitions to range of reversed torsion	Shear stress limit in cylinder repetitions to mean range of reversed bending	Shear stress limit in cylinder to ultimate tensile strength
			Reversed bending		Range of reversed torsion	Range of stress due to pressure in cylinders			
			Range in longitudinal direction	Range in transverse direction					
Mild Steel	3,880	1,780	±2,250	±1,900	±1,360	0 to 1,340	0·495	0·324	0·345
En 25T	8,600	4,040	±4,930	±4,320	±3,000	0 to 2,870	0·48	0·310	0·330
En 25V	10,300	4,620	±6,170	±5,060	±3,650	0 to 3,630	0·50	0·323	0·353
En 40S	8,300	2,930	±5,000	±4,050	±2,870	0 to 2,970	0·52	0·328	0·355
En 56C	7,100	2,770	±4,030	±3,770	±2,340	0 to 2,550	0·545	0·327	0·358
Stainless	5,900	620	±3,900	±3,790	±1,840	0 to 1,670	0·455	0·217	0·283
En 31	17,200 to 19,300	11,300	±6,800	±4,480	±5,100	0 to 3,470	0·34	0·308	0·18 to 0·20
Maraging	17,900	7,580	±5,770	—	±3,860	0 to 3,470	0·45	0·30	0·194

All stresses in bars

6.2.3 *Correlation of Results*

In the original discussions between Morrison and the representatives of I.C.I., before the work at Bristol started, it was thought that there might be a useful correlation with repetitions of torsion, seeing that in both cases the loading is essentially one of shear, although the cylinders have a hydrostatic tension superimposed thereon. It is known from the results of other kinds of fatigue testing that shear stress plays a considerable part in initiating fatigue damage, but tensile stress is also concerned in this. The results of the tests on the 6 diameter ratios in the En 25T steel were therefore replotted, first on the basis of the maximum tensile stress (i.e. the tangential stress at the inner surface) and secondly on the basis of maximum shear stress. The results, shown here in Figs. 6.5 and 6.6, leave little doubt as to the superiority of the latter, at any rate for this material. Thus, if we consider the specimens of each diameter ratio that remained un-

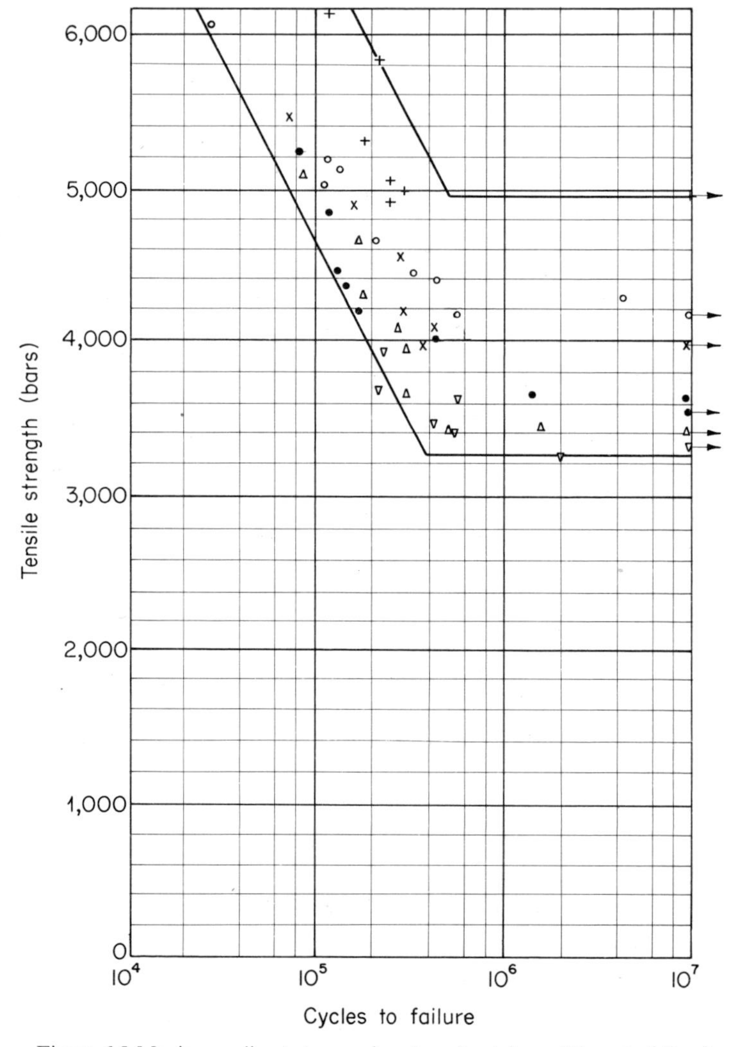

Figure 6.5 Maximum direct stress v log. (number of repetitions to failure) B.S.En 25

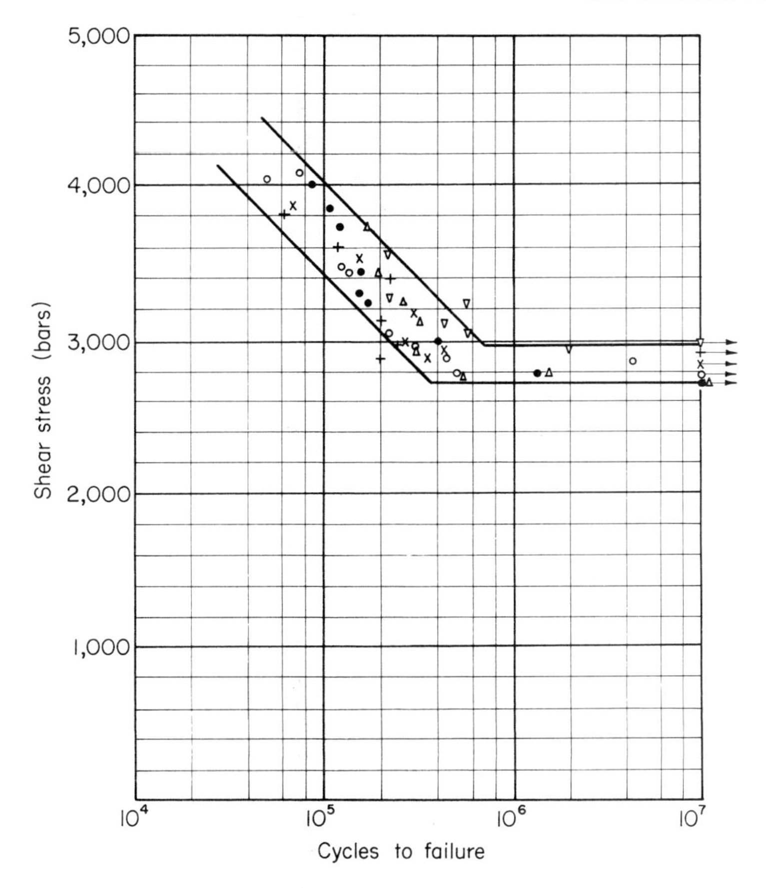

Figure 6.6 Maximum shear stress v log. (number of repetitions to failure) B.S.En 25

broken after 10 million repetitions, their tensile stresses ranged from 3,340 to 4,980 bars, or very nearly $\pm 20\%$ above and below the mean value of 4,160 bars. The shear stresses, on the other hand, ranged from 2,700 to 3,000 bars, or $\pm 5\frac{1}{4}\%$ above and below the mean value of 2,850 bars.

Similar correlation was found in most of the other materials tested in the form of simple cylinders, and for the steels (other than an austenitic stainless material) it appears that the limiting maximum shear stress in the cycle which can be endured indefinitely is about one third of the ultimate tensile strength, at least up to values of the latter of 10,000 bars. This is, of course, completely empirical, but it has been found to hold for at least 5 different steels.

6.2.4 *Effect of Varying the Mean Stress*

All the earlier work at Bristol was concerned with cycles in which the pressure rose to a specified maximum value and then dropped almost to zero, thus making the mean stress half the maximum, and also half the range of stress. Later however a simple method was found for varying both the mean stress and the range. The results of the tests thus carried out are given by Burns and Parry[(6.3)], who also discuss them from the designer's point of view. Incidentally, their paper gives a good bibliography of other published work in this field; see also Morrison, Crossland and Parry(6.4) and Davidson, Eisenstadt and Reiner(6.5).

For the material En 25 (in the condition of 8,500 bars U.T.S.), the apparent fatigue limit conditions, i.e. the repeated stress systems which are likely to be endurable for indefinitely long periods, are probably best represented by the diagram given in Fig. 6.7, although there appears to be some influence from the direct stresses, depending on whether they are tensile or compressive. At the time of writing, however, this phenomenon has not been fully elucidated, and a diagram such as Fig. 6.7 is probably the best guide yet available.

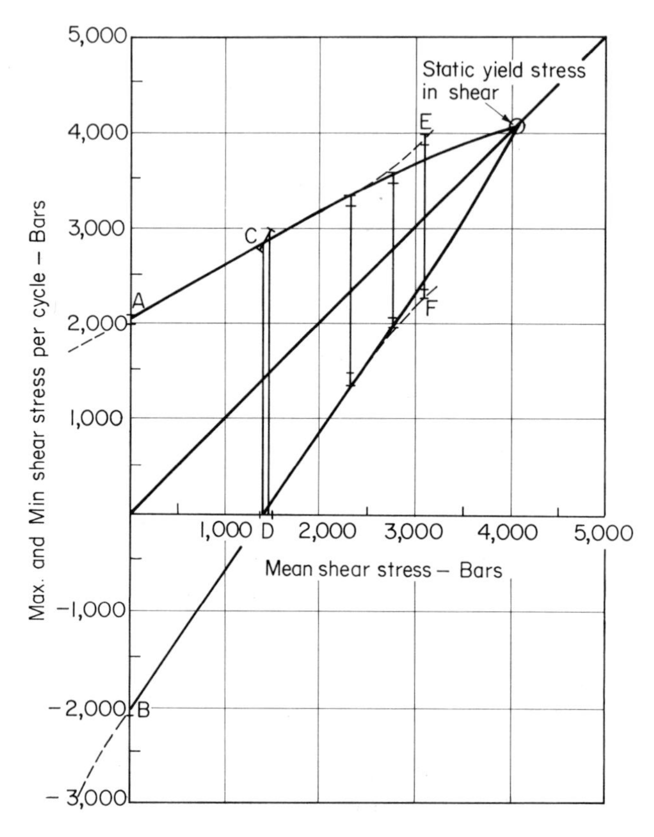

Figure 6.7 Diagram showing effect of mean stress on endurance of B.S.En 25 steel

The ranges of stress are plotted with the maximum and minimum values on each side of the mean stress point, all stresses being shear and the values at the inner surface. Curves drawn through these points are nearly straight and meet near the static yield stress in shear. It is not clear what significance this has, especially as the maximum and minimum points for the highest mean stress, *E* and *F*, suggest that the range is there tending to increase, as indicated by the broken lines.

At the lower end of the diagram there may be some doubt about the points *A* and *B*, since they were obtained by testing a series of duplex cylinders, the inner component being the actual specimen. In this way it was possible to get close to a zero mean stress, but—as has been explained in Chap. 3, Section 3.2—it is impossible to avoid setting up some longitudinal stresses in the shrinking operation, and this may affect the contact pressure and hence the stresses acting when the cyclic pressures are applied.

The points *C* and *D* represent the bulk of the early Bristol experiments, the points of all the specimens which withstood 10^7 applications of pressure lying within the two close parallel lines. Some attempts have been made to find how the bounding lines would run if projected further to the left, i.e. to negative values of the mean stress. So far, however, the behaviour in this range is not consistent enough to mark on the diagram, but it appears that the bounding curves diverge more rapidly than would straight lines; i.e. the upper becomes concave and the lower convex when viewed from above. This is a further indication of the strengthening effect of compressive stresses, although more research is needed before this can be fully formulated.

6.2.5 *Effects of Surface Treatment*

It has already been stated that honing without stress-relieving increases the endurance of cylinders to fatigue, and that this was explained by the setting up of quite large compressive stresses in the bore layers. We can therefore expect that any other treatment which introduces compressive stresses will be similarly beneficial.

Nitriding is one such treatment. Normally used to put a very hard surface on a fairly ductile core, it may also prove a useful way of prolonging the life of such parts as compressor cylinders, where the hardness is also valuable for wear resistance. With a simple cylinder the endurance appears to be raised by some 30% above its value for the honed and stress-relieved specimens, see Parry.[6.6]

Autofrettage also has a very beneficial effect, especially where stress-raisers, such as radial holes, are present. The probable explanation of this is that any localised stress, whether due to locked-in stresses resulting from heat-treatment, cold work, or bad machining, or to irregularities of shape such as holes, screw threads, bad welds, etc., or to slag inclusions or other flaws in the metal, will cause preferential yielding which then redistributes the stress and smooths out the peak values. Thus conditions favouring fatigue damage at relatively low applied stresses are removed.

The results of this will be considered further in the next section, but it is worth noting here that this phenomenon provides a good reason for applying overpressure tests to containers which are likely to undergo fluctuating stress loading.

6.2.6 *Effects of Small Radial Holes*

It is often desirable to drill small holes through the thick walls of high pressure vessels for inserting thermocouples or other control devices. This inevitably weakens the vessel, but until the problem was studied experimentally at Bristol there was no clear idea as to the extent of this weakening. It is common practice to put a strap round the section which has been drilled, thereby providing external support for the closure or connection to the hole, but again no one knew how effective this was in overcoming the effects of the hole.

The Bristol team investigated this problem and their first paper on it was published in 1959.[6.7] They produced an analytical solution, by assuming that the diameter of the hole was negligible compared with that of the cylinder, and this suggested that the maximum shear stress would be raised by a factor of 2.5.

Their tests were carried out with specimens of the En 25T steel, having a main bore of 1 in and an outside diameter of $2\frac{1}{4}$ in. Three holes running right through the specimen from one side and out the other were drilled in each, one of $\frac{3}{16}$ in diameter and two of $\frac{1}{8}$ in. The holes were closed either by a steel ball supported by an external strap, or by a plug screwed into the wall itself. These are shown diagrammatically in the inserts on Fig. 6.8, which also gives the results. Incident-

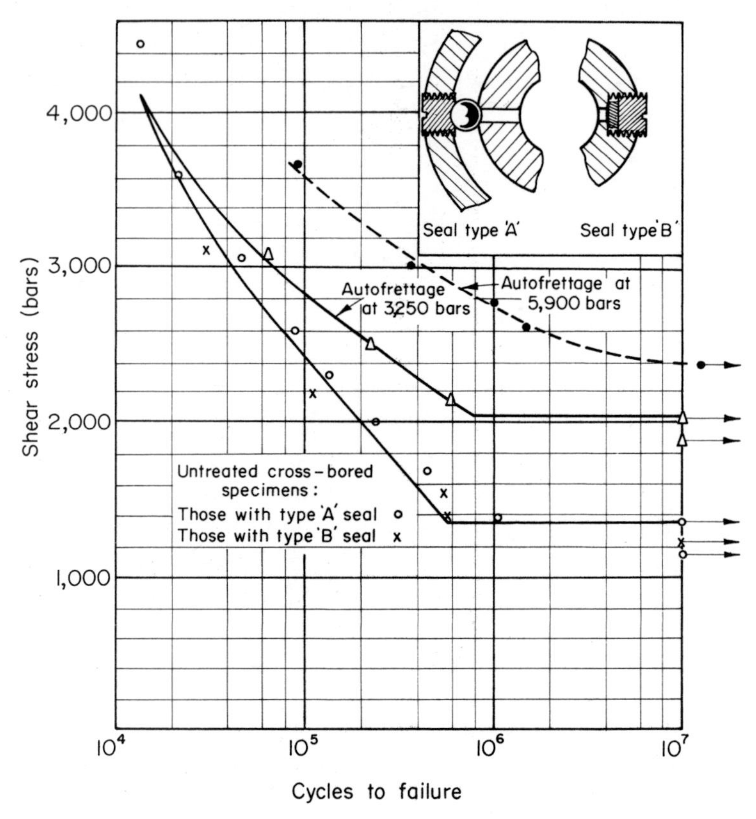

Figure 6.8 Effect of small radial holes on fatigue strength of cylinders of B.S.En 25 steel; also effects of autofrettage (from Ref 6.6)

ally, the bores of these specimens were not lapped since the side holes were obviously much greater stress-raisers than any local roughnesses left by the machining operations.

These results were remarkably consistent, as this figure shows, and it should be noted that the fatigue cracks invariably began at the intersection of the hole with the main bore, and there was no indication that they were more liable to start at a $\frac{3}{16}$ in hole than at the smaller $\frac{1}{8}$ in holes; nor did the method of supporting the side bore closures appear to have any effect.

The stresses plotted in Fig. 6.8 are the maximum shear stresses set up in the wall of a corresponding undrilled cylinder. Thus the ratio of the fatigue limit shear stress as shown in Fig. 6.6 to that in Fig. 6.8 should give the stress concentration factor resulting from the side bore holes. As will be seen, this is appreciably less than the theoretical value of 2.5, being about 2.1. Incidentally this has been confirmed by Dr. Fessler of Nottingham University in the course of an investigation of a somewhat similar problem by means of photo-elastic techniques.[6.8] It appears also that, if the intersecting holes can have the sharp edges removed, i.e. if smooth rounded corners can be produced at these edges, the stress-raising effects can be considerably reduced. Mönch and Ficker[6.9] suggest that this reduction might be as much as 30%.

This smoothing would be prohibitively difficult to achieve with $\frac{1}{8}$ in holes intersecting a 1 in bore, but in large vessels, even if it had to be done by hand, the small extra cost would be a small price to pay for greater safety and increased life. There is however the possibility that overstrain might appreciably reduce the resulting stress concentrations, and this was also investigated at Bristol. The upper full line curve shows the effect of a series of tests on specimens which had previously been subjected to a static pressure of 3,250 bars, which would just about cause incipient overstraining in the undrilled inner surface, and therefore—presumably—considerable overstrain at the side bore intersections. The fatigue limit has thus been raised by at least 50%. Overstrain to the geometric mean radius was also tried, using an initial static pressure of 5,900 bars, and the fatigue test series is shown by the broken line and by the black points in Fig. 6.8. In this case a further improvement has been brought about, raising the fatigue limit by at least 80%.

6.2.7 *Other Materials*

As we have seen, most of the Bristol work so far published has been concerned with various types of steel. A few other materials have, however, been tested in the form of simple cylinders, of which the most important are tabulated in Table 6.2.

None of these three materials gave very consistent results. In particular there was no clear indication of a "knee" in any of the curves, so that one cannot speak of a fatigue limit; instead the figures in the last column relate to the shear stresses in cylinders to withstand 10^7 repetitions of pressure.

For the titanium the shear stress was appreciably different for the two diameter ratios tested. This may be explained by the probability that this material has an endurance limit well above its elastic range, and in consequence there is always some autofrettage occurring during the initial pressure cycles. The two curves shown in Ref. (6.2) appear to be straight over the last decade shown and they would meet if produced for about another decade, i.e. to about 10^8 repetitions. The shear stress would then be down to 1,500 bars approximately, which corresponds well with the static limiting elastic shear stress.

The beryllium copper results were disappointing on account of the scatter of the points, but this was traced to considerable variations in grain size from place to place along the bar from which the specimens were taken. This is a frequent trouble with this material for reasons which appear not to be fully understood.

The aluminium alloy also had certain peculiarities. In the first place the slope

Table 6.2
(All stresses in bars)

Material	*Analysis*	*Heat treatment*	*Ultimate tensile strength bars*	*Range of shear stress in cylinder bars for 10^7 repetitions*
Aluminium alloy (DTD 364)	4·2% Cu, 0·8% Si, 0·49% Mg, 0·77% Mn, 0·37% Fe, 0·012% Ti	Solution treated at 512°C. Aged for 5 hours at 175°C	5,030	0 to 1400
Titanium (commercially pure)	0·055% C, 0·04% Fe, 0·002% H_2 – rest Ti	Annealed at 700°C for ½ hr	4,130[1]	varied between 0 to 1740 and 0 to 1970
Beryllium copper	1·8% Be, 0·23% Co, 0·05% Fe, rest Cu	Solution treated at 800°C, water quenched, then precipitation hardened at 315°C for 3 hours	11,600	0 to 2,080

[1] At a strain rate of 0·0003 per second

of the lines on the shear stress versus logarithm of numbers of repetitions was much flatter than in any of the other materials tested. In spite of considerable scatter there can be no doubt about this; for instance lowering the shear stress (and therefore also the pressure) by 20% corresponds with a 60-fold increase in the endurance. The other peculiarity is the type of fracture. Evidently this is a material in which the crack spreads almost instantaneously and in every broken specimen the failure ran right through the wall in an axial-radial plane and even went through the enlarged ends.

6.3 Other Reported Experiments

Several investigators have studied the effect of high stress fatigue leading to failure after numbers of repetitions much fewer than in any of the Bristol results. For instance, Austin and Crossland[6.10] describe a machine capable of applying repetitions of pressures up to about 6,000 bars. This was applied to both plain and cross-bored cylindrical specimens of steel similar to the En 25T, both in analysis and mechanical properties. The results lay very close to the backward extrapolation of the mortal range lines obtained by Morrison et al as already described.[6.2] With diameter ratios of 1.4 and 1.8 this was remarkably consistent up to within a few per cent of the pressure causing overstrain throughout the whole thickness of the cylinder walls. Even then the endurance was nearly 10,000 repetitions. Very slight increases of pressure then caused failure after considerably fewer repetitions, and in one experiment a 1.8 ratio specimen endured 250 cycles of a pressure of 5,100 bars, which was within 7% of its Ultimate Bursting Pressure.

Austin, Reiner, and Davidson[6.11] describe some rather similar experiments with actual gun barrels of 105 mm and 175 mm calibre, both before and after the rifling had been machined in. Their material was similar to SAE 4330, and compared with the En 25T, contained rather less nickel (about 2·2 %) and rather more chromium (0·95 %) with the extra addition of about 0·1 % of vanadium; the ultimate strength, however, was considerably greater lying between 12,500 and 13,300 bars. The results were reasonably consistent, but after careful analysis and statistical appraisal they concluded that the maximum direct stress (i.e. the hoop stress at the bore) provided a better parameter for comparing the behaviour of different diameter ratios than the maximum shear stress. It is possible that this may be a correct law for the harder materials although the fact that the shear comparison was preferred in virtually all the Bristol work, ranging over a number of materials of widely varying strength, should not be forgotten. It should also be noted that the specimens in this work had their end forces externally supported so that the axial stress σ_z would have been substantially zero.

Jones and Tomkins[6.12] investigated the types of fatigue produced by repetitions of pressure in very thick-walled cylinders, i.e. up to a diameter ratio of 7, in low carbon steels. In most of their experiments the material was grossly overstrained and they were able to correlate the endurance in terms of the cyclic external tangential strain. They also showed that, under certain circumstances, the strain cycle would settle down to a constant value after about 2,000 repetitions, but would then start to rise again shortly before fracture. The particular importance of this work would seem to be the high values of the diameter ratio to which it has extended; unfortunately, however, being confined to mild steel—due no doubt to the pressure ceiling of about 5,000 bars of their apparatus—the results would seem to be of limited value, but it is to be hoped that these workers will find a way to raise the pressure and apply the technique to stronger materials, since the correlation by means of strain could be very valuable in dealing with limited life problems in ultra-high pressure studies.

Morrison, in some of his early work as reported in Ref. (6.1), suggested that the surprisingly low endurance of cylinders to repetitions of internal pressure when judged by the results of other fatigue tests might be explained if the oil transmitting the pressure in some way attacked the metal surface, and he mentions the possibility of increasing the endurance by protecting this surface by plating, or by lining with non-metallic sheaths of various kinds. This was studied by Frost and Burns[6.13] who subjected specimens of the En 25T material to a static internal pressure with a superimposed "push-pull" fluctuating axial load. In some experiments thin rubber sheaths were used to protect the bores, and various liquids were used to transmit the pressure, both with and without this protection. The conclusion was, however, that no effect on the fatigue life could be attributed to any attack on the highly stressed metal by these liquids.

6.4 Design Procedure

The available data for dealing with problems of design involving cyclic fluctuations of pressure is still scanty, and the work at Bristol described above still seems the best guide. For repetitions of pressure in which the minimum in each cycle is zero and for ferritic or martensitic steels with tensile strengths not exceeding 10,000 bars (145,000 lbf/in^2) a limiting maximum shear stress of about 33% of the tensile ultimate seems a reasonable design basis, although if the material is notably different, either in analysis or thermal history, from those listed in Table 6.1 it would be prudent to allow a rather larger factor of safety, since the 33% rule is purely empirical.

With other kinds of pressure cycle, i.e. where the mean shear stress is not half the maximum, the only guide would appear to be the diagram shown in Fig. 6.7. This was obtained from tests on the En 25 material only, but similar tests on mild steel gave comparable results. In all such cases it would be desirable to take account of the uncertainties by allowing rather generous factors of safety.

The suggested procedure is illustrated below by a few examples.

Example 1*a*. A ram pump is required to pump oil in one stage from atmospheric pressure to 500 bars. The cylinder is to have a bore of 10 cm and is to be made from En 25T material with a tensile strength of 8,500 bars. What should be the thickness of the cylinder wall?

Here the material specified is similar to that tested very completely in the Bristol work, and as the cycle is similar to that of most of their experiments we suggest that a safety factor of 33% would be adequate.

Then if we take a value of 2,800 bars as the fatigue limit in shear for this material, the design working shear stress must not exceed 2,800/1·33 = 2,100 bars. The maximum pressure in each cycle is 500 bars, so that, if K is the diameter ratio, we must have, from eq. (2.29),

$$2{,}100 = 500 \frac{K^2}{K^2 - 1}$$

from which $K = 1{\cdot}146$. Thus the outside diameter would be 11·46 cm and the wall thickness 0·73 cm.

There would probably be other reasons for making the cylinder wall thicker than this, and therefore in this particular problem fatigue is unlikely to be the governing consideration so far as the cylinder is concerned.

Example 1*b*. Here the requirement is the same as in the previous example, but the material specified is a medium carbon steel to B.S.S. En 10, having an ultimate tensile strength of 7,000 bars.

Here we have no direct experimental determination of the endurance to repetitions of pressure to help us, and we must therefore resort to the empirical rule which suggests that the limiting shear stress is 33% of 7,000, or 2,333 bars. However, in view of the uncertainty introduced by this empiricism it will be prudent to use a rather higher factor of safety, say 50%.

Thus the shear stress we can allow is 2,333/1·5 or 1,560 bars. Then the diameter ratio K will be given by:

$$1{,}560 = 500\frac{K^2}{K^2-1}$$

from which $K = 1{\cdot}213$. Thus the outside diameter would be 12·13 cm and the wall thickness 1·07 cm.

Example 2. A compressor is required to raise a hydrocarbon gas from a suction pressure of 500 bars to a delivery of 1,500 bars. The bore of the cylinders is 7·5 cm and they are to be made from steel to B.S.S. En 56C. What should be the wall thickness?

Here we have a material for which the endurance to repetitions of the full pressure falling to zero once a cycle has been measured, but in this case the lower pressure is not zero but $\frac{1}{3}$ of the maximum. Some uncertainty is therefore introduced, and we must work by analogy with the En 25 material to which Fig. 6.7 applies. There we see that with a mean shear stress of 2,200 bars the range of endurance is between 3,300 and 1,100 which is the same ratio as that required in this problem. It is however for a stronger material, having a shear fatigue limit of about 2,870 bars compared with 2,550 for the En 56C. We assume then that the corresponding figures will be reduced pro rata so that the range of shear stress that can be allowed is from a maximum of:

$$3{,}300 \times (2{,}550/2{,}870) = 2{,}920 \text{ bars}$$

Again we must allow for the uncertainty of this correlation with a slightly higher factor of safety, say 50%, and we can therefore design for a maximum shear stress of 2,920/1·50 = 1,950 bars.

The diameter ratio K is then given by:

$$1{,}950 = 1{,}500\frac{K^2}{K^2-1}$$

from which $K = 2{\cdot}082$. Thus the outside diameter would be 15·615 cm and the wall thickness 4·06 cm.

6.5 Retarding Effect of Compressive Stresses

Several of the more successful manufacturers of high pressure piston compressors have found by experience that considerable benefits are to be obtained in places where large and unanalysable stresses may be expected, as in the heads and valve passages for instance, by applying compressive stresses by external means. Some examples of this are mentioned in.[6.14] In some cases it is possible to make the pressure itself supply some of the compressive effects. Fig. 6.9 is an illustration from an I.C.I. patent[6.15] which is more or less self-explanatory; the valves are mounted within an enclosed space which can either be allowed to fill with gas from the delivery system by drilling a small duct as shown by the broken lines

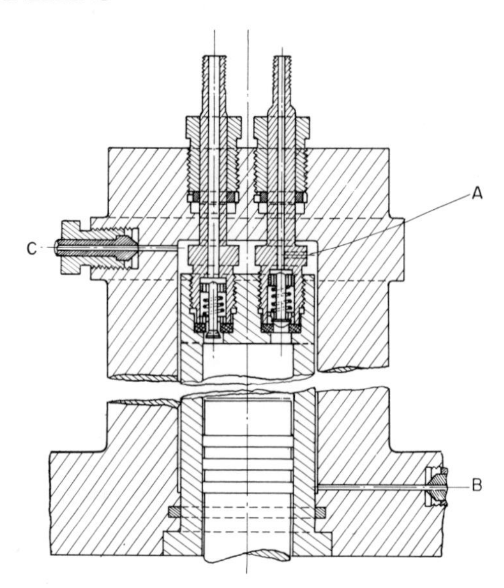

Figure 6.9 Method of applying compressive stress to the valves in the head of a compressor (*ICI design, from British Patent No.* 1,027,934)

at *A*, or it can be supplied with high pressure fluid (which can also be used for cooling), through the duct *B* and out through *C*. In this way the outer jacket has only to resist a steady pressure, while the actual cylinder wall takes the fluctuating pressure, but as this is mainly setting up compressive stresses the walls can be made comparatively thin. A further advantage of this design is that the outer jacket would act as a safety protection should either the valves or the inner cylinder fail.

REFERENCES

6.1. MORRISON, J. L. M., CROSSLAND, B. and PARRY, J. S. C., *Proc. Inst. Mech. Eng.*, **170,** 697, 1956.
6.2. MORRISON, J. L. M., CROSSLAND, B. and PARRY, J. S. C., *Proc. Inst. Mech. Eng.*, **174,** 95, 1960.
6.3. BURNS, D. J. and PARRY, J. S. C., *Proc. Inst. Mech. Eng.*, **182,** Pt 3C, Conference on High Pressure Engineering, London, Sept. 1967, Paper No. 5, 72.
6.4. MORRISON, J. L. M., CROSSLAND, B. and PARRY, J. S. C., *Trans. Am. Soc. Mech. Eng.*, **82** (Series B), 143, 1960.
6.5. DAVIDSON, T. E., EISENSTADT, R. and REINER, A. N., *J. Basic Eng., Trans. Am. Soc. Mech. Eng.*, **85** (Series D), 555, 1963.
6.6. PARRY, J. S. C., *Proc. Inst. Mech. Eng.*, **180,** Pt 1, 387, 1965–6.
6.7. MORRISON, J. L. M., CROSSLAND, B. and PARRY, J. S. C., *J. Mech. Eng. Sci.*, **1,** 207, 1959.
6.8. FESSLER, H., Private communication to authors.
6.9. MÖNCH, E. and FICKER, E., *Konstruktion,* **5,** 168, 1959.
6.10. AUSTIN, B. A. and CROSSLAND, B., Conference on Machines for Material and Environmental Testing, Manchester, Sept. 1965, Paper No. 31, *Proc. Inst. Mech. Eng.*, **180,** Pt 3A, 134, 1965–6.
6.11. AUSTIN, B. A., REINER, A. N. and DAVIDSON, T. E., *Proc. Inst. Mech. Eng.*, **182,** Pt 3C, Conference on High Pressure Engineering, London, Sept. 1967, Paper No. 35, 91.
6.12. JONES, P. M. and TOMKINS, B., *Proc. Inst. Mech. Eng.*, **182,** Pt 3C, Conference on High Pressure Engineering, London, Sept. 1967, Paper No. 30, 81.
6.13. FROST, W. J. and BURNS, D. J., *Proc. Inst. Mech. Eng.*, **182,** Pt 3C, Conference on High Pressure Engineering, London, Sept. 1967, Paper No. 29, 65.
6.14. MANNING, W. R. D., *Bulleid Memorial Lectures,* Nottingham University, 1963, III, 13.
6.15. British Patent No. 1,027,934.

7 Materials and Working Stresses

7.1 General Considerations

The factors which must be taken into account in deciding the most suitable material, technically and economically, are legion, even for the simplest high pressure component. In dealing with vessels and larger items of equipment attempts have been made to optimise the choice by means of computers, and Witkin[7.1] has described some of the methods used by the National Forge Co. of Irvine, Pa. (U.S.A.) for dealing with forged vessels. Luft[7.2] goes further and shows how this can be extended to vessels made by shrink construction or by autofrettage, although in this reference he is mainly concerned with simplifying the stress calculations. It is evident, however, that complicated decisions of this kind can be greatly assisted by such aids, although many of the problems can be adequately dealt with much more simply.

Usually the conditions of service, e.g. temperature, pressure cycles, corrosive environment, local design requirements and codes, etc., will first have to be taken into account, and then the most economical means of construction, e.g. fabrication by welding, forging, autofrettage, etc. can be considered. In the lower ranges of pressure (i.e. not exceeding say 1,000 bars) we have ample experience of successful designs and a number of materials have been fully proved; above that pressure level, however, and particularly for higher temperatures, the problems facing the designer are by no means fully catered for by previous experience, and the successful solution will depend greatly on the engineer's judgement.

The various national Codes for pressure vessels, some of which are enforceable by the laws of the countries concerned, differ considerably as to the stress levels permitted and will be considered in more detail in Section 7.2. They are primarily related to simple cylinders, but as the required factors of safety (all based on the maximum tensile stress) vary between 1·5 and 1·7 times the "minimum yield stress", and some refer also to the ultimate tensile strength, the task of the designer may be considerably complicated. There is at the time of writing a committee working on international standards for the I.S.O., but so far their results have not been published. It is to be hoped that they will be related to shear data, at any rate for the higher pressures and thicker walls, since experiments have shown conclusively (see Chapter 2, Section 2.5) that the limit of elastic behaviour is determined by shear rather than direct stresses.

It should also be noted that not all these Codes specify an upper limit of pressure or diameter ratio beyond which they do not apply, and there would seem to be a real danger of designers using them for conditions in which the assumptions they are based on are no longer valid.

Where creep conditions are encountered it is clear that the design must be based on a limited life, but this is seldom expressly stated in these Codes. The same may also apply where repetitions or fluctuations of pressure are liable to cause fatigue damage, although it would appear that most steels have a limiting value of shear stress below which the endurance of a vessel is very large, if not infinite.

Such considerations emphasise the importance of regarding *all* pressure vessels' lives as finite, even though many are amply safe for a working period far longer than is ever likely to be required. The authors feel that the idea of unending service for plant is a little too readily accepted and – as we shall see – the whole philosophy of design needs re-thinking when we come to very high pressure operation.

In what follows we describe the sort of properties which are most suitable in materials for the different kinds of high pressure service and give a few examples. It would be quite impossible in such a work to be at all comprehensive, owing to the very large and rapidly increasing number of possible materials. The preparation of, for instance, steels by modern methods of purification, de-gassing, etc. is briefly considered, since this is very important in a fully satisfactory material for high pressure service. The need for expert inspection and testing of equipment at various stages of its manufacture from the raw material to the finished machined job is also briefly discussed. Essentially, this chapter can only be a very brief summary of some of the requirements for obtaining the right materials for a successful high pressure unit, but we believe it may be of service, especially to those with no previous experience in such work. Basically, success in this field, as in many others of modern technology, requires the understanding cooperation of a number of professional skills, a point to be remembered by those contemplating short cuts to achievement.

7.2 Factors of Safety in Cylinders

7.2.1 *Existing Codes.*

Table 7.1 shows the factors of safety recommended by the various Codes, for conditions where creep due to temperature will not be encountered, and it will be noted that they all relate to direct stresses. In some instances the method of calculating the stresses in the walls is not specified, although there are at least three ways in which this can be done, namely the "Thin Cylinder" approximation, the Lamé relation (which is the most realistic and accurate) and the so-called "Mean Diameter Formula", which – though wholly empirical – is surprisingly accurate even up to diameter ratios as great as 1·5, at which the error is no more than 4%.

It will be noticed that in some cases reference is made to the Ultimate Tensile Stress as well as the Proof Stress, and this will be considered later in the dis-

Table 7.1

PRESSURE VESSEL CODES FOR NON-CREEP CONDITIONS

Country	*National Pressure Vessel Code*	*Body responsible for Code*	*Legal standing*	*Design stress (tensile)*
France	Réglementation des Appareils à Pressions de Gaz	Service des Mines	None	$\frac{1}{3}\times$ U.T.S. at working temperature
West Germany	A.D. Merkblätter	Technische Überwachungs-verein	Enforced	0·2% proof stress at working temperature ÷ 1·5
Italy	A.N.C.C. Code	Associazone Nazionale Controllo Combustione	Enforced	0·2% proof stress at working temperature ÷ 1·5
Japan	Unfired Pressure Vessel Code	Japanese Industrial Standards Committee	None	0·2% proof stress at working temperature ÷ 1·7
Sweden	Tryckkarlsnormer	Swedish Pressure Vessel Committee	Mandatory in Practice	0·2% proof stress at working temperature(1) ÷ 1·5
United Kingdom	B.S. 1515	British Standards Institution	None in 1968	U.T.S. at room temperature ÷ 2·35 or 0·2% proof stress ÷ 1·5 (whichever is lower)
United States of America	A.S.M.E. Code – Section VIII	American Society of Mechanical Engineers	Enforced in some States	U.T.S. at working temperature ÷ 4, or 0·2% proof stress at working temperature ÷ 1·6 (whichever is lower)

NOTE that all Codes are based on the maximum tensile stress in the walls of the vessel, and in several the word "yield" is used which we have here interpreted as the 0·2% proof stress.

(1) The working temperature here is to be interpreted as 150°C or higher.

cussion of brittleness and the risks of brittle fracture; all we need say at this stage is that it represents a precaution which is probably unnecessary having regard to the methods likely to be adopted anyway in the selection and inspecting of materials for equipment of this kind.

Where the Codes are based on Yield or Proof Stress data, it will be seen that a safety factor of 1.5 is now becoming fairly general, although the Americans and the Japanese still prefer a higher factor (the latter as much as 1.7†).

7.2.2 *Shear Basis for Design*

The alternative procedure for the design of elastic cylinders which the authors are recommending is based on the maximum shear stress in the walls, as cal-

† This figure appears to refer to steels and a lower factor (1·5) is allowed for aluminium.

culated on the assumptions of the Lamé analysis. According to this the working pressure p_W and the allowable shear stress τ_{al} are related by:

$$\tau_{al} = p_W \frac{K^2}{K^2-1} \tag{7.1}$$

where K is the diameter ratio. Moreover, we suggest that ideally this value for the shear stress should be derived from torsion tests, but as the great preponderance of the available information has been obtained from tensile testing it is essential to have a correlation between comparable "yields" or proof stresses in these two forms of test. It appears that this can best be done by means of the Maxwell (von Mises) criterion, according to which the critical shear stress τ^* is related to the corresponding critical tensile stress σ_T^* by:

$$\tau^* = \sigma_T^*/\sqrt{3} \tag{7.2}$$

although some caution should be exercised when applying this to very high strength materials, e.g. with U.T.S. values in excess of say 16,000 bars (232,000 lbf/in²), or to non-ferrous metals, where a divisor of 2 instead of $\sqrt{3}$ may be closer.

7.2.3 *Method of Comparing Different Design Bases*

It is desirable for the problems we are now concerned with to have some parameters for comparing materials of widely different strengths, as applied to containing widely different pressures, and one such is the dimensionless ratio of working pressure to tensile proof stress, which we will denote S. Then:

$$S = \frac{p_W}{\sigma_T^*}$$

and this will be related to K, or to the ratio:

$$Y = \frac{\text{wall thickness}}{\text{internal diameter}}$$

The appropriate relations for the differently based designs are given below in §7.2.5.

7.2.4 *Specification of Yield*

Before proceeding further we must consider what value to assign to σ_T^*. In most of the older Codes this is referred to as the Yield Stress, but, since we are seeking a comprehensive relation applicable to as many as possible of the materials likely to be used in high pressure equipment, this ought to cater for the three kinds exemplified in Fig. 2.9, p. 31. The true elastic limit is not easy to measure, even in a tensile test, and it is seldom included among the mechanical properties called for in manufacturers' specifications. There are clearly two yield values in Fig. 2.9a, but the higher is very sensitive to the test procedure and will be greatly reduced if the tensile pull is accompanied by the slight bending effects which are liable to occur with routine measurements in an ordinary test-house. Thus the lower, or secondary, yield is usually preferred for room temperature tests. The older idea of yield is associated with steelyard beam testing machines, which seldom applied a truly axial pull, and which therefore underestimated the primary yield, although the actual discontinuity was so abrupt that the beam immediately dropped onto its supports. Moreover, unless the weight on the beam was at once run back so as to decrease the load (not by any means an easy piece of manipulation), the secondary yield was completely obscured. These difficulties have led to the increasing use of the idea of Proof Stresses, the strain most used being 0·2%. This will in general be the same as the secondary yield stress in materials behaving as in Fig. 2.9a, but it will also give reproducible results for the other two kinds shown in that diagram. Thus we suggest that σ_T^* should be interpreted as the 0·2% Proof Stress for all materials in what follows.

7.2.5 *Comparison of Designs*

The most "advanced" of the Codes call for a factor of 1·5 on the Yield Stress, which we will here assume to be the 0·2% Proof Stress. For the shear basis which we are advocating, the possibility of using a rather lower factor seems reasonable, assuming that great care has been taken in the selection, testing, and inspection of the material. The figure of 1·33 or $\frac{4}{3}$ is therefore put forward. The allowable shear stress τ_{al} would then be $\tau^*/1{\cdot}33$ or having regard to eq. (7.2):

$$\tau_{al} = \frac{\sigma_T^*}{1{\cdot}33\sqrt{3}}$$

whence, by substituting also into eq. (7.1):

$$K = \sqrt{\left(\frac{1}{1-2{\cdot}309S}\right)} \qquad (7.3)$$

For comparison, according to the Thin Cylinder Formula (which assumes that the tangential stress does not vary across the wall thickness):

$$K = 1+1{\cdot}5S \qquad (7.4)$$

using the 1·5 factor on the tensile 0·2% Proof Stress. The more correct Lamé relation gives, correspondingly:

$$K = \sqrt{\left(\frac{1+1{\cdot}5S}{1-1{\cdot}5S}\right)} \qquad (7.5)$$

and the Mean Diameter Formula, which is specified in the leading British Code, B.S. 1515,† and according to which the allowable tangential stress σ_{al} is given by:

† The actual formula for the thickness t of cylindrical walls as given in B.S. 1515, Part 1: 1965 (p.13) is:

$$t = \frac{pD_i}{2fJ-p}$$

where p is the working pressure, f the allowable stress, D_i the inside diameter, and J a joint factor to take account of weld imperfections, etc. If there are no welds or other seams, this can clearly be given the value of unity, and the relation then becomes identical with eq. (7.6), i.e. with the Mean Diameter Formula.

$$\sigma_{al} = p_W \frac{K+1}{2(K-1)} \tag{7.6}$$

Again on the 1·5 safety factor basis this leads to:

$$K = \frac{1+0{\cdot}75S}{1-0{\cdot}75S} \tag{7.7}$$

The comparison is shown in Fig. 7.1 for the four above relations, in which values of S are plotted as abscissae against ordinates of Y, where:

$$Y = \frac{t}{D_i} = \frac{1}{2}(K-1) \tag{7·8}$$

As will be seen, the shear stress basis allows appreciably lighter vessels in the lower ranges, although it actually becomes more conservative when S exceeds about 0·2 and Y about 0·18.

7.2.6 *Comparison of Results*

The resulting plots of S and Y for the two main criteria of design, i.e. maximum direct stress and maximum shear stress, are shown in Fig. 7.1, using a safety factor of 1·33 for the former and of 1·5 for the latter, in which the three ways of calculating the stress are plotted separately. It is at once clear that, for the left-hand half of the diagram, there is little to choose between the methods. However, the danger of using the Thin Cylinder Formula for cases where S exceeds about 0·14 is clearly shown, since this has no upper limit for S and is therefore obviously invalid. On the other hand, an infinite value of K is required by the other relations, when $S = 0{\cdot}43$ for the shear basis, when $S = 0{\cdot}67$ for the tension basis using the Lamé relation, and when $S = 1{\cdot}33$ for the formula used in B.S. 1515.

For the lower values, i.e. for the bottom left-hand parts of the curves in Fig. 7.1, the pressures are likely to lie below those normally covered by the considerations of this book and might call for walls so thin as to be unsatisfactory for other reasons, i.e. stability. Even up to a value for S of 0·05, the corresponding values of t/D_i are virtually coincident for all three methods of calculating the tangential stress, although the shear stress basis allows an appreciably thinner wall. Again, however, this is in a range where the construction would normally be by welding plate and some allowance for joint weakness would be required. But this is in the region below the main considerations of this book. For values of S above say 0·02, the shear stress design basis with the lower factor of safety appears reasonable, but it would require special care in the material selection and in the subsequent workmanship.

Examples. As higher pressures are considered requiring higher strength materials, the situation changes as is seen from Table 7.2, in which the values of $Y = t/D_i$ are plotted for various values of S according to the various theories. As an example, considering the steel tested by Crossland and Bones[7.3] which was found to have a value of approximately 7,450 bars for σ_T^* to be used for a cylinder

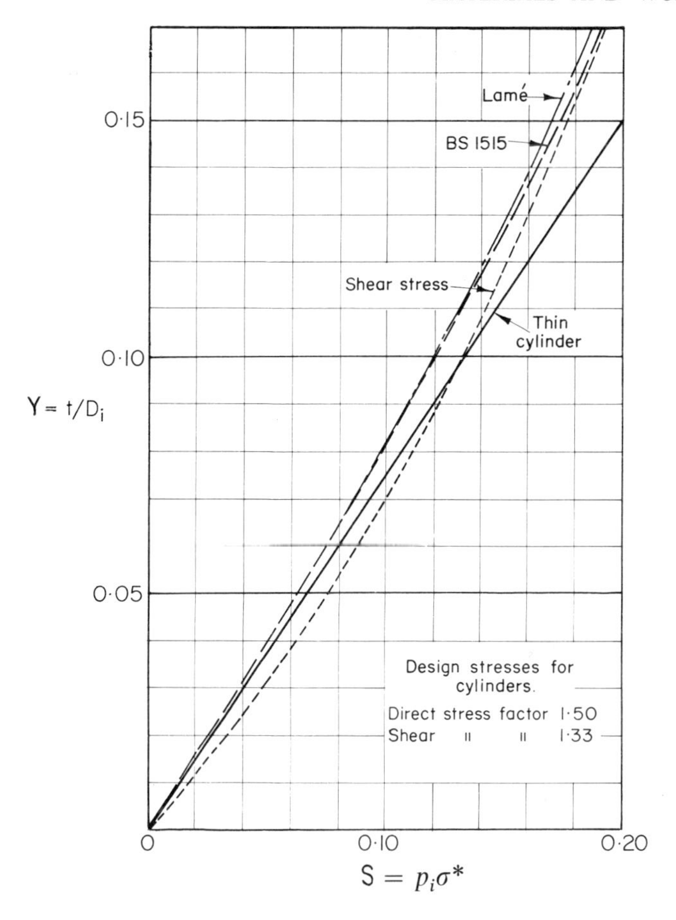

Figure 7.1 Comparison of design formulae for cylinders

Table 7.2

Values of $Y = t/D_i$

Design basis	*Safety factor*	$S = 0{\cdot}05$	$S = 0{\cdot}10$	$S = 0{\cdot}15$	$S = 0{\cdot}20$	$S = 0{\cdot}25$	$S = 0{\cdot}30$
Thin Cylinder Formula	1·5	0·0375	0·0750	0·1125	0·1500	0·1875	0·2250
Lamé Equation	1·5	0·0390	0·0826	0·1285	0·1814	0·2415	0·3119
B.S. 1515	1·5	0·0389	0·0811	0·1268	0·1765	0·2308	0·2903
Shear basis	1·33	0·0315	0·0701	0·1185	0·1816	0·2690	0·4020

to contain 1,000 bars, we have a value for S of 0·134, which would give a corresponding value for Y of about 0·10 for both the shear criterion and the tensile method when calculated by the Thin Cylinder Formula, and of about 0·127 by either of the other two relations. Thus, if required for a tubular reactor of 20 cm bore (8 in approximately) the wall thicknesses would be 2·0 and 2·54 cm respectively, giving diameter ratios of 1·20 and 1·254. This would seem reasonable, but the question of safety margins will be considered later (Section 7.6.1).

It is interesting to note the results if these systems are used to design a reactor to be made from the same material to work at twice that pressure, i.e. 2,000 bars when S becomes 0·268. The corresponding wall thicknesses are shown in Table 7.3. Clearly the use of the Thin Cylinder Formula is no longer permissible.

Table 7.3

Designs for 2,000 bar Reactor of 20 cm Bore

Design basis	*Safety factor*	*Wall thickness (cm)*	*Diameter ratio*
Thin Cylinder Formula	1·5	4·02	1·40
Tensile stress (Lamé)	1·5	5·30	1·53
B.S. 1515	1·5	5·00	1·50
Shear basis	1·33	6·20	1·62

According to the experiments of Crossland and Bones[7.3] a cylinder of 1·40 diameter ratio (i.e. $Y = 0{\cdot}20$) made from this material initially overstrained at 2,070 bars and finally burst at almost exactly 3,000 bars.

7.3 The Danger of Brittle Failure

A number of instances have been reported in which pressure vessels and components have failed with catastrophic suddenness and sometimes with the high speed projection of large fragments in all directions. Some of these occurred after a very short working life, and a few even on test. Some were in large installations designed for operation at moderate pressures, such as the drum of the 190 bar working pressure boiler at Cockenzie[7.4] which failed when under test as the pressure reached 275 bars, while others were in smaller units designed for considerably higher pressures, such as the vessel examined by Saibel[7.5] which failed on test at about 3,000 bars as the result of a 360° circumferential crack which caused the heavy closure at one end to break away and penetrate the wall of a building.

The most disconcerting feature in nearly all such cases is that the failures occurred in material which could reasonably be assumed to be ductile; also the pressures rose slowly and there could be no question of explosive or shock effects. It is, of course, well known that ductility in metals is very much a function of temperature, and this would no doubt explain some of these failures, for instance those in surface pipe lines in cold climates. There remain, however, a number for which this was evidently not the cause, and we must refer therefore to the newly emerging science of Fracture Mechanics.

It is doubtful if this has yet reached the state where all these past failures can be explained and all future ones prevented, but it is clear that many, if not all, were due to the tendency of cracks to spread extremely rapidly when they exceeded a certain critical size, and this is a phenomenon which Fracture Mechanics is particularly concerned with. The cracks may have been present in the first place, due perhaps to defects in welding, or have been caused by the onset of fatigue at localised stress concentrations in otherwise sound material. But this is evidently not the whole story, for there are plenty of examples of fatigue cracks slowly spreading right through thick walls without resulting in any sort of catastrophic break-up (see Chapter 6).

Several special forms of test have been devised in attempts to elucidate this phenomenon; most of them are elaborate and costly, and it is doubtful if any are wholly satisfactory as yet. Probably the best is that devised under the auspices of the American Society for Testing Materials and described in their special publication.[7.6] Essentially this consists in making a bend test-piece in the form of a rectangular bar with a notch across one side perpendicular to the axis. This is first subjected to some form of bending fatigue until a slight crack is started at the base of the notch; it is then tested under static loading to determine at what stress the crack begins to spread uncontrollably. The results are given in terms of the product of the stress and the square root of the crack length, described as "the stress intensity factor of the elastic strain field at the crack tip", and usually denoted K_{1c}. An inconvenient aspect of this test is that the size of the test piece has to be related to the static strength of the material, and while this is of quite reasonable dimensions for very high tensile steels such as those of the maraging type, it becomes very cumbersome for the hardened and tempered low alloy steels with U.T.S. in the range, say, 7,500 to 11,000 bars (about 110,000 to 160,000 lbf/in^2) when the cross-section called for may have an area of the order of 60 cm^2 (about 10 in^2).

Fortunately it appears that, at any rate in the range just mentioned, the variations in K_{1c} value correspond with changes in the values obtained in either the Charpy or the Izod tests fairly adequately. In fact, as Fig. 7.2 shows, in the case of a nickel/chrome/molybdenum/vanadium steel, the Charpy test appears even more sensitive by showing a fall in impact resistance in the range of tempering temperatures between 250° and 500°C which does not appear in the corresponding K_{1c} curve, although it should be noted that the Charpy value never falls below about 13 ft lb, while the U.T.S. of the material is at 17,500 bars (250,000 lbf/in^2) or over.

Thus, provided material with a good impact resistance as measured by the Charpy or Izod test is obtained, and provided it is reasonably homogeneous and directionally uniform in its mechanical properties, brittle fracture is unlikely unless the design or manufacture is faulty, and it is important to realise that deficiencies in these requirements can be just as dangerous as the use of inherently brittle material. Saibel[7.5], for instance, found that the mechanical properties of the failed vessel he examined were very poor in the transverse direction compared with those taken longitudinally.

Geometry can also play an important part in accentuating the effects of fatigue. We know from the work of Morrison and his school at Bristol (discussed

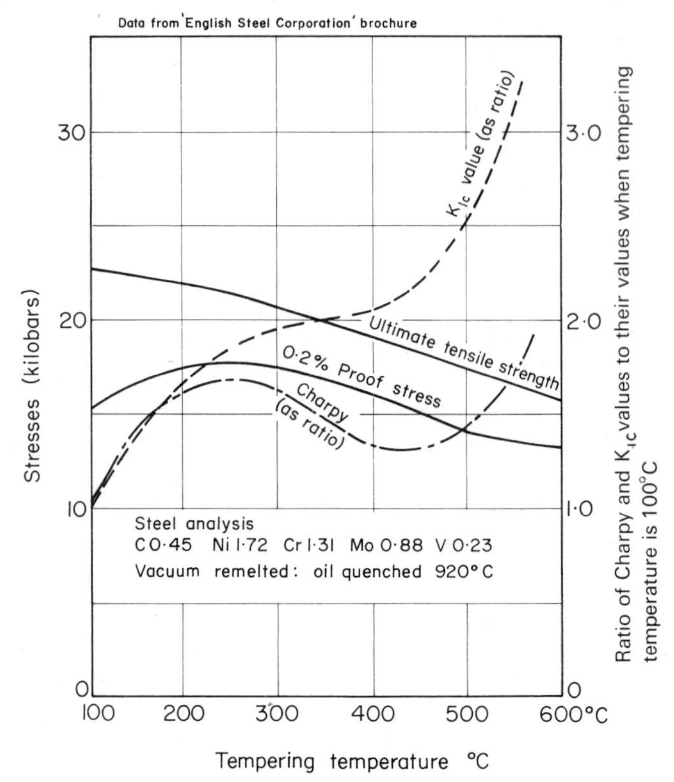

Figure 7.2 Comparison of ASTM crack propagation index (K_{1c}) with Charpy value for varying tempering temperatures in nickel-chrome-molybdenum-vanadium steel *(English Steel Corporation Ltd)*

in Chapter 6) that very large cracks can develop before actual separation takes place if the shape of the component is such that the cracks can spread without running into other areas of high local stress. On the other hand, if there are several stress-concentrating shape changes close to one another and a fatigue crack starts in their neighbourhood, the result is likely to be serious, leading to rapid crack propagation and rupture.

We can sum up this Section by saying that many factors may contribute to the brittle failure of a high pressure component, of which the material quality is only one. Especially important from the designer's point of view is to be sure of the temperature range within which it will work, since an abnormally cold winter, for instance, may chill it down so that its impact resistance is much impaired. Again, too rapid starting up, even of components normally operated at high temperatures, may lead to trouble as the result of differential expansion stresses. The avoidance of stress-raising changes of shape, e.g. sharp corners, especially in counter-bored holes, or abrupt changes of section, is very important, as is care in all fitting and assembly operations.

So far as material is concerned, it is important not only to select the appropriate material, but also to ensure that one gets what has been specified. This is no reflection on the steel manufacturer, but a warning that troubles have often arisen because he was not taken into the customer's confidence to the extent of knowing what was going to happen to the component to be made from the plate, bar, or forging which he had been asked to supply. In particular, directional uniformity is extremely important, especially when very high strength materials are chosen, and the supplier should certainly be made aware of this, since it affects the positions from which the test pieces are to be taken.

In fact, brittle failure is certainly something to beware of, but when its possibility is recognised and appropriate steps are taken to deal with it, such as ensuring good impact resistance in terms of say Izod or Charpy figures, it need not cause undue anxiety.

Some Codes, for instance the A.S.M.E. Section VIII, include both the proof stress and the U.T.S. – each with an appropriate factor – so that materials for which these quantities are close to one another are not worked up to the stress which would be permitted by the proof stress alone. The reason behind this is the belief that such materials are potentially brittle and certainly with the older, more conventional types, this is often true. But with maraging steels, for instance, it appears that there is a very great elastic strength, after which deformation occurs at nearly constant shear stress until fracture occurs. This is particularly marked in a torsion test. So far as the danger of brittle fracture is concerned however, the protection of special care in manufacture, followed by careful impact tests, is probably a better insurance than specifications based on both proof stress and ultimate strength.

The effect of this type of behaviour, i.e. extensive straining at constant stress very near the U.T.S., on the ultimate bursting pressure is not known because no bursting tests have yet been reported, but on theoretical grounds it would appear unlikely that they could withstand a steady pressure greater than:

$$\frac{\sigma_T^*}{\sqrt{3}} \times \ln K^2$$

and in that case the margin of safety when working to eq. (7.3) may not be sufficient. This is clearly a field where more experimental work is urgently required.

7.4 Creep Conditions

7.4.1 *General*

As has been explained in Chapter 5, design in such circumstances should be based on an acceptable working life, but it is doubtful if the data yet available for any single material are sufficient for this to be satisfactorily used at the present time; and certainly this could not be done in general terms for high pressure components in all the materials likely to be specified. Design must therefore be based on tensile creep data, although the quantities commonly used are more often than not themselves the result of considerable extrapolation.

Among the specifications of the national Codes for high temperature conditions at least four quantities are used, namely:

Average stress for rupture after 10^5 hours at working temperature, denoted σ_{Ru}
Average stress for 1% creep after 10^5 hours at working temperature, denoted σ_c
Average stress for 1% creep after 10^5 hours at working temperature $+15°C$, denoted σ_{c+15}
Average stress for 0·2% Proof Stress at working temperature, denoted σ_T^*.

In addition the French Code apparently still bases its creep philosophy on designing to keep the maximum tensile stress down below one third of the short-time ultimate strength at the working temperature.†

The first two of the above were discussed in Chapter 5, where it was explained that there was increasing preference amongst the leading authorities for the stress to cause rupture after 10^5 hours, mainly because there was less scatter in the results of creep rupture tests of shorter duration, and consequently the necessary extrapolations could be made with more confidence. There was also the important work of Chitty and Duval[7.7] in which the Mean Diameter Formula was used to correlate the maximum tension in the cylinder with that in a simple tensile specimen; and when these were the same, the life of the cylinder was much the same as that of the specimen.

It should be noted that, in each case, the actual figure to be taken for the stress is an average of all available for the material at the temperature in question. Usually the spread of such figures is not more than about 10% on either side of the mean, but sometimes this may be twice as wide and the usual practice then is to take the lowest figure and multiply it by 1·25. Several of the national Codes specify for the working stress:

$$\sigma_{Ru}/1{\cdot}5$$

including the German T.U.V.,‡ the Italian A.N.C.C., and B.S. 1515. On the other hand, the A.S.M.E. in their Section VIII specifies the value of σ_c without any safety factor, but includes also a quarter of the U.T.S. at the working temperature, and the 0·2% (short time test) Proof Stress divided by 1·6. σ_c is also permitted by B.S. 1515.

Having regard to the above, and particularly to the work of Chitty and Duval (loc. cit.) we feel that a factor of 1·5 on the creep rupture stress should suffice for quite thick-walled cylinders, e.g. at least up to a diameter ratio of 1·5 if the Mean Diameter Formula is used, since the correlation these authors investigated was carried up to a ratio of 1·7.

The fact remains, however, that the available information on the actual mechanism of cylinder creep under internal pressure is still very meagre and we are

† It may be noted in passing that the earlier British Standard Specification (B.S. 1500), which is still in use in some quarters, was based on the same idea although it was a little more conservative, specifying a quarter of the U.T.S. at the working temperature.

‡ This actually specifies the lower of $\sigma_{Ru}/1{\cdot}5$ and $\sigma_{c+15}/1{\cdot}5$, although it appears that the former is usually taken owing to the difficulty of satisfactorily determining the latter.

not in a position to change the conventional design procedure on the basis of what has been published so far. On the other hand work on this subject is continuing in Belfast and elsewhere, and particular attention is being given to the direct study of creep under shear stress using hollow cylinders under torsion. Crossland[7.8] has reported that the strain which can be endured before rupture in torsion is vastly greater than it is in tension. This, together with the known fact that the stress distribution in the cylinder wall is completely altered by the creep process (see Fig. 5.4), is a clear indication that the present position is far from satisfactory, and we must hope that it may be possible before long to derive a method somewhat along the lines proposed in Section 5.3.4.

7.4.2 *Temperature Variations*

In practice many high pressure components are operated at varying temperatures, and the safe designed life span may be considerably affected thereby. The usual sequence of working times interspersed with periods when the plant is allowed to cool for cleaning, etc. probably will do little more than increase the life by the aggregate cool time, but the situation may be very different if the temperature control allows any parts to get hotter than the designed temperature. Even a few degrees for a few minutes will have its effect, and it is most important to design for the maximum likely temperature.

7.4.3. *Special Materials*

The quest for safe working at higher and higher temperatures has led to the extensive use of various non-ferrous materials for creep-resistant pressure containers, especially nickel-based alloys of the Nimonic type. Some of these are mentioned in the lists given below, but it must be emphasised that progressive metallurgical changes may be involved with these—as indeed they may be eventually with almost any material—and it may not be possible to meet demands for long life high pressure equipment for service, say, above 700°C.

In practice, some control can be kept of the extent to which creep has progressed, by periodically measuring the external diameter. For instance, if the rate of increase is found to be accelerating, this should at once be regarded as a danger signal. As operations of this kind are usually only on a small research scale, the fact that the service life of the equipment is likely to be a matter of considerable uncertainty need not perhaps be too serious, and in any case such apparatus is likely to be small and reasonably easy to guard, so that failure should not cause injury to personnel or damage to property. These considerations are however typical of what one must always expect when working near the limit of available knowledge and experience, and the region in which high pressures at high temperatures are called for is one where the demands are coming close on the heels of the technical know-how for making the containers.

7.5 Preparation and Testing of Material

7.5.1 *General*

This is of course a very big subject and clearly we can only touch on certain aspects of it here. Normally high pressure equipment will be built from material, whether ferrous or non-ferrous, which has been specially handled by the manufacturer with full knowledge of any special requirements. The material actually chosen should obviously be the cheapest that will suffice, but cheapness in this sense must be decided with due regard to the methods of manufacture. For instance, with "solid drawn" tubing it may pay to use a less strong material and make the walls thicker, if by so doing the drawing process is made easier, more reliable, and less costly.

With steel the manufacturing process may, in very general terms, be divided into the following stages:

Melting
Ingot treatment
Rolling or forging
Heat treatment
Inspection and testing
Fabrication and machining

7.5.2 *Melting*

This involves the correct feeding of the various ingredients into the furnace so that the analysis of the resulting ingot will be to specification; there is also the question of what kind of furnace to use, but the special high strength materials are now usually electrically melted, either by an arc process or by induction. It is at this stage that small traces of highly injurious substances may be picked up, and it is of course the manufacturer's job to see that this is kept under control.

7.5.3 *Ingot Treatment*

This covers the controlled solidification and cooling of the material during and after pouring into the ingot mould, and also the removal of the "discards", which may amount to nearly half the total material cast. The British Admiralty, for instance, used to specify the removal of at least 40% of the material at the top of the ingot and 5% at the bottom in steel for big guns, and proportionally similar removal of doubtful material is desirable where forgings for high pressure vessels are involved. Modern methods of "de-gassing", either in the ladle or during pouring into the ingot mould, may be highly advantageous, by removing a considerable proportion of the dissolved gases, particularly hydrogen which is known to be one cause of subsequent internal cracking, and the extra cost of these is only about 5%.

For the highest quality materials, e.g. for very high pressure work involving very thick walls and extensive autofrettage of very high strength materials, the introduction of further treatment by vacuum re-melting may be advisable, although this will probably double the cost. It consists in preparing from the remainder of the original ingot a rough machined solid cylinder, to which is welded a long electrode. The whole is then inserted in a water-jacketed copper chamber (the electrode passing out through a gland) which is connected to high capacity vacuum pumping equipment. The insulated electrode with the solid cylinder of material attached is next lowered to the bottom of the chamber and an arc is struck there, which fuses the material under high vacuum and allows the melt to resolidify quickly as its heat is removed by the water cooling. In this way a new ingot is formed, in which the crystal size and the directional properties are both remarkably uniform. Many of the non-metallic inclusions are decomposed, and the dissolved quantities of hydrogen, oxygen, and nitrogen greatly reduced. As has been remarked, this is an expensive process and can only be recommended for components to work under very severe conditions, but it is probably worth while where material with tensile strength in excess of 12,500 bars (180,000 lbf/in^2) is required.† See also Leach [7.12] on developments in vacuum treatment.

7.5.4 *Rolling or Forging*

Here we are concerned with converting into plate or bar for fabrication by welding, or with forging into the approximate shape for machine finishing; these are processes which need great skill and experience, especially where alloy steels are concerned. In forgings for large vessels it can be a great advantage if "hollow forging" can be applied. This is usually done by removing the central core of the ingot and supporting it on a massive mandrel; forging then takes place between the press head and the mandrel, thus squeezing the metal out circumferentially and "working" it in the most favourable direction, i.e. to resist tangential stress. This is particularly valuable with materials which may tend to retain some directional effects from the varying crystal size within the original ingot, but it can only be carried out if the final minimum internal diameter is at least 30 cm and even then if the overall length is no more than about 2 metres; otherwise the bending resistance of the mandrel is insufficient to sustain the required press load. The possibility of being able to hollow-forge large pressure vessels is a matter that should be borne in mind at the design stage. It may sometimes be worthwhile in the long run to make a vessel with a hole right through it, and thus to accept the extra complication of joints and couplings at each end, for the sake of improving the metal of the walls and so obtaining material which, if it has any directional strength variations, will be strongest in the direction of greatest load. There is, incidentally, the further advantage in this design that all the material which had been in the central core of the ingot is thus discarded, and if there are impurities or serious zones of weakness in a piece of material of this sort they are most likely to be near the centre line.

7.5.5 *Heat Treatment*

This is the process of achieving the specified condition of strength and ductility and may take a number of forms. With low or medium carbon steels (i.e. $C < 0{\cdot}4\%$) this consists of simple "normalising", or heating to about 950°C and allowing to cool in air.

The low alloy, high tensile steels of the kind mentioned in Section 2.7, with U.T.S. in the range 8,000 to 11,000 bars (say 115,000 to 160,000 lbf/in^2), are generally used in a hardened and tempered condition, which means heating to somewhere in the region of 850 to 920°C (the temperature depending on the com-

† This process is also known as Consumable Electrode Melting.

position), followed by plunging into oil or water to cool quickly. The material is then extremely hard and brittle, but by tempering, i.e. heating again, usually in the range 500 to 650°C, the strength is reduced and good ductility and impact resistance obtained.

A similar process may be used to achieve considerable ductility allied with great strength, e.g. 15,000 bars and over, with some of the more recently developed steel compositions. The problem then is the final machine finishing, and the usual practice is to do most of the machining while the material is in a moderately soft condition, leaving only a small allowance for final grinding or machining after re-treating to the high strength and hardness. Some distortion may then occur, especially in the quenching operation, and enough material has to be left to enable the part to be trued up by its removal.

From this point of view, precipitation hardening materials have great advantages. Examples of these are the maraging steels, various high strength aluminium alloys, and beryllium copper. The basic idea is that the material is first heat-treated to bring it into the condition of a solid solution, in which it is reasonably soft and suitable for machining. It is then given a treatment at a relatively moderate temperature, between 400 and 500°C for most of the steels, which precipitates one or more phases to produce very much greater hardness and strength, virtually without any distortion.

In the case of austenitic stainless steels the heat treatment necessary to give the best corrosion resisting properties is by what is usually known as "fully softening" and consists in water quenching from about 1100°C. The material is then extremely tough and has great capacity for work-hardening, but it has only moderate tensile strength – U.T.S. normally about 4,500 to 7,000 bars (65,000 to 100,000 lbf/in^2) – and very low Proof Stress, of the order of 1,800 bars (26,000 lbf/in^2). However, precipitation hardening varieties of austenitic steels are now becoming available.

All heat treatment operations need skill and experience, for there are many subtleties which can cost the novice dearly. For instance, the time taken for heat to penetrate to the centre of a large mass is considerable and there is a tendency for the inner parts not to get hot enough, especially with materials which suffer deterioration if kept hot for too long. It is like cooking a joint of meat: if it is a big joint, the outside tends to be overcooked while the inside is still raw. Fortunately in the heating prior to quenching there is usually a fair amount of latitude so far as time at temperature is concerned, and it is safer to "soak" the material for an ample period of time. The hardening effect of the quench, whether in oil or water, is to a considerable extent a function of the rate of cooling, and this is clearly dependent on the mass of metal being treated, particularly its greatest thickness; the heat from inside can only be removed by conduction through the outer layers. In consequence very large masses can never be given properties as good as those of smaller pieces of the same material. This also explains the desirability of removing as much metal as possible before heat treatment, and emphasises the need for the test specimens to be taken from the forging *after* heat treatment so that they will have been subjected to the same mass effects, i.e. rates of heating and cooling, and under no circumstances should the test pieces be separated first. A good way of providing for this, where two or more large pieces are being forged as one lump, is to leave enough material for a disc to be cut out between the two when the heat treated forging is finally parted, the specimens to be machined from this disc.

7.5.6 *Inspection and Testing*

The object here is to ensure that the material, whether as a forging or as a batch of rolled plates or bars, is within specification in respect of mechanical properties, and that it is as far as possible completely sound and free from flaws or other defects. The mechanical properties are of course determined directly from specimens cut out from a suitable disc as mentioned above, but as a check it is sometimes necessary to cut off a small piece on which an indentation hardness test can be made. As is well known, the hardness is in direct relation to the U.T.S. for nearly all ferrous metals, and this enables the effect of the heat treatment to be checked; if necessary, it can then be repeated.

The chief anxiety regarding the quality of a large piece of metal lies in the possibility of internal cracks, or concentrations of non-metallic inclusions, or—in the case of plates—of laminating effects. Modern techniques of ultrasonic inspection go a long way towards reassuring the customer that none of these is present, but probably the best safeguard is in the manufacturing which, with modern methods of control, can do much to prevent these troubles from arising.

It is however sometimes necessary to use for high pressure components material whose history is not known. In such conditions tensile and impact specimens should always be cut out, preferably in the transverse direction (i.e. parallel to a diameter in the case of round bar) and tested; the uniformity of properties can be partially checked by indentation hardness measurements at positions well apart from one another, e.g. at each end of a length of bar, and if these give widely different results the material should be regarded as highly suspect. Another useful guide to the quality of steel is the sulphur print,† which is simple to carry out and will show if there is a concentration of sulphur-containing substances, usually an indication of slag inclusions and of inadequate discarding from the ingot. If a large number of unevenly distributed patches of sulphur are revealed, it would be safer not to use such material for highly stressed components.

7.5.7 *Fabrication and Machining*

The manufacture of pressure vessels by rolling plate into cylindrical form and welding along longitudinal seams is a major industry, but such vessels are usually covered by insurance and made in conformity with established codes, which specify sizes, materials, welding methods, etc. We shall not therefore consider this aspect of high pressure engineering in any detail, especially as the working pressures involved seldom exceed 300 bars (say 4,500 lbf/in^2). The possibility of

† This requires a smooth and fairly well polished surface and consists in taking a piece of soft bromide photographic paper, immersing it for a few minutes in 2% sulphuric acid, draining it and laying it uniformly over the surface, which must be free from all oil or grease. After 3 minutes the paper is removed, washed, and fixed in a hypo solution in the usual way. The presence of sulphur in the steel shows up as dark brown patches on the paper.

using such manufacturing techniques for much higher pressures is however worth a brief consideration, since modern welding techniques can comfortably handle plate up to at least 15 cm (6 in) thick, and of high-tensile composition.

The limiting factor then is usually the bending into cylindrical form, which, for final diameter ratios above 1·2, becomes more expensive than starting with a forging. On the other hand, the welding-in of connections is being used increasingly even in the range of working pressures above 1,000 bars. This calls for the use of materials which are reasonably easy to weld, and for stress-relieving heat treatment afterwards wherever possible. Long pipe lines can be joined by resistance butt welding in which squared-off sections are forced against one another while a high current is passed between them. This raises the metal at the interface to welding temperature and, when properly controlled, ensures a very good joint.

Machining of high pressure components does not, in general, call for any special comment except to say that the materials involved are often harder than machine operators without experience of this technique are used to; and in consequence care should be exercised when giving such work to those new to it. In particular, there is a tendency to produce sharp changes of section when cutting internal threads and these can, of course, produce stress concentrations and lead to cracking if the load fluctuates and causes fatigue damage.

The finishing of internal surfaces by grinding or honing has much to recommend it for very high pressure parts such as the cylinders of intensifiers. Not only does it produce a very good surface with materials that would be difficult to finish with a tool, but also it develops a high residual compression stress in the surface layers, thereby increasing the fatigue resistance.

7.6 Some Suitable Materials

The following notes and tables describe a few steels and other metals which have been used successfully in high pressure engineering. They are divided into groups according to the usage to which they could be safely put, having regard to the design and safety considerations described herein. The analysis figures are for typical materials, and we have, where possible, included the Code numbers of the nearest national specifications.

7.6.1 *Steels for Fully Elastic Cylinders under Static Conditions*

Here we are concerned with temperatures between 0°C and about 250°C, i.e. conditions where there is no marked embrittlement due to low temperature and no detectable high temperature creep. Also we are thinking in terms of a monobloc diameter ratio of about 1·5, which – on the basis of a 33% safety factor over the Shear Proof Stress in the wall according to Lamé's theories – corresponds to a ratio S (of working pressure to 0·2% Proof Stress of material) of about 0·25.

The analysis of these materials together with some notes relating to their heat treatment etc., is given in Table 7.4, while characteristic mechanical properties are given in Table 7.5. The static test figures, i.e. 0·2% Proof Stress and Ultimate Tensile Strength are given in bars and in lb force/in^2 in brackets underneath.

Table 7.4

Serial No.	Material	Analysis % by Weight											Heat Treatment	B.S. Code	S.A.E. Equivalent
		C	Si	Mn	S	P	Ni	Cr	Mo	Co	Va	Ti			
C.1	Carbon steel	0·28	0·12	0·82	0·033	0·035							Normalised from 900°C	En 4	1026
C.2	1% Nickel Mn	0·37	0·22	1·53	0·040	0·042	0·81						Normalised from 900°C	En 12	
C.3	Chrome molybdenum	0·39	0·30	0·77	0·043	0·039		1·10	0·27				Oil-hardened from 870°C Tempered at 650°C	En 19	4140
C.4	Stainless iron	0·23	0·38	0·18	0·017	0·024	0·32	12·78					Oil-hardened from 960°C Tempered at 710°C	En 56C	51410
C.5	Nitriding steel	0·25	0·19	0·45	0·014	0·015	0·22	3·27	0·51				Oil-hardened from 910°C Tempered at 625°C	En 40S	
C.6	Gun steel	0·30	0·15	0·66	0·015	0·013	2·55	0·58	0·59				Oil-hardened from 850°C Tempered at 660°C	En 25T	
C.7	ditto: lower tempering temperature	0·32	0·21	0·60	0·034	0·025	2·68	0·68	0·52				Oil-hardened from 850°C Tempered at 585°C	En 25V	
C.8	ditto: higher carbon	0·40	0·20	0·68	0·020	0·019	2·52	0·70	0·49				Oil-hardened from 860°C Tempered at 525°C	En 26	
C.9	"NCMV"	0·44	0·64	0·45	0·006	0·008	1·79	1·50	0·89		0·23		Oil-hardened from 920°C Tempered at 610°C	English Steel Corpn. special	
C.10	ditto: special heat treatment	0·45	0·75	0·44	0·008	0·012	1·72	1·31	0·88		0·23		ditto with refrigeration at −75°C and tempering at 250°C	"	
C.11	Maraging	0·027	0·10	0·05	0·010	0·004	18·08		4·80	7·16		0·46	Solution-treated at 820°C Aged at 480°C	E.S.C.'s Nimar	

Table 7.5

Serial No.	Material	0·2% Proof stress bars (lbf/in^2)	Ultimate tensile stress bars (lbf/in^2)	Elongation % on $4 \times \sqrt{A}$	Reduction in area %	Impact resistance		Safe Working Pressure of K = 1·54 cylinder bars	Estimated probable bursting pressure of 1·54 cylinder	Notes
						Izod ft lb	Charpy ft lb			
C.1	Carbon steel	2,500 (35,000)	5,050 (72,500)	25		>20		600	2,000 [2]	
C.2	1% Ni/Mn	4,100 (59,000)	6,150 (89,200)	20		20		1,000	2,600	
C.3	Chrome molybdenum	4,950 (71,500)	7,970 (115,000)	22		40		1,200	3,400	Suitable for tube drawing
C.4	Stainless iron	5,040 (73,000)	7,080 (102,500)	26·5	60	30		1,200	3,100	Suitable for tube drawing
C.5	Nitriding steel	6,800 (98,500)	8,300 (120,000)	>20		40–45		1,700	3,500	Useful for fatigue resistance
C.6	Gun steel	7,450 (107,500)	8,450 (123,000)	21·5	50	38		1,850	3,500 [2]	As used in reactor discussed in §2.7
C.7	ditto: lower tempering temperature	9,100 (132,000)	10,250 (149,000)	15	35	30		2,250[1]	3,650	
C.8	ditto: higher carbon	11,100 (161,000)	12,300 (179,000)	12	26		17	2,700[1]	no available	
C.9	"NCMV"	13,800 (200,000)	15,200 (220,000)	18·2	54	23		3,400[1]	shear data	Material made by E.S.C. Sheffield
C.10	ditto: special heat treatment	17,800 (258,000)	21,400 (310,000)	12·5			15	4,400[1]	for esti-	ditto fully fully hardened
C.11	Maraging	17,000 (246,000)	18,900 (274,000)	7·2 transverse 10·7 longitudinal	15·3 transverse 41·8 longitudinal		18	4,200[1]	mating	E.S.C. "Nimar 110"

[1] Calculated only: no experience with containing pressure with these materials.
[2] Figures from actual tests to destruction.

The column of Safe Working Pressures is based on the shear stress with a factor of safety of 1·33, i.e. from eq. (7.3), and should be taken as an indication only. It will be seen, however, that the first 5 materials listed would each give an overall factor of safety of more than 2, as referred to the Ultimate Bursting Pressure of a cylinder, while C.6, which has been extensively used for such work, is known to have a safety margin only just smaller. The other materials are largely untried and should only be specified therefore for service where adequate precautions can be taken in the event of unexpected failure.

It is interesting to note that the "NCMV" steel appears at least equal to the maraging steels and its much lower content of expensive alloying elements should make it much cheaper at the re-melted ingot stage, assuming that both would need that treatment. The subsequent refrigeration treatment at $-75°C$ after quenching would, however, be expensive. On the other hand, the possibilities of such materials for containing pressures above 10 kilobars, with or without autofrettage, appear considerable.

7.6.2 *Corrosion-Resisting Steels*

Material No. C.4 in Tables 7.4 and 7.5 has considerably greater corrosion resistance than ordinary carbon steels, or in fact than any of the low alloy high tensile materials such as Nos. C.5 to C.10 inclusive. It is hardly to be considered as a true "stainless" material, but it is favoured by some tube manufacturers for its cold-drawing qualities and is valuable from that point of view; and its chemical resistance adds to this an appreciably longer life under mildly corrosive conditions. Incidentally, in the form of 3 to 1 diameter ratio tubing in small sizes it has given long service at pressures up to 4,000 bars, an indication of its suitability in conditions where autofrettage is necessary. Higher proportions of chromium, up to 20%, increase this corrosion resistance, but only to a small extent, and have very little effect on the mechanical properties.

More recently, however, various precipitation-hardening alloys of similar composition have been developed. One such, known as "Chromar D.70", made by the English Steel Corporation at River Don Works in Sheffield, is included as No. S.1 in Tables 7.6 and 7.7. As will be seen, it can develop very high strength, but the small range between its 0·2% Proof Stress and the U.T.S. shows that it has very little capacity for work-hardening and would not therefore be very suitable for autofrettage applications.

The austenitic stainless alloys are much better from the point of view of corrosion resistance, but in their best condition for that, i.e. fully softened, they have relatively low strength and a very small elastic range. On the other hand, their capacity for work-hardening is very great, but this is to some extent at the expense of chemical resistance; i.e., the resistance is lowered by overstraining. They are of course expensive because of the high proportion of costly alloying elements, and for many applications it may pay to use them only in the form of relatively thin liners, as discussed in Section 3.4.

Precipitation-hardening can also be produced in austenitic materials of this kind, leading to great increases in strength, but again at the expense of some reduction in chemical resistance. An example is given at No. S.3.

Table 7.6

Corrosion Resistant Steels

Serial No.	*Material*	*Analysis % By Weight*											*Heat Treatment*
		C	*Si*	*Mn*	*S*	*P*	*Ni*	*Cr*	*Mo*	*Co*	*Va*	*Ti*	
S.1	Preciptn. Hardened High Chrome	0·02	0·08	0·05	0·005	0·004	4·25	11·95	4·50	13·20		0·40	Water quenched from 865°C Refrigerated at −75°C Aged at 450°C
S.2	Austenitic 18/8/Ti	0·06	0·70	1·56	0·013	0·034	8·6	18·2				0·34	Water quenched from 1,100°C
S.3	Preciptn. Hardened Austenitic Steel:[1]	0·03	0·52	1·37			25·9	15·5	1·20		0·32	2·28	Quenched from 980°C Aged at 700°C

[1] Details supplied by English Steel Corp. for their "Rex 559" material which also contains a small quantity of Boron.

Table 7.7

Corrosion Resistant Steels

(Stresses in bars with corresponding lbf/in^2 in brackets)

Serial No.	*Material*	*0·2% proof stress Bars (lbf/in^2)*	*Ultimate tensile strength Bars (lbf/in^2)*	*Elongation % on $4\sqrt{A}$*	*Reduction in area %*	*Impact resistance*		*Safe working pressure of K = 1·54 cylinder Bars*	*Estimated probable bursting pressure (K = 1·54) Bars*	*Notes*
						Izod ft lb	*Charpy ft lb*			
S.1	Precipitation-hardened high chrome	14,400 (209,000)	14,800 (216,000)	10·5	51	20		3,700[2]	6,400	"E.S.C. Chromar D.70"
S.2	Austenitic 18/8/Ti	2,280 (33,000)	5,920 (85,700)	55	74	> 100		700	1,930[1]	Welding quality
S.3	Precipitation-hardened austenitic A.286	7,000 (101,000)	11,200 (162,000)	30	49·7	66		1,800[2]	3,100	See note 1, Table 7.6.

[1] Result of actual test to destruction in small specimens.
[2] Calculated only: no experience of containing pressure with these materials.

7.6.3 *Materials for High Temperature Service*

Here we are primarily concerned with creep resistance, i.e. with stress to cause rupture in 100,000 hours, which we are taking as the criterion for lack of other data. Where very high temperature service is contemplated the effects of oxidation and scaling of exposed surfaces may be important, but – beyond this warning of the possibility – we shall not consider the matter further.

The highest creep resistance is undoubtedly to be got by using nickel-based non-ferrous alloys of the Nimonic type, but many of the alloy steels which are stable at the temperatures concerned have good high temperature properties.

The design of cylinders is here based on an allowable tangential stress of:

$$\frac{\text{Average stress to cause rupture in 100,000 hours}}{1{\cdot}5}$$

since this is preferred by the German, Italian and British (B.S. 1515) codes, and the correlation is that suggested by Soderberg.[7.9] According to this the stress is related to the pressure by:

$$(\sigma_\theta)_{al} = \frac{\sqrt{3}}{2}\,\frac{K+1}{2(K-1)}\,p_w \tag{7.9}$$

from which it will be seen that this is substantially the same as the Mean Diameter Formula of eq. (7.6) except for the factor $\sqrt{3}/2$ which is introduced to allow for the stress variation across the wall.

Table 7.8 summarises this data for various temperatures and for ferrous materials. A mild carbon steel is also included for comparison.

Example 1. Suppose that a reaction vessel is required to work at 500°C with a steady internal pressure of 600 bars. If the required internal diameter is 30 cm what will be the necessary wall thickness?

Referring to Table 7.8 the two materials for which the creep rupture data is given for this temperature are the En 17 and the En 40C, the carbon steel being then too weak. We will consider these in parallel hereunder.

	En 17	En 40C
Stress for creep rupture in 10^5 hours at 500°C	850 bars	1,580 bars
Thus, on basis of 1·5 factor, working stress	$\frac{850}{1{\cdot}5} = 567$ bars	$\frac{1{,}580}{1{\cdot}5} = 1{,}053$ bars
Then, from eq. (7.9)	$567 = \frac{\sqrt{3}}{2}\,\frac{K+1}{2(K-1)} \times 600$	$1{,}053 = \frac{\sqrt{3}}{2}\,\frac{K+1}{2(K-1)} \times 600$
Therefore	$260\frac{(K+1)}{(K-1)} = 567$	$260\frac{(K+1)}{(K-1)} = 1{,}053$
Whence	$K = 2{\cdot}69$	$K = 1{\cdot}66$
Hence the outside diameter required is	$30 \times 2{\cdot}69 = 80{\cdot}70$ cm	$30 \times 1{\cdot}66 = 50$ cm
and the wall thickness is	25·35 cm	10 cm

Table 7.8
High Temperature Steels
(Stresses in bars with corresponding lbf/in² in brackets)

Serial No.	*Material*	*British Standard equivalent*	*Temperature/Stress to Rupture in 10^5 hours*						
			450°*C*	500°*C*	550°*C*	600°*C*	650°*C*	700°*C*	750°*C*
H.1	0·3% Carbon steel	En 4	1,030 (15,000)	370 (5,400)	165 (2,400)	90 (1,300)			
H.2	0·5% Molybdenum steel	En 17	1,440 (21,000)	850 (12,300)	390 (5,700)	210 (3,050)	125 (1,800)	55 (850)	
H.3	$2\frac{1}{2}$% Cr, 1% Mo steel	En 40C		1,580 (23,000)	900 (13,100)	530 (7,700)	310 (4,500)	115 (1,700)	
H.4	25/20 Cr/Ni austenitic				1,410 (20,500)	1,030 (15,100)	535 (7,800)	315 (4,600)	145 (2,100)
H.5	18/12 Cr/Ni with Cb and Ta				2,550 (37,000)	1,480 (21,500)	920 (13,400)	575 (8,400)	375 (5,600)

These materials will be only marginally different in cost, so that the En 40C will evidently be the better and more economical.

Example 2. A tube system is required to carry 1,000 bars at 600°C. What should be the diameter ratio?

Here either of the two austenitic materials, serial nos. H.4 and H.5, could be used, and it can be assumed that the latter would be 15% more expensive than the former.

	H.4	H.5
Stress for creep rupture in 10^5 hours at 600°C	1,030 bars	1,480 bars
Thus on basis of 1·5 factor working stress	$\frac{1{,}030}{1\cdot5} = 687$ bars	$\frac{1{,}480}{1\cdot5} = 987$ bars
Then from eq. (7.9)	$687 = \frac{\sqrt{3}}{2} \frac{K+1}{2(K-1)} \times 1{,}000$	$987 = \frac{\sqrt{3}}{2} \frac{K+1}{2(K-1)} \times 1{,}000$
or	$433 \frac{(K+1)}{(K-1)} = 687$	$433 \frac{(K+1)}{(K-1)} = 987$
Whence	$K = 4\cdot410$	$K = 2\cdot563$

Now assuming that we are here concerned with a long pipe system so that ends and fittings form a small part of the total cost, it is clear that this cost will be proportional to the weight, which in turn will be proportional to the cross-sectional area (the overall length being the same in each). Thus the area a is given by:

$$a = \frac{\pi}{4}(D_o^2 - D_i^2) = \frac{\pi}{4} D_i^2 (K^2 - 1)$$

and therefore, if the costs are denoted C_{H4} and C_{H5} respectively:

$$\frac{C_{H4}}{C_{H5}} = \frac{K_{H4}^2 - 1}{(K_{H5}^2 - 1) \times 1\cdot15} = 2\cdot88$$

Thus the additional weight which would be required if the cheaper material were used would far outweigh the extra cost of the more creep-resistant material.

7.6.4 *Materials for Low Temperatures*

The importance of cryogenic work in modern technology has called for materials capable of carrying high stress at temperatures of only a few degrees absolute. The usual effect of cooling metals is to increase their static strength, but at the same time to reduce their ductility and their impact resistance – in some cases – almost to zero. Thus the chief consideration is this latter, and in recent years a large amount of work has been done by the manufacturers of metals for such service to determine the variation in, for instance, Charpy value with falling temperature, and figures of this kind are usually to be found in their catalogues.

From many points of view the ordinary austenitic stainless steels in their fully softened condition are very well suited for such work, as was shown by Colbeck, MacGillivray and Manning[7.10]. For example, a typical steel of this kind containing 0·11 % carbon, 0·25 % silicon, 0·49 % manganese, 0·019 % sulphur, 0·005 % phosphorous, with 18·3 % chromium and 10·05 % nickel, after water quenching from 1,100°C had a U.T.S. at room temperature of 6,400 bars (93,000 lbf/in^2) with an elongation of 54·0 % and a reduction in area of 74 %. At −180°C, however, the U.T.S. value had risen to no less than 15,400 bars (223,000 lbf/in^2) while the elongation remained at 45·7 % and the reduction of area was still well over 50 %. Moreover, the Izod impact tests were 117 ft lb at room temperature and 98 ft lb at −180°C and in each case the specimen was only bent and not broken.

A cheaper steel, containing 9 % nickel, is being increasingly used in low temperature applications although it requires rather special heat treatment involving two separate tempering operations at different temperatures. In the U.S. it is designated A-353 and it is expected to be covered shortly by a B.S. Specification. A typical example is LTN 9, manufactured by Messrs. Firth-Brown Ltd. of Sheffield. It has a Charpy value of at least 30 ft lb at −200°C and is easily weldable provided suitable precautions are observed. A typical analysis is: C 0·07 %, Si 0·23 %, Mn 0·64 %, Ni 9·08 %. Its 0·2 % proof stress at room temperature is 7,000 bars (102,000 lbf/in^2), but this rises to about 10,000 bars at −196°C; correspondingly, the U.T.S. rises from 7,850 bars at room temperature to about 11,600 bars (167,000 lbf/in^2) at −196°C.

7.6.5 *Non-Ferrous Metals*

The use of these in high pressure service is comparatively small, being chiefly confined to instruments and other auxiliary fittings. On the other hand, the demand by the chemical and metallurgical industries for increasing temperatures at high pressures is already approaching the limits to which ferrous alloys† can be used, and the most likely alternatives are those based on nickel, chromium or cobalt, although these—as well as being very expensive—are difficult to produce with their best heat-resisting properties except in small pieces.

We give below, in Table 7.9, a few typical non-ferrous alloys for general high pressure service with some of their mechanical properties; then in Tables 7.10 and 7.11 are added some of the latest temperature-resisting materials. With the latter the creep strengths are given in terms of the stress to give a total strain of 1 % in 10^4 hours, since the figures for rupture in 10^5 hours are too small to be of much use. It is not surprising therefore that, when dealing with temperatures above 700°C, highly stressed equipment must be considered in terms of relatively short lives (10^4 hours is a year and seven weeks) and even if one again used a factor of safety of 1·5 on the basis of the Soderberg formula, eq. (7.9) it is doubtful if a safe working life of much more than a year could be achieved.

† The term is used here to cover those containing more than 50 % of iron.

Table 7.9

Non-Ferrous Metals for General High Pressure Service

Serial No.	*Material*	*Approximate analysis*	*0·2% proof stress Bars (lbf/in²)*	*U.T.S. Bars (lbf/in²)*	*Heat treatment*
NF.1	Phosphor bronze	6 to 8% Sn, 0·3% P, rest Cu	2,000 (29,000)	3,090 (44,800)	Hot rolled
NF.2	Aluminium alloy	4·0% Cu, 0·49% Mg, 0·80% Si, 0·37% Fe, 0·77% Mn, 0·012% Ti	4,650 (67,500)	5,070 (73,400)	Annealed at 512°C Aged 5 hours at 175°C
NF.3	Beryllium copper	1·8% Be, 0·23% Co, 0·05% Fe, remainder Cu	9,100 (132,000)	11,600 (168,000)	Annealed at 800°C Aged 3 hours at 315°C

Table 7.10

High Temperature Materials (Non-Ferrous). Figures taken from Buchter[7.11]

Serial No.	*Material*	*Analysis % by Weight*												*Notes*
		C	*Si*	*Mn*	*Cr*	*Ni*	*Co*	*Mo*	*Ti*	*Cb*	*W*	*Al*	*Fe*	
NF.4	Nimonic 80	0·05	0·50	0·70	20	76			2·3			1·0	0·5	Developed by the International Nickel (Mond) Co. Ltd. in conjunction with Henry Wiggin and Co. Ltd.
NF.5	Nimonic 90	0·08	0·40	0·50	20	58	16		2·3			1·4	0·5	
NF.6	Inconel X	0·04	0·30	0·70	15	73			2·5	0·9		0·09	7·0	
NF.7	Refractaloy 26	0·05	0·70	0·70	18	37	20	3·0	2·8			0·20	18	
NF.8	S–590	0·40	0·70	1·5	20	20	20	4·0		4·0	4·0		24	
NF.9	L–605	0·12	1·0	1·5	20	10	51				15		1·0	

Table 7.11

High Temperature Materials (Non-Ferrous). Figures taken from Buchter[7.11]

Serial No.	*Material*	*Stress bars (lbf/in^2) to cause 1% strain or rupture in 10^4 hours at temperatures °C*				
		700	750	800	850	900
				Rupture		
NF.4	Nimonic 80		1,950 (28,000)	1,070 (15,000)	340 (5,000)	
NF.5	Nimonic 90		2,550 (37,000)	1,700 (25,000)	800 (11,600)	400 (5,800)
				1% Strain		
NF.6	Inconel X	3,100 (45,000)	2,140 (31,000)	1,380 (20,000)	900 (13,000)	
NF.7	Refractaloy 26	2,490 (36,000)	1,860 (27,000)	1,250 (18,000)		
NF.8	S–590	1,510 (22,000)	1,180 (17,000)	830 (12,000)	600 (8,500)	350 (5,000)
NF.9	L–605		1,800 (26,000)	930 (13,500)	650 (9,400)	480[1] (7,000)

[1] L–605 has a value for 1% strain in 10^4 hours of about 270 bars at 1,000 C.

REFERENCES

7.1. WITKIN, D. E., *Industrie Chimique Belge*, **31**, 1250, 1966.
7.2. LUFT, G., *Industrie Chimique Belge*, **31**, 1129, 1966.
7.3. CROSSLAND, B. and BONES, J. A., *Proc. Inst. Mech. Eng.*, **172**, 777, 1958.
7.4. Report of Board of Enquiry on "Brittle Fracture of a High Pressure Boiler Drum at Cockenzie Power Station", South of Scotland Electricity Board, January 1967.
7.5. SAIBEL, E., *Ind. Eng. Chem.*, **53**, 975, 1961.
7.6. American Society for Testing Materials, *Special Technical Publication* No. 410, 1967.
7.7. CHITTY, A. and DUVAL, D., Joint International Conference on Creep 1963–4, Paper 2, *Proc. Inst. Mech. Eng.*, **178**, Pt 3A.
7.8. CROSSLAND, B., private communication, 1968.
7.9. SODERBERG, C. R., *Trans. Am. Soc. Mech. Eng.*, **64**, 769, 1942.
7.10. COLBECK, E. W., MACGILLIVRAY, W. E. and MANNING, W. R. D., *Trans. Inst. Chem. Eng.*, **11**, 89, 1933.
7.11. BUCHTER, H. H., *Apparate und Armaturen der Chemischen Hochdruck-technik*, Springer-Verlag, Berlin/Heidelberg/New York 1967, 647.
7.12. LEACH, J. C. C., Proc. Symposium on Chemical Engineering in Iron and Steel Industry: Inst. Chem. Eng. (London) Mar. 1968, 65.

PART II

PUMPS AND COMPRESSORS

Introduction to Part II

Of the various machines required for the production of high pressures the compressor is the most important since all industrial high pressure processes such as the manufacture of ammonia, methanol, urea and low density polyethylene (to mention the major ones) involve the compression of a gaseous working fluid. In the laboratory or on the pilot-plant scale intensifiers, membrane and reciprocating compressors are used. For large industrial applications the reciprocating machine formerly held the field but for certain processes in the medium high pressure range it has, within the past decade, become possible to use the centrifugal or turbo-compressor for plants above a certain minimum size.

Where liquids at high pressure are involved, reciprocating – plunger – pumps are still the most widely used, though membrane or diaphragm pumps are sometimes used in the laboratory for medium high pressures (of the order of 350 bars). The role of the liquid pump is far less than that of the compressor and it has not experienced the spectacular development which the latter has undergone, but the same principles which govern the successful design of a compressor for high and extremely high pressures are equally applicable to pumps.

The high pressure centrifugal pump has so far found very limited application in chemical plants, mainly because in modern processes the volumes to be handled are generally below the limit for which this type of pump is suitable.

In addition to the considerations which govern the design of the high pressure compressor from the mechanical point of view, and which have been covered in Part I, a knowledge of thermodynamic principles is equally essential since at high pressures gases no longer obey the so-called perfect gas laws. Accordingly we start in Chapter 8 with a brief introduction to the thermodynamics of real gases. It is assumed that the reader already has some knowledge of the laws governing the behaviour of perfect gases and thermodynamic principles applied to such gases.

Thermodynamic charts with pressure in kg(force)/cm^2 are available for a wide range of gases and are adequate for most problems in the design of high pressure compressors: the difference between the two units of pressure – the bar and the kgf/cm^2 is less than 2%. The work unit, metre kilogram force (m kgf) is equal to 9·807 joules or 7·233 ft lbf.

Chapter 9 deals with various aspects of the general design and arrangement of high pressure reciprocating compressors and discusses the mechanical design of some of the more important elements. It is followed (Chapter 10) by illustrations of some modern high pressures and hypercompressors. From the large number of examples of their machines submitted by the various manufacturers it has been possible to include only a few, and the authors wish to thank all the firms for generously placing so much information at their disposal.

Chapter 11 is concerned with the high pressure centrifugal compressor, a comparatively recent development, and Part II concludes with a brief chapter devoted to liquid pumps and intensifiers.

Principal Notation used in Part II

A	1/427 kcal/m kgf, mechanical equivalent of heat
c_p, c_v	Specific heat capacity at constant pressure, constant volume (kcal/kg degC)
G	Mass (kg)
I	Enthalpy or total heat (kcal)
M	Molecular weight
N	No. of mols of mol. wt M in mass of gas G ($G = NM$)
N, N_{is}, N_{ad}	Power, isothermal power, adiabatic power (kW)
P, p	Absolute pressure
Q	Heat quantity (kcal)
R	Gas constant (m kgf/kg deg)
r	Compression ratio
$\mathscr{R}$	Universal gas constant (m kp/kmol deg)
ρ	Density, mass per unit volume
S, s	Entropy (kcal/deg; kcal/kg deg)
T	Absolute temperature
u	Internal energy
V	Volume
v	Specific volume (m^3/kg)
W, W_{is}, W_{ad}	Work in mechanical units (m kgf), isothermal, adiabatic
x_i	Mol. concentration of the ith component of a gas mixture
Z	Compressibility factor or p–v deviation for a real gas
γ	Adiabatic index
n	Polytropic index

8 Pressure, Volume, Temperature Relationships

8.1 General

The simple relation $pv = RT$ derived by combining Boyle's Law and Charles' Law is only an approximation to the behaviour of actual or, as we shall call them, real gases. It can be regarded as a limiting condition which all gases approach as the pressure is lowered. A gas which behaves exactly in accordance with the above relation is said to be an ideal or perfect gas. There are other ways of defining an ideal gas, for example by Boyle's Law and Joule's Law, which states that the energy of a gas is a function of its temperature only, and also by Boyle's Law combined with the condition that such a gas would exhibit a zero Joule-Thomson effect. It can be shown that all three definitions are equivalent.

If, in

$$pv = RT \tag{8.1}$$

where p is the pressure in kgf/m^2 (9·807 N/m^2 where N, the newton, is the unit of force) and v is the specific volume in m^3, we put $v = V/G$, where V is the volume occupied by G kg of the gas,

$$pV = GRT \tag{8.2}$$

Since, by Avogadro's Law, in an ideal gas a given volume contains the same number of molecules irrespective of the nature of the gas, if M is the molecular weight of the gas and V the molar volume (22·4 m^3) at atmospheric pressure ($p = 1{\cdot}033 \times 10^4$ kgf/m^2) and 0°C ($T = 273$°K),

$$MR = 843 \text{ m kgf/kmol degC}$$

$MR = \mathscr{R}$ is the universal gas constant: $\mathscr{R}$ is the same for all gases.

As already mentioned, equations (8.1) and (8.2) only hold for real gases at very low pressures, that is as $p \to 0$. Many equations of state or relations between the variables p, v, T have been devised in an attempt to express the condition of the gas over the entire pressure and temperature range. All are of more or less complex form involving a number of constants which must be determined from experimental data and the equations are accordingly empirical. For a discussion of the more important of these equations reference should be made to any standard text on thermodynamics.[8.1] We shall only mention here one of these equations—the van der Waals' equation—which we shall have occasion to discuss later.

The use of a complicated equation of state for calculations relating to high pressure is, in general, not practicable since it cannot be reduced to a form convenient for introduction into compressor theory. It is therefore simpler to introduce a "correction" factor Z into eq. (8.1) or (8.2).

Thus for a real gas

$$pv = ZRT \text{ or } pV = ZGRT \tag{8.3}$$

Z is defined as the real gas factor or "pv deviation". In the literature it is sometimes referred to as the compressibility factor. $Z = 1$ for a perfect gas and also for real or actual gases as $p \to 0$.

Compressibility data are frequently expressed as the ratio of the value of pv at a given temperature to the value of the same product at atmospheric pressure (760 mm Hg) and 0°C, $p_N v_N$ where the suffix N refers to the above standard conditions. The compressibility factor so defined is expressed by

$$K = pv/p_N v_N$$

The relation between Z and K can be shown to be given by

$$Z = KT_0/T$$

As already mentioned, for real gases the ideal gas law only holds in the limit as p approaches zero. For most gases, at 0°C the limiting value of pv, that is $p_0 v_0$, differs from $p_N v_N$ by a negligible amount although in some cases the difference may be as much as 4 per cent. Its value for a few of the commoner gases is given in Table 8.1.

Table 8.1

Gas	$p_0 v_0$ ($p_0 \to 0$)
A	1·0009
H_2	0·9993
N_2	1·0004
O_2	1·0009
NH_3	1·015
CO	1·0005
CO_2	1·0070
CH_4	1·0024
C_2H_4	1·0078

8.2 Gas Mixtures

The definition of an ideal gas implies that in a mixture of ideal gases each constituent behaves as if it were an ideal gas occupying the whole volume. Thus if P, V denote the total pressure and total volume of the mixture, P_i, V_i the corresponding partial pressure and partial volume of the ith component, then

$$P = \Sigma P_i \text{ (Dalton's Law) and } V = \Sigma V_i \text{ (Avogadro's Law)}$$

Dalton's Law does not hold for a mixture of real gases. The determination of P_i or V_i requires a knowledge of the equation of state for the gas in question, in one case in the form which gives P_i as an explicit function of the mol. concentration x_i and total volume V at the temperature T, that is $P_i = \phi(x_i, V, T)$, and in the other case where V_i the total volume of the constituent is expressed explicitly in terms of x_i, P, T or $V_i = \psi(x_i, P, T)$.

Analogous to eq. (8.3) for a single real gas we can write, for a mixture of gases,

$$PV = ZGRT = ZG\frac{\mathscr{R}}{M}T \tag{8.4}$$

For N mols $G = NM$ and

$$PV = ZN\mathscr{R}T \tag{8.5}$$

where V is the total volume of the mixture, and for the ith component of the mixture

$$P_iV = Z_iN_i\mathscr{R}T \tag{8.6}$$

Denoting the volume, mass and mol. concentrations by c, g and n respectively, we can deduce from equations (8.5) and (8.6)

$$c_i = \frac{Z_i}{Z}n_i = \frac{Z_i}{Z}g_i\frac{M}{M_i}; \quad n_i = g_i\frac{M}{M_i} \tag{8.7}$$

and for the specific volume

$$v_i = \frac{Z_i}{Z}\frac{M}{M_i}v \tag{8.8}$$

Since $c_i = V_i/V$; $g_i = G_i/G$; $n_i = N_i/N$,

$$\Sigma c_i = 1; \Sigma g_i = 1; \Sigma n_i = 1 \tag{8.9}$$

From eq. (8.7)

$$n_i = \frac{Z}{Z_i}c_i, \quad c_i = \frac{Z_i}{Z}n_i$$

so that

$$\Sigma c_i = \frac{1}{Z}\Sigma Z_in_i$$

But $\Sigma c_i = 1$, so that for the compressibility of the real gas mixture:

$$Z = \Sigma Z_in_i \tag{8.10}$$

Reference to equations (8.5) and (8.6) shows that we have defined the compressibility factors of the mixture and the individual components all referred to the same temperature and total volume. It can easily be shown that this implies an additive law for the pressures exerted by each constituent although, as pointed out above for actual gases, $P \neq \Sigma P_i$ since we are dealing with real gases.

The following relations for the properties of the gas mixture can be derived simply from eq. (8.7) for the gas mixture:

Specific volume	$v = \Sigma g_i v_i$	(8.11)
Molecular weight	$M = \Sigma c_i M_i$	(8.12)
Gas constant	$R = \Sigma c_i R_i$	(8.13)
Density	$\rho = \Sigma c_i \rho_i$	(8.14)

It is possible to define the compressibility factor in other ways, for example by reference to the total pressure. This method implies an additive volume law for the total volume from the partial volumes of the separate components. The pressures of the mixture deduced from these two ways of defining the compressibility factor of the mixture would, in general, differ from each other. In what follows we shall define Z in accordance with eq. (8.10).

In considering problems involving compressibility a useful technique, and one which is particularly so when dealing with gas mixtures, is based on what is known as the law of corresponding states. According to this law, which is only an approximation, all gases behave alike and have the same thermodynamic and other properties, including compressibility factor under the same "reduced conditions". These reduced conditions are defined by the ratios of p, v, T to the corresponding values, for the gas in question, at the critical point, viz. p_c, v_c, T_c. These ratios, referred to as reduced pressure p_r, reduced temperature T_r and reduced volume v_r, are given by

$$p_r = p/p_c;\ T_r = T/T_c;\ v_r = v/v_c \tag{8.15}$$

This means that if the compressibility factor Z is plotted with the reduced pressure as abscissa at various reduced temperatures, the group of curves so obtained applies to all gases. Such a generalised compressibility chart is shown in Fig. 8.7, p. 178. For most engineering applications it can be used with sufficient accuracy: with some gases however—for example ammonia—experimental data on compressibility show fairly wide deviations from the corresponding values obtained from the chart.

It has been found that the use of the general compressibility chart can be extended successfully to gas mixtures provided that instead of the true critical point data we introduce a "pseudo-critical" temperature and "pseudo-critical" pressure calculated by the simple linear mol. fraction rule from the individual critical temperatures and pressures. A method for calculating the compressibility factor for gas mixtures directly from molecular and p, v, T data of the pure constituents, which—at least up to moderate pressures—gives values in very close agreement with experimentally determined values, has been described by Kaske.[8.2]

As stated on p. 159, to represent the p, v, T behaviour of a gas accurately requires a complex equation involving a number of constants and even then any such equation only holds over a limited range. Many of these equations have been developed primarily to provide a means of deriving the thermodynamic properties of a particular gas so that, as would be expected, the various equations differ greatly in the accuracy with which they represent the behaviour of other gases. One of the most widely—and indeed one of the earliest—equations of state is that due to van der Waals:

$$\left(P+\frac{a}{v^2}\right)(v-b) = RT \tag{8.16}$$

which contains only two constants a and b and is therefore useful for preliminary calculations where the available data is limited and the order rather than an accurate result is aimed at. The van der Waals equation correctly predicts the general form of the pv isotherms at temperatures above the critical and the critical phenomenon itself. That it is only an approximation can be seen at once from the fact that it gives a value of 0·375—the same for all gases—for the compressibility $p_c v_c/RT_c$ at the critical point and a value of 3·375 for the ratio of the temperature at the Boyle point $[\partial(pv)/\partial p]$ to the critical temperature T_c. The actual values for real gases differ considerably in both cases from the above predicted values and moreover vary from gas to gas.

8.3 Thermodynamic Considerations

We shall next state (but without attempting any detailed consideration or analysis of them) some of the more important thermodynamic conceptions and relationships which enter into the design of compressors and particularly those for high pressures.

These relationships follow from the First and Second Laws of Thermodynamics. The first of these postulates the equivalence of heat and work as forms of energy and the relation between the two is expressed simply by the equation

$$Q = AW$$

where A is the proportionality factor between heat Q and work W. If the work is mechanical work $1/A$ is referred to as the mechanical equivalent of heat and has the value 427 m kgf/kcal. We shall regard as positive heat Q which is supplied from an external system or source to the "working substance"—for example gas which is expanding in a heat engine or being compressed in a compressor—and the work W as positive if it is work obtained from the working substance and used externally. Conversely $-Q$ and $-W$ will denote heat given up by, or withdrawn from, and work done from an external source on, the working substance.

The First Law can then be expressed in the form

$$Q = u_2 - u_1 \pm AW \tag{8.17}$$

where u_2 and u_1 denote the internal energy of the working substance in the final and initial states, Q the heat supplied to, and W the work done by it. Throughout the present treatment suffixes 1 and 2 will be used to denote the initial and final states of the substance respectively.

The pressure P, specific volume v, and temperature T are functions of the state of the fluid, as are also the total heat or enthalpy I and the entropy s to be referred to later. This means that in any change of state each of these quantities changes by a definite amount which depends only on the initial and final states and not at all on how the change was effected. Mathematically this is expressed by saying that each of these five quantities is a perfect differential. On the other hand Q and W depend on the particular way or path in which the change of state is brought about and are, therefore, not perfect differentials.

Accordingly, if we know the relation between any two of these functions of state and a third, this is sufficient to determine, uniquely, all the other functions. Even when the particular form of the equation of state is not known we can deduce a general relation between the variables involved. The derivation of these relationships is given in any standard text on thermodynamics such as Ref. 8.1 already quoted.

Reverting to eq. (8.17) for an infinitesimal change of state:

$$dQ = du \pm A\,dW$$

or

$$\int dQ = \int du \pm A \int_{v_1}^{v_2} P\,dv \qquad (8.18)$$

where we again emphasise that whilst du is the differential of a function u of the variables determining the state, for example $u = \phi(v, T)$ or $u = \xi(P, v)$, dQ and dW are not differentials of functions Q, W as such functions of the state do not exist.

In the case of a compressor where external work is done on the substance we must take the negative sign in eq. (8.18) and we then have

$$\int dQ = \int du - A \int_{v_1}^{v_2} P\,dv \qquad (8.19)$$

We proceed to consider the application of this equation to certain changes of state which are of most interest in the present context. These are changes which occur at (*a*) constant pressure, (*b*) constant temperature, (*c*) changes in which there is no heat flow to or from the working substance during the change and (*d*) changes accompanied by heat flow between the working substance and its surroundings. Initially we confine ourselves to the case of a perfect gas.

8.3.1 *Changes at Constant Pressure, i.e. P = constant*

Changes of this type are also known as isobaric. With P constant, eq. (8.19) becomes, on integration:

$$\int dQ = \int du + AP \int_{v_1}^{v_2} dv$$

$$Q = (u_2 - u_1) + AP(v_2 - v_1)$$

and with

$$Pv = RT,\ (u_2 - u_1) = c_v(T_2 - T_1)$$

$$Q = (c_v + AR)(T_2 - T_1)$$

$$= c_p(T_2 - T_1) \qquad (8.20)$$

Of the heat supplied Q, the quantity $c_v(T_2-T_1)$ is the amount used to raise the temperature of the gas from T_1 to T_2; the term $AP(v_2-v_1) = (c_p-c_v)(T_2-T_1)$ represents that part of Q which is converted into external work.

8.3.2 *Constant temperature, T = constant*

A change of state in which the temperature remains constant is referred to as an isothermal change. Since T = constant, u must also remain constant so that $u_2 = u_1$ and we have from eq. (8.17)

$$Q = \pm AW_{is}$$

Thus when a perfect gas is compressed isothermally the whole of the work done on the gas reappears as heat which must be removed immediately it is produced. The value of W_{is} is given by

$$W_{is} = -\int_{v_1}^{v_2} P\,dv \text{ in mechanical units}$$

and since in an isothermal change Pv = constant

$$W_{is} = P_1 v_1 \log_e P_2/P_1 \text{ m kgf/kg} \tag{8.21}$$

In compressor problems it is usually more convenient to consider the work involved in compressing a mass G kg having a volume $V_1 = Gv$ at the initial condition. If P_1 is expressed in kgf/cm^2, V_1 in m^3/h the isothermal power in kilowatts (kW) to compress the gas is

$$N_{is} = 0{\cdot}0627\, P_1 V_1 \log_{10} P_2/P_1 \tag{8.22}$$

8.3.3 *Change of state with $Q = 0$.*

A change of state in which there is no interchange of heat between the working substance and its surroundings is known as an adiabatic change. Such a change can be achieved if the substance is perfectly insulated from its surroundings or if the change is brought about so rapidly that there is no time for any transfer of heat to take place.

From eq. (8.7) with $Q = 0$,

$$u_2 - u_1 = \pm AW$$

Hence all the work done by the substance is done at the expense of its internal energy and (in compression) the work done on the substance all goes to increase its internal energy.

The fundamental eq. (8.18) takes the form:

$$dQ = c_v\,dT + AP\,dv$$

For an ideal gas in which $Pv = RT$ we can derive the P–v relation for an adiabatic change. It is given by:

$$Pv^\gamma = \text{constant} \tag{8.23}$$

where $\gamma = c_p/c_v$, the ratio of the specific heat at constant pressure to the specific heat at constant volume.

$$W_{ad} = -A\int_{v_1}^{v_2} P\,dv \tag{8.24}$$

the negative sign indicating that—for compression—work is done on the gas. From equations (8.24) or (8.23) we obtain:

$$W_{ad} = \frac{P_1 v_1}{\gamma - 1}\left[\left(\frac{P_2}{P_1}\right)^{(\gamma-1)/\gamma} - 1\right] \text{ m kgf/kg} \tag{8.25}$$

and for the power, if V_1 is in m^3/h and P_1 in kgf/cm^2:

$$N_{ad} = 0{\cdot}0281\,\frac{P_1 v_1}{\gamma - 1}\left[\left(\frac{P_2}{P_1}\right)^{(\gamma-1)\gamma} - 1\right] \text{ kW} \tag{8.26}$$

8.3.4 *Change of State with $Q \neq 0$*

In actual machines in which a change of state involving compression or expansion is carried out, isothermal or adiabatic conditions are never realised though they may be closely approached. Accordingly there is always some interchange of heat between the working substance and its surroundings. A general change of this type is known as polytropic and the P–v relation during the change can be expressed by Pv^n = constant. The index n is obviously greater than unity corresponding to an isothermal change of state and may be less or greater than the adiabatic index γ.

It follows that for a polytropic change of state, eq. (8.26) is replaced by

$$N_{pol} = 0{\cdot}0281\,\frac{P_1 v_1}{n - 1}\left[\left(\frac{P_2}{P_1}\right)^{(n-1)/n} - 1\right] \tag{8.27}$$

To determine Q—the amount of heat exchange between the working substance and the surroundings—we proceed as follows.

Differentiating the polytropic equation Pv^n = constant:

$$nP\,dv + v dP = 0$$

and from the equation of state for ideal gases $Pv = RT$:

$$P\,dv + v\,dP = R\,dT$$

so, by subtraction,

$$(n-1)\,P\,dv = -R\,dT$$

and from eq. (8.18)

$$dQ = du + AP\,dv$$

$$= c_v\,dT - \frac{AR}{n-1}\,dT$$

$$= \left\{c_v - \frac{c_p - c_v}{n-1}\right\} dT$$

$$dQ = c_v\,\frac{n-\gamma}{n-1}\,dT \tag{8.28}$$

or, if we write

$$c_n = c_v \frac{n-\gamma}{n-1}$$

$$dQ = c_n dT$$

c_n has accordingly the significance of a specific heat. If during the change of state c_n is constant, we have:

$$Q = c_n(T_2 - T_1) \tag{8.29}$$

If n is less than γ, that is $1<n<\gamma$, dQ is negative which means that during the change of state heat must be withdrawn from (for compression) and supplied to (for expansion) the working substance. If n is $>\gamma$, c_n is positive and heat must be supplied during compression and withdrawn during expansion.

So far we have considered the relationship between heat and work during an isolated or single operation involving a change of state. To repeat the process or cycle the working substance must be brought back to its initial state. However it is clear that to obtain any useful effect – work or heat – we must choose a different path for the return from state 2 to the initial state 1 from that followed in the change from state 1 to state 2. In this manner we have a cyclic or closed process in one part of which heat must be supplied to, and in the other withdrawn from, the working substance. Since in the cycle as a whole there is no change in internal energy the net work done by the medium is the difference between the two heat quantities mentioned above.

An example of such a cyclic process is that which occurs in the cylinder of a compressor, the P–V or "indicator" diagram for which is shown in Fig. 8.1. For simplicity we assume that the cylinder has no clearance volume – the effect of which will be considered later. We again restrict ourselves for the present

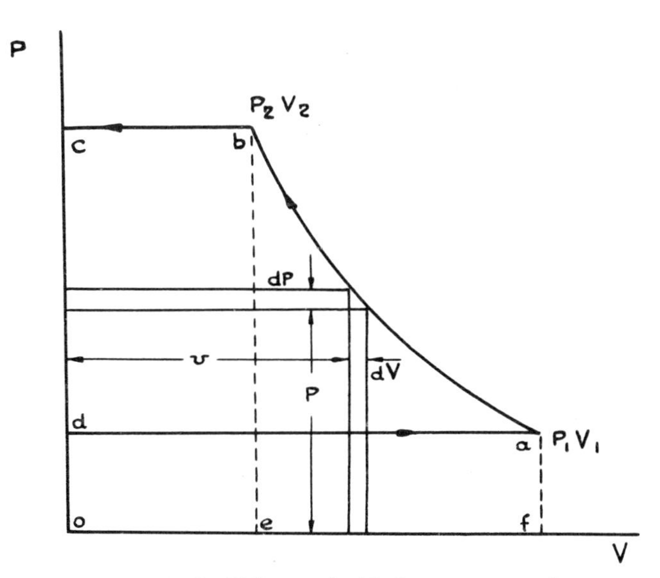

Figure 8.1 $P-V$ diagram for ideal compressor cycle

to an ideal gas as the working fluid and assume further that all stages are carried out in a reversible manner in which there is at all times equilibrium between the working substance and its surroundings. *da* and *bc* represent the stages in which gas is drawn into and discharged from the cylinder, in each case at constant pressure. Accordingly, they correspond to pure displacements in which there is no actual change of state. The net work of the cycle is given by the area *dabc* and in line with our convention regarding signs is to be taken as negative for compressive work.
Hence from eq. (8.19),

$$Q = (u_2 - u_1) - A\int_{v_1}^{v_2} P\,dv$$

From Fig. 8.2,

$$A\int_{v_1}^{v_2} P\,dv = A\int_{P_1}^{P_2} v\,dP + AP_1v_1 - AP_2v_2$$

so that

$$Q = (u_2 + AP_2v_2) - (u_1 + AP_1v_1) - \int_{P_1}^{P_2} v\,dP \tag{8.30}$$

$(u_2 + AP_2v_2)$ and $(u_1 + AP_1v_1)$ depend only on the final and initial states of the working substance and not at all on the way in which this is brought about: they are defined as the total heats or enthalpies I_2, I_1 corresponding to the final and initial states respectively.
Hence eq. (8.30) becomes:

$$Q = I_2 - I_1 - A\int_{P_1}^{P_2} v\,dP \tag{8.31}$$

For an infinitesimal closed process of the kind here considered:

$$dQ = dI - Av\,dP \tag{8.32}$$

The integral

$$A\int_{P_1}^{P_2} v\,dP$$

represents the total external work of the cycle and it should be noted that only in the case where the working substance is taken into, and discharged from, the cylinder at constant pressure is

$$\int_{1}^{2} v\,dP \text{ equal to } \Sigma P\,dv$$

taken over the whole process.

For a cycle in which compression is carried out isothermally $I_2 = I_1$ and eq. (8.31) reduces to

$$Q = -A\int_{P_1}^{P_2} v\,dP = -AW_{is}$$

as shown earlier. Likewise for an adiabatic compression cycle, $Q = 0$ and

$$I_2 - I_1 = A\int_{P_1}^{P_2} v\,dP = AW_{ad} \tag{8.33}$$

which shows that the whole of the external work is used in raising the total heat of the working substance; in other words the work is obtained simply as the difference between the final and initial values of the enthalpy.

For polytropic compression eq. (8.31) cannot be reduced to a simplified form.

8.4 Application to real gases

An expression which represents the P–v–T behaviour of a gas is referred to as an equation of state. A large number of such equations have been proposed varying not only in their complexity but also in the range of their validity and degree of accuracy throughout this range. One such equation which has been widely used is the van der Waals equation already mentioned, eq. (8.16).

We have seen that the external work in a compression cycle is given by $-\int v\,dP$ (in mechanical units). For the special case of isothermal compression this is also given by $\int P\,dv$ as can be seen at once by differentiating the equation $Pv = ZRT$. Accordingly if we can transform the equation of state to express P explicitly in terms of v, or v explicitly in terms of P we can evaluate the corresponding integral and so determine the isothermal work W_{is}^1 for the real gas. For the van der Waals equation the result is easily found to be:

$$W_{is}^1 = RT\ln\frac{v_2 - b}{v_1 - b} - a\left(\frac{1}{v_1} - \frac{1}{v_2}\right)$$

An equation which has the same general form as that of van der Waals but involves an additional "constant" is the equation due to Clausius:

$$\left\{P + \frac{a}{(v+h)^2}\right\}(v-b) = RT$$

which also lends itself easily to the calculation of W_{is}^1. This latter equation agrees very well with observed data on carbon dioxide and predicts the critical phenomena of this gas better than the equation of van der Waals.

For the reasons already mentioned it is better to approach the calculation of the isothermal work for real gases from the standpoint of the compressibility

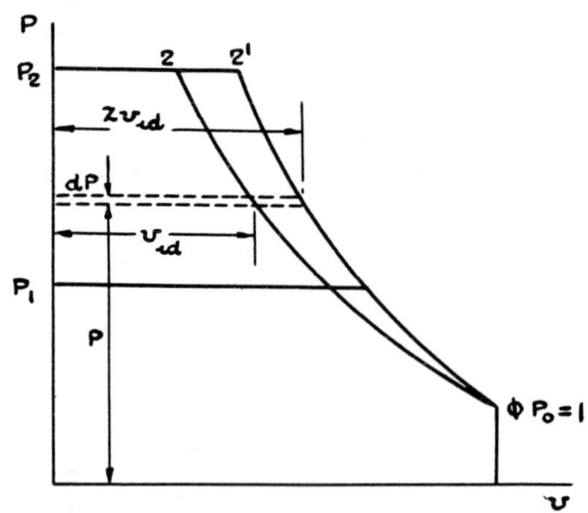

Figure 8.2 $P-v$ diagram for isothermal compression of real and ideal gas

deviations from the ideal gas. Fig. 8.2 shows the P–V diagram for the isothermal compression of a real gas in which the compression curve is represented by 1, 2′ and the corresponding curve for the ideal gas is 1, 2. If Z denotes the compressibility factor at pressure P we have:

$$Zv_{id} = v_{id} + (Z-1)v_{id}$$

where v_{id} is the ideal volume.

The isothermal work for the cycle, W_{is}^{1} is given by

$$W_{is}^{1} = \int_{P_1}^{P_2} v_{id}\, dP + \int_{P_1}^{P_2} (Z-1)v_{id} \quad dP \quad \text{m kgf/kg}$$

If v_0 is the specific volume at a pressure $P_0 = 1\ \text{kgf/cm}^2$ and at suction temperature:

$$Pv_{id} = P_0 v_0 = v_0$$

Substitution in the above gives:

$$W_{is}^{1} = v_0 \left[\log_e P_2/P_1 + \int_{P_1}^{P_2} \frac{(Z-1)}{P} \, dP \right] \quad \text{m kgf/kg}$$

which we can express in the form:

$$W_{is}^{1} = v_0 \log_e P_2/P_1 + v_0 \left[\int_{P_o=1}^{P_2} \frac{Z-1}{P} dP + \int_{P_o=1}^{P_1} \frac{Z-1}{P} dP \right] \text{m kgf/kg} \quad (8.34)$$

or with

$$\phi_P = \int_{P_o=1}^{P=P} \frac{Z-1}{P} dP$$

$$W_{is}^{1} = v_0 \log_e P_2/P_1 + v_0 \left(\phi_{P_2} - \phi_{P_1}\right) \quad \text{m kgf/kg} \quad (8.35)$$

If V_0 is the volume compressed in m^3/h measured at $1\ \text{kgf/cm}^2$ and suction temperature we have, corresponding to eq. (8.21)—which, as shown earlier, also gives the external power for an isothermal compression cycle for an ideal gas—for a real gas:

$$N_{is}^{1} = 0{\cdot}0627\ V_0 \log P_2/P_1 + 0{\cdot}0281\ V_0\,(\phi_{P_2} - \phi_{P_1}) \tag{8.36}$$

$$N_{is}^{1} = 0{\cdot}0627\ V_0 \left[\log P_2/P_1 + \frac{1}{2{\cdot}301}(\phi_{P_2} - \phi_{P_1})\right] \tag{8.37}$$

The first term in eq. (8.37) gives the theoretical isothermal power for the ideal gas and the second term the additional isothermal power due to the deviation from the ideal gas law. Values of $(Z-1)/P$ can be calculated and plotted against P and the integral determined graphically.

Linnartz[8.3] has given a simple "rule of thumb" formula for estimating quickly the isothermal power required at the coupling. He assumes an overall "isothermal efficiency" of 62·7 per cent—a figure which takes into account some additional losses, for example those in the compressor delivery piping. Using the above notation, Linnartz's formula for the isothermal power at the coupling for a real gas is:

$$W_{is-c}^{1} = \frac{V_0}{10} \log P_2/P_1 + 10\% \quad \text{kW}$$

For a compressor cycle with *adiabatic* compression the adiabatic external work for the ideal gas is, as already indicated, given (in mechanical units) by:

$$W_{ad} = -\int_{P_1}^{P_2} v\,dP = -v_1 P_1^{1/\gamma} \int_{P_1}^{P_2} P^{-1/\gamma}\,dP$$

which reduces ultimately to:

$$W_{ad} = -\frac{\gamma}{\gamma-1}\ P_1 v_1 \left[\left(\frac{P_2}{P_1}\right)^{(\gamma-1)/\gamma} - 1\right] \quad \text{m kgf/kg} \tag{8.38}$$

and if V_0 again denotes the volume compressed in m^3/h measured at 1 kgf/cm^2 and suction temperature we have for the adiabatic power

$$N_{ad} = 0{\cdot}0281\ V_0\ \frac{\gamma}{\gamma-1}\left[\left(\frac{P_2}{P_1}\right)^{(\gamma-1)/\gamma} - 1\right] \quad \text{kW} \tag{8.39}$$

Comparison of eq. (8.38) with eq. (8.27) shows that the adiabatic work for the complete cycle is γ times the work involved in the change of state from 1 to 2 along the adiabatic curve.

For the real gases a close approximation to the external work is obtained by multiplying eq. (8.38) by the factor $\sqrt{(Z_2/Z_1)}$, but since account must be taken of Z_1 in calculating the suction volume we have, as a close approximation—the error is, in general, less than 1 per cent—for real gases:

$$W_{ad}^{1} = \sqrt{(Z_1 Z_2)}\,W_{ad}; \quad N_{ad}^{1} = \sqrt{(Z_1 Z_2)}\,N_{ad} \tag{8.40}$$

The external work W_{ad} can also be expressed in the form:

$$W_{ad} = \frac{\gamma}{\gamma-1} R(T_2 - T_1) \tag{8.41}$$

where the temperature T_2 after compression is given by:

$$T_2 = T_1(P_2/P_1)^{(\gamma-1)/\gamma} \tag{8.42}$$

If the compression is *polytropic* with index n we need only replace γ in the expressions (8.38)–(8.42) to obtain the corresponding expressions for power etc.

For many gases including air, nitrogen, hydrogen, carbon monoxide and ammonia synthesis gas (a 3 to 1 mixture of H_2 and N_2), Z_1 and Z_2 at the temperatures encountered in compressors are $>1{\cdot}0$. For CO_2 the compressibility factor is $<1{\cdot}0$ for pressures below about 600 kgf/cm^2 and in the lower pressure range, which is of interest for the intermediate stages of high pressure CO_2 compressors, it varies widely. For example, at 50°C, $Z \approx 0{\cdot}34$ at 120 kgf/cm^2 and 0·85 at 60 kgf/cm^2. Propane, ethane and ethylene behave in a manner generally similar to CO_2 at low and medium pressures, i.e. $Z<1{\cdot}0$, but increase rapidly for pressures above about 400 kgf/cm^2. Ethylene, for instance, has a compressibility factor of 1·0 at approximately 400 kgf/cm^2 and 50°C; the corresponding value at 3,000 kgf/cm^2 is 5·0.

The P–V diagram with which we have been concerned so far enables only the changes in pressure and volume occurring in the cycle to be followed. If we wish to obtain information on the heat quantities and temperature changes involved in the various stages of the cycle as well we must make use of a temperature–entropy or total heat–entropy diagram (if one is available); if not, recourse must be made to calculation to prepare such a diagram and we shall indicate below how this can be done using compressibility data.

Entropy—a term introduced by Clausius—may be defined in various ways; one definition is expressed in the statement: if in a reversible change a substance takes in a quantity of heat dQ at a temperature T, the ratio dQ/T is called the increase in entropy of the substance. Entropy is a property of the substance in the same way that P, v, T, u and I are properties, which means that the change in its value between two states is independent of the process joining them but *only* in a reversible process is it equal to $\int dQ/T$. Mathematically dQ/T is a perfect differential. We shall use s to denote the specific entropy (kcal/kg degC) so that we can write:

$$ds = dQ/T \tag{8.43}$$

It should be borne in mind that although dQ/T is a perfect differential, dQ alone is not, since Q itself is not a property of the substance.
Integrating the above expression:

$$\Delta s = s_2 - s_1 = \int_1^2 dQ/T \tag{8.44}$$

The above definitions apply only to changes which are completely reversible. If the process does not satisfy this condition Δs or the change in entropy is not equal to $\int dQ/T$. When a substance has passed through a complete cycle its entropy is the same at the end as at the beginning and this is true whether the cycle is reversible or irreversible. Thus in both cases $\Delta s = 0$, but for the irreversible cycle $\int dQ/T \neq 0$; it has, in fact, a negative value. Thus in the present context we can regard entropy as a means of enabling us to express the Second Law of Thermodynamics in analytical form and so providing a measure of the degree of irreversibility of the process or processes.

Since from eq. (8.43) $T\,ds = dQ$ for a reversible operation, in the temperature–entropy diagram, which has temperature as ordinate and entropy as abscissa, the area under a path on the diagram represents the heat received by the system when that path is followed reversibly but not – as emphasised above – if the process is irreversible. Similarly the area within any closed path gives the net heat received by the system in passing reversibly through the corresponding cycle and by the First Law the same area represents the net-external-work done by the system.

We shall now see how some of the changes of state referred to above are represented on the temperature–entropy, T–s, diagram.

Fig. 8.3 shows the temperature–entropy diagram for an isothermal change in a real gas for which the enthalpy I decreases with increasing pressure at constant temperature, i.e. $I_2 < I_1$. For a compression change of state the heat to be removed is represented by the area $a12ba$ below the constant temperature line. The area $a1dca$ is the heat which has to be removed due to the decrease in total heat, viz. $-(I_2 - I_1)$, and the difference between the two areas $cd12bc$ is the heat equivalent of the external work. For an ideal gas the area $a1dca$ vanishes and the external work to compress the gas is given by $a12ba$.

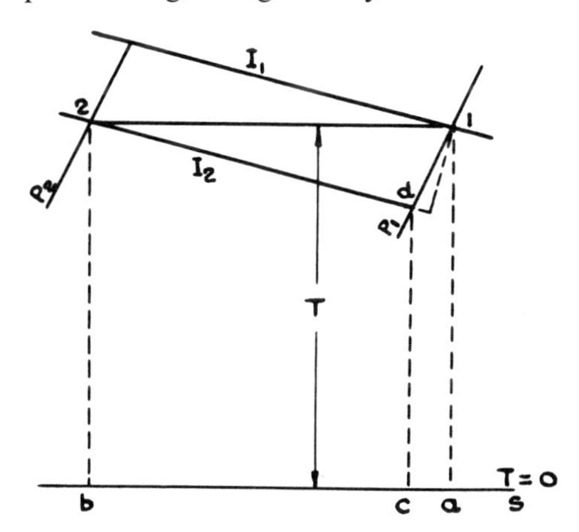

Figure 8.3 Isothermal change of state for a real gas in $T-s$ diagram

For an adiabatic change of state, Δs and therefore Q are both zero so that from eq. (8.33):

$$AW_{ad} = I_2 - I_1$$

that is, the adiabatic work is obtained from the T–s diagram simply by reading off the difference in enthalpy between the final and initial states. Such a change is shown in Fig. 8.4 in which, in this case, we assume that at constant temperature enthalpy increases with pressure. The heat equivalent of the external work is represented by the area $a2cda$. If after compression the gas is cooled at constant

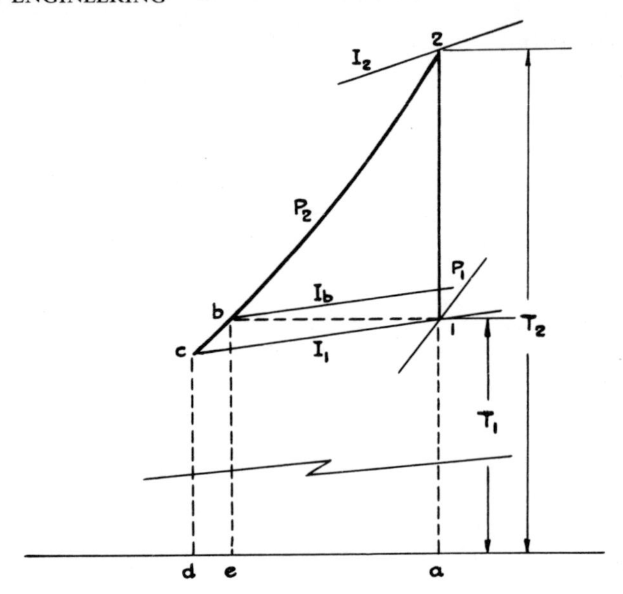

Figure 8.4 *T–s* diagram for adiabatic change of state in a real gas

pressure to suction temperature the heat to be removed is given by the area *a2bea*. The former is greater than the latter by the area *ebcde* which corresponds to the increase in enthalpy at constant temperature due to the increase in pressure from P_1 to P_2.

For a polytropic change of state, represented in Figs. 8.5 and 8.6, the total heat to be removed due to the work of compression is given by the area *a*1*2ha*, the increase in heat content $I_2 - I_1$ by *h2cdh* and the external work by the area *a*1*2cda*. The latter is greater than the heat to be removed in cooling to suction temperature by the amount $I_b - I_1$, the increase in enthalpy due to the increase in pressure from P_i to P_2.

So far we have confined our attention to the heat and work involved in a change of state—isothermal, adiabatic or polytropic compression—followed by

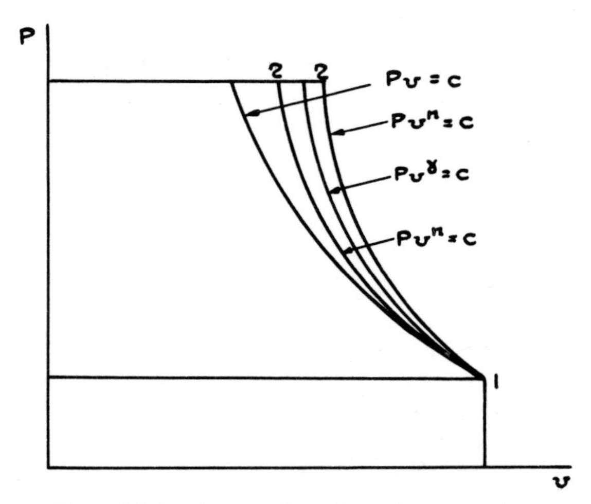

Figure 8.5 *P–v* diagram for polytropic change of state

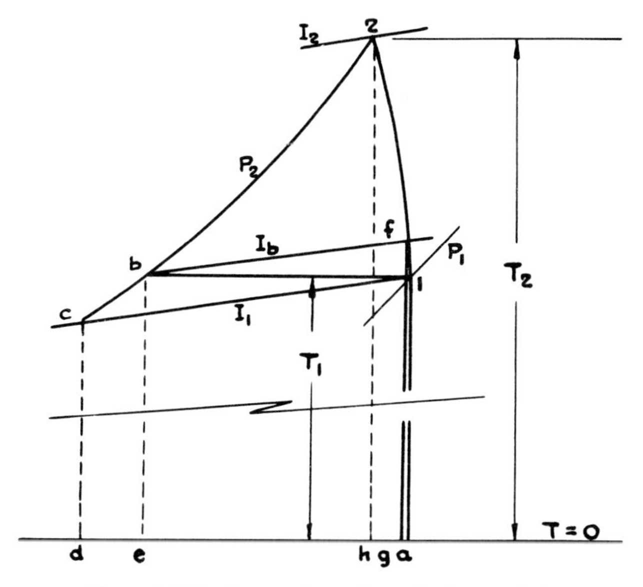

Figure 8.6 T–s diagram for polytropic change of state

cooling of the compressed gas at constant pressure, conditions which would apply in a compressor cylinder without clearance. The effect of this factor will be considered later (Chapter 9).

Temperature–entropy diagrams are at present available for a very limited number of gases and gas mixtures which have so far proved of industrial importance. The variety of gases and the pressure range for which such diagrams will be needed in the future will almost certainly increase and it is therefore of interest to sketch briefly the method by which the values of entropy and enthalpy can be calculated from compressibility data.

From the definition of entropy:

$$ds = dQ/T \tag{8.45}$$

and from eq. (8.32)

$$ds = [dI - Av\,dP]$$

Since I is a function of the variables P and T,

$$dI = \left(\frac{\partial I}{\partial P}\right)_T dP + \left(\frac{\partial I}{\partial T}\right)_P dT$$

and substituting in eq. (8.45):

$$ds = \left(\frac{\partial I}{\partial T}\right)_P \frac{dT}{T} + \frac{1}{T}\left[\left(\frac{\partial I}{\partial P}\right)_T - Av\right] dP \tag{8.46}$$

Similarly since s is also a function of P and T,

$$ds = \left(\frac{\partial s}{\partial T}\right)_P dT + \left(\frac{\partial s}{\partial P}\right)_T dP \tag{8.47}$$

so that by equating the coefficients of corresponding terms in equations (8.46) and (8.47):

$$\left(\frac{\partial s}{\partial T}\right)_P = \frac{1}{T}\left(\frac{\partial I}{\partial T}\right)_P; \quad \left(\frac{\partial s}{\partial P}\right)_T = \frac{1}{T}\left[\left(\frac{\partial I}{\partial P}\right)_T - Av\right] \tag{8.48}$$

It was pointed out on p. 172 that s is a perfect differential, so that:

$$\frac{\partial^2 s}{\partial T \partial P} = \frac{\partial^2 s}{\partial P \partial T}$$

Performing the appropriate differentiations on the two expressions in eq. (8.48) we obtain, after some simplification:

$$\left(\frac{\partial I}{\partial P}\right)_T = -AT\left(\frac{\partial v}{\partial T}\right)_P + Av \tag{8.49}$$

and from equations (8.48) and (8.49):

$$\left(\frac{\partial s}{\partial P}\right)_T = -A\left(\frac{\partial v}{\partial T}\right)_P \tag{8.50}$$

Finally, since $dQ = Tds = c_p dT$,

$$\left(\frac{\partial c_p}{\partial P}\right)_T = -AT\left(\frac{\partial^2 v}{\partial T^2}\right)_P \tag{8.51}$$

We thus see that values of the entropy and total heat (enthalpy) can be calculated from the first differential coefficient of the volume with respect to temperature at constant pressure. To obtain values of the specific heat at constant pressure we require the second differential coefficient of v with respect to T at constant P. Unfortunately, as mentioned earlier, none of the equations of state so far proposed is accurate over a sufficiently wide range for the exact determination of $(\partial v/\partial T)_P$ and $(\partial^2 v/\partial T^2)_P$ so that recourse must be had to a graphical approach using experimental P, v, T data. However the relation between v and T at constant pressure is almost linear so that the degree of accuracy in graphically finding the tangent to the v–T curve is very low, and even lower for the second derivative.

A number of methods for avoiding this difficulty have been proposed depending on separating the real gas relation $Pv = ZRT$ into two terms, one of which expresses the relation for an ideal gas and the other the deviation from this state. Accordingly we can write the general relation for a real gas, $Pv = ZRT$, in the form:

$$v = RT\left[\frac{1}{P} + \frac{Z-1}{P}\right]$$

from which, by differentiation and substitution in eq. (8.49) we can derive:

$$\left(\frac{\partial I}{\partial P}\right)_T = -ART^2 \left\{\frac{\partial}{\partial T}\,\frac{Z-1}{P}\right\}_P \tag{8.52}$$

and by integration:

$$I_{P_1} - I_{P_1} = -ART^2 \frac{\partial}{\partial T}\left[\int_{P_1}^{P_2} \frac{Z-1}{P}\, dP\right]_P \tag{8.53}$$

Similarly from eq. (8.50) we obtain:

$$\left(\frac{\partial s}{\partial P}\right)_T = -AR\left[\frac{\partial}{\partial T}\left\{T\,\frac{(Z-1)}{P}\right\}_P + \frac{1}{P}\right] \tag{8.54}$$

$$= -AR\left[T\left\{\frac{\partial}{\partial T}\,\frac{(Z-1)}{P}\right\}_P + \frac{Z-1}{P} + \frac{1}{P}\right] \tag{8.55}$$

and thus by integration:

$$(s_{P_2} - s_{P_1}) = -AR\left[\left\{\frac{\partial}{\partial T}\left(T\int_{P_1}^{P_2} \frac{Z-1}{P}\right) dP\right\}_P + \ln\frac{P_2}{P_1}\right] \tag{8.56}$$

or alternatively:

$$(s_{P_2} - s_{P_1}) = -AR\left[T\left(\frac{\partial}{\partial T}\int_{P_1}^{P_2} \frac{Z-1}{P}\, dP\right)_P + \int_{P_1}^{P_2} \frac{Z-1}{P}\, dP + \ln\frac{P_2}{P_1}\right] \tag{8.57}$$

The last term in eq. (8.57), viz. $AR \log_e P_2/P_1$ gives the change of entropy in an ideal gas and the first two terms the change due to the P–v deviation in a real gas; the first term, as can be seen from eq. (8.52), is the entropy change arising from the change in enthalpy whilst the second term

$$-AR\int_{P_1}^{P_2} \frac{Z-1}{P}\, dP$$

gives the change resulting from the additional isothermal work.

If we denote, for simplicity, the first term in eq. (8.56) by $(\Delta s)_{P_1}^{P_2}$, the expression can be evaluated from the tangents to the curves

$$T\int \frac{Z-1}{P}\, dP.$$

In practice, assuming that P–v deviations or compressibility data are available for a range of isotherms, for example at 25°C intervals, for each temperature we determine $(\Delta s)_{P_1}^{P_2}$ for a series of intervals $P_1 - P_2$ and then the summation of these values from $P_1 = 1$ kgf/cm^2 to P, viz.

$$\sum_{P_1=1}^{P} (\Delta s)_{P_1}^{P_2}.$$

If the resulting values are plotted against P for the various temperatures, Δs corresponding to each pressure and these temperatures can be read off, and the actual entropy difference, for the T–s diagram, calculated from eq. (8.55).

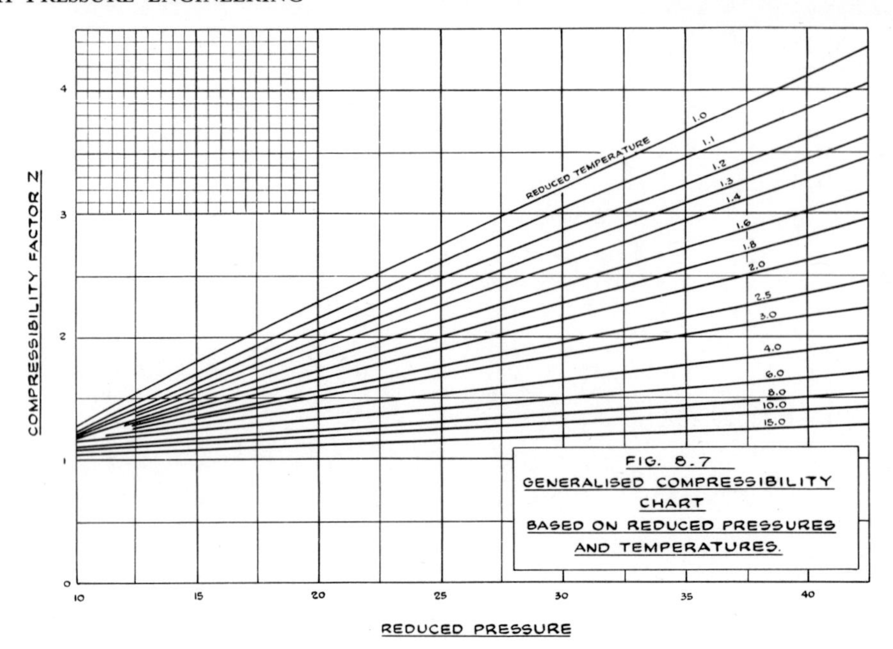

FIG. 8.7
GENERALISED COMPRESSIBILITY CHART
BASED ON REDUCED PRESSURES AND TEMPERATURES.

REFERENCES

8.1. SPALDING, D. B. and COLE, E. H., *Engineering Thermodynamics,* 2nd Ed., Edward Arnold, London 1966.
LAY, J. E., *Thermodynamics,* Pitman, London 1965.
FROELICH, F., *Kolbenverdichter,* Springer Verlag, 1961.

8.2. KASKE, G., Calculation of real gas factors for gaseous mixtures, *Chem.-Ing.-Tech.,* No. 10, 1060–63, Oct. 1966 (in German).

8.3. LINNARTZ, H., Practical experience in the selection of high pressure compressors, *Chem.-Ing.-Tech.,* No. 3, 237–241, 1962 (in German).

9 Mechanical Design

9.1

In the preceding pages we have considered, on a theoretical basis, the changes of state which occur in an ideal compressor cylinder, that is one without clearance volume and in which P, v and T remain constant during the suction and discharge parts of the cycle. We shall now examine more closely the conditions which occur in an actual compressor cylinder and formulate the factors which have to be taken into account in deciding the principal design parameters of a high pressure compressor.

9.2 Effect of Clearance Volume

Fig. 9.1 shows how the ideal diagram of Fig. 8.2 is modified by the clearance volume in the cylinder. The compression (1–2) and re-expansion (3–4) curves, are assumed to be polytropic and the pressures during the suction and discharge strokes to be constant, that is we ignore, for the present, the losses due to flow resistance which occur as the gas enters and leaves the cylinder, also the heating up of the fresh gas during the suction stroke and the leakage losses past the piston and through the valves and glands.

With these assumptions we define the theoretical volumetric efficiency η_v of the cylinder as the ratio of the volume of gas drawn into the cylinder to the volume swept by the piston (piston displacement). From Fig. 9.1,

$$\eta_v = v/v_s$$

If re-expansion follows the polytropic relation $PV^m =$ constant, we find, for ideal gases:

$$\eta_v = 1 - c\left[\left(\frac{P_2}{P_1}\right)^{1/m} - 1\right] \tag{9.1}$$

The clearance volume c, expressed as a percentage of the stroke volume, varies with the size of the cylinder and the velocities which can be allowed through the

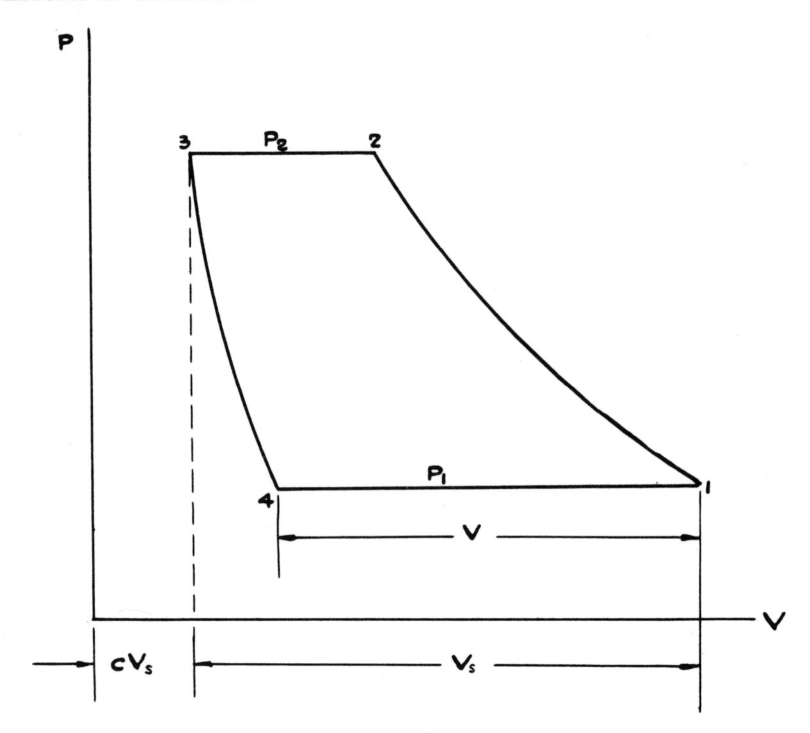

Figure 9.1 Ideal pressure–volume diagram for compressor cylinder with clearance

valves etc. For low pressure cylinders where, with due account to the density of the gas, significantly higher valve velocities can be allowed than for high pressure cylinders, an average value of c may be taken as 10% ($c = 0{\cdot}10$), a figure which may increase to as much as 20% ($c = 0{\cdot}20$) or more for cylinders to operate at very high pressures.

In developing a new design of cylinder c is of course unknown and a value must be assumed on the basis of the nearest similar design. Later the actual clearance volume of the proposed cylinder must be calculated and checked against the assumed value so that, if necessary, adjustments to the design can be made.

For real gases, and particularly at high pressures, account must be taken of the deviation from the ideal gas law. If $r = P_2/P_1$ is the ratio of compression (and re-expansion) and Z_1 and Z_2 the compressibilities at suction and discharge conditions respectively, the volumetric efficiency can be obtained from the approximate expression:

$$\eta_v = 1 - c\left[\frac{Z_1}{Z_2}\,(r^{1/m} - 1)\right] \tag{9.2}$$

The index of re-expansion m varies with the pressure and the speed of the compressor. For diatomic gases it usually lies within the range 1·2 to 1·3 for low and medium pressures, and within the range 1·25–1·35 for moderately high pressures. With increase of speed, m also increases and tends to approach the adiabatic index γ. It should be noted that, at high pressures, γ can vary considerably with the pressure and temperature.

In practice, the actual volumetric efficiency is, for the reasons mentioned above, less than the value obtained from eq. (9.1) or (9.2) and various empirical modifications have been proposed to include the appropriate correction. One such general expression is:

$$\eta_v = K - c\left[\frac{Z_1}{Z_2}(r^{1/m} - 1)\right] \tag{9.3}$$

where K lies between 0·95 and 1·0 and is frequently taken as 0·95, which is equivalent to assuming a figure of 5% for the losses due to heating up of the fresh gas and to valve and piston ring leakage. In large, well maintained machines, this leakage loss is usually small, not exceeding 1%.

A second expression which differs only slightly from eq. (9.3) is

$$\eta_v = 0{\cdot}95\left[1 - c\left\{\frac{Z_1}{Z_2}(r^{1/m} - 1)\right\}\right] \tag{9.3a}$$

which corresponds to the assumption of a value of rather more than 5 per cent for the losses referred to above.

Where the nominal suction pressure is atmospheric, the actual pressure in the cylinder at the end of the suction stroke will generally be slightly below atmospheric, and only reach this value again a short distance from the dead centre on the compression stroke. This represents a further loss in volumetric efficiency of 2% to 3%.

Another, and perhaps more instructive, way of considering the effect of clearance and the other factors mentioned above is from the standpoint of the amount of gas discharged from the cylinder, since this is what primarily interests the user. Accordingly, we define a discharge or output factor f where f is the ratio of the actual volume of gas (corrected to suction conditions), discharged in a given time, say per hour, to the piston displacement over the same period. Thus

$$f = G_d V_1/60nV_s$$

where G_d is the mass of gas discharged in kg/h and v_1 is the specific volume at suction temperature (temperature in the suction branch) and at the mean pressure during the suction stroke and n is the speed in rev/min.

The output of a compressor is frequently specified in terms of the volume of gas V_o measured at standard pressure P_o (1·0133 bar) and temperature T_o. In the United Kingdom, T_o is taken as 293 K (20°C), but in Europe the standard temperature is 273 K (0°C). If, for any stage of the compressor, the actual suction pressure and temperature are P_s and T_s, then the effective volume of gas to be discharged, corrected to suction conditions P_s, T_s is:

$$V_{eff} = V_o \frac{P_o}{(P_s - xP')} \frac{T_s}{T_o} \frac{Z_s}{Z_o} \tag{9.4}$$

where P' is the saturation pressure at suction temperature T_s, x the degree of saturation and Z_s the compressibility under suction conditions. We then have, from the definition of f:

$$V_s = V_{eff}/f. \tag{9.5}$$

The factor f cannot be calculated. It depends on the volumetric efficiency, the size and type of compressor, and on the gas which is being compressed. Thus f is significantly lower for refrigeration compressors, which in general have uncooled cylinders.

It is convenient to include in f the losses due to valve and piston ring leakage, which in a well maintained compressor will normally be small. Table 9.1 gives empirical values of f/η_v for diatomic gases for compression ratios from 2·5 to 5·0. The higher value is applicable to large diameter cylinders, the lower value to cylinders of small diameter.

Table 9.1

Compression ratio	f/η_v
2·5	0·97–0·95
3·0	0·95–0·94
3·5	0·94–0·925
4·0	0·93–0·92
4·5	0·925–0·90
5·0	0·91–0·89

A close approximation to the mean values of f/η_v over the range $2{\cdot}5<r<5$ is given by the linear relation:

$$f/\eta_v = 1{\cdot}025 - 0{\cdot}025r$$

from which

$$f = (1{\cdot}025 - 0{\cdot}025r)\eta_v \tag{9.6}$$

where η_v is obtained from eq. (9.2).

The piston displacement V_s can then be calculated from equations (9.5) and (9.6) with V_{eff} given by eq. (9.4). With V_s in m^3/h, the stroke s in m and the speed n in rev/min, the required piston area A in cm^2 is:

$$A = 10{,}000V_s/60sn \tag{9.7}$$

It is evident from the various expressions for the volumetric efficiency that a small clearance is necessary for a high volumetric efficiency, and in modern compressors the valves contribute a significant percentage of the clearance volume. For a high efficiency it is desirable that the speed of the gas through the valve should be as low as possible which implies a large area of flow through the latter, a requirement which conflicts with that of aiming at a small clearance, and indeed in high pressure cylinders the two cannot be completely reconciled.

Experience has shown that the following "rule of thumb" expression gives a good approximation to the piston displacement V_s which should be provided for a specified output V measured at standard pressure and temperature.

$$V_s \approx 1{\cdot}4 \text{ to } 1{\cdot}5\ V \tag{9.8}$$

This relation is useful in enabling the designer to form a quick mental picture of a compressor to meet a specified duty. It can be applied with suitable allowances for pressure, suction temperature, and compressibility to the higher pressure stages, but is not recommended for use at the highest pressures.

It is important that the compressor manufacturer should be able to estimate the overall volumetric efficiency as closely as possible in order to meet the specified duty within the guarantee. With the tendency today to allow no negative tolerance on the output – particularly in process compressors – he must be able to design his machine with confidence, with no more than a margin of 3% to 5% over the specification figure.

9.3 Compressor Speed

The speed (rev/min) is one of the most important parameters in the design of a compressor. Indeed it affects most, if not all, of the factors which determine the size of the machine – overall volumetric efficiency, indices of compression and re-expansion, number and size of cylinders, etc. – as well as having an influence on the arrangement so far as the method of balancing is concerned, the selection of valves and the most suitable type of drive.

The choice of speed is governed by several considerations of which the chief are:

(1) Increase of speed reduces the dimensions and weight of the compressor, and in particular the weight of the reciprocating parts.

(2) Increase of speed favours the adoption of direct drive by a high speed motor. Until comparatively recently the use of a high speed – e.g. 1,500 rev/min – motor in conjunction with a reduction gear was not regarded with confidence, although double reduction gears in combination with a turbine prime mover have, in one particular installation, been in satisfactory use for over forty years. In recent years there has been renewed interest in the high speed motor/reduction gear drive, particularly among Continental manufacturers, and there is no doubt that the ease and thoroughness with which the torsional analysis of such a system can be carried out with the aid of a computer has contributed greatly to the readier acceptance of this form of drive. Hitherto the spur gear, usually with helical or double helical teeth, has been favoured, but in the future it is likely that the epicyclic (planetary) gear will be more extensively used, even for the highest powers.

(3) Increase of speed also results in higher velocities through the valves, and therefore higher valve losses. Moreover, with the reduction in cylinder diameter it becomes virtually impossible, in the limit, to provide sufficient space for the valves. To avoid excessive loss through the valves a limit is imposed on the mean piston velocity. Too high a mean speed is also undesirable for other reasons, for example increased rate of piston ring and cylinder wear, although with the continual developments which have taken place in

the quality of lubricating oils and in the surface treatment of piston rings this is no longer a major factor in limiting the mean piston speed. In modern high pressure compressors the latter is of the order of 4·0 m/s and up to the present has seldom exceeded 4·2 m/s. It is of interest to mention that mean piston velocities as high as this latter value were being used over 30 years ago, but it would be wrong to conclude that this means that no advance has been made in the intervening years. Progress has been in the ability to operate for considerably longer periods without the need to renew the piston and gland rings.

(4) As a corollary to (1), increase of speed increases the inertia forces. The latter increase as the square of the compressor speed, and must obviously be less than the gas forces on the piston.

The above observations on compressor and mean piston speeds apply to normal lubricated machines where no special problems arise due to the nature of the gas. They are not applicable to non-lubricated compressors nor to compressors for exceptionally high pressures, which will be dealt with in a later section. Even for lubricated machines there is a tendency at pressures in the range 700–900 bars to reduce the mean piston speed below 4 m/s.

During the past decade the reciprocating compressor for high pressures has faced increasing competition from the centrifugal type – which will be dealt with in a later chapter. If it is successfully to resist further inroads into what, until now, has been regarded as its exclusive province a new approach will be needed to many aspects of the design of the reciprocating compressor which has fundamentally changed little over the past half century. Compared with the attention given to other forms of prime mover, in particular the internal combustion engine, the high pressure reciprocating compressor has received only scant treatment.

9.4 Some Mechanical Factors

Before proceeding to describe some typical modern high pressure compressors it may be useful to discuss, from the design point of view, a few of the more critical components which are still – all too often – responsible for shutting down the machine. Many, if not most, of these "failures" can be attributed to some form of stress raiser. In some instances, these cannot be avoided, as, for example, the transverse bores for suction and delivery passages and valve pockets in cylinder heads or the holes for lubrication purposes through cylinder walls, but in others modern techniques of manufacture can eliminate them entirely or reduce them to quite harmless proportions. Much can be done also by attention to design, particularly where fatigue is an important factor.

The screw threads on connecting rod bolts and piston rods are an illustration of details where modern manufacturing methods – thread rolling – can produce a far superior component than the best machined and ground bolt, and moreover one that is better suited to the severe operating conditions to which these bolts are subjected, and at a fraction of the cost. Yet it remains true that only relatively few compressor makers have so far adopted this technique, probably because they feel that the small number of failures does not justify the installation of equipment to roll threads of perhaps (if they wish to roll the threads of piston rods as well) up to 150 mm or more diameter. But the breaking of a connecting rod bolt is usually accompanied by disastrous consequences which can put a machine out of commission for several weeks or even months, so that with the present-day tendency to operate plants without any standby machines or equipment, there is likely to be increased pressure from the users of high pressure compressors for the adoption of any technique which reduces the possibility of component failure.

Connecting rod bolts do not always fail in the thread. Sometimes breakages due to corrosion fatigue or fretting corrosion (this latter a comparatively rare phenomenon) occur in the shank. The design of the bolt to give maximum elasticity and its location in the connecting rod and yoke are important and it is advantageous to use a form of nut which provides a more uniform distribution of stress along the depth of the thread.

The importance of the elasticity of the bolts can be seen most clearly from the deformation diagram, Fig. 9.2, for the bolts and connecting rod. Referring to Fig. 9.2a, and assuming all deformations to be elastic, the initial extension of the bolts is represented by OB and the corresponding tightening load by AB. The compression of the connecting rod end (including any shims between the two halves) under the action of the same tightening load AB is represented by CB.

With the notation of Fig. 9.2a we can write:

$$\frac{F_i}{\lambda_b} = \tan\alpha = C_b \text{ and } \frac{F_i}{\lambda_c} = C_c = \tan\beta \tag{9.9}$$

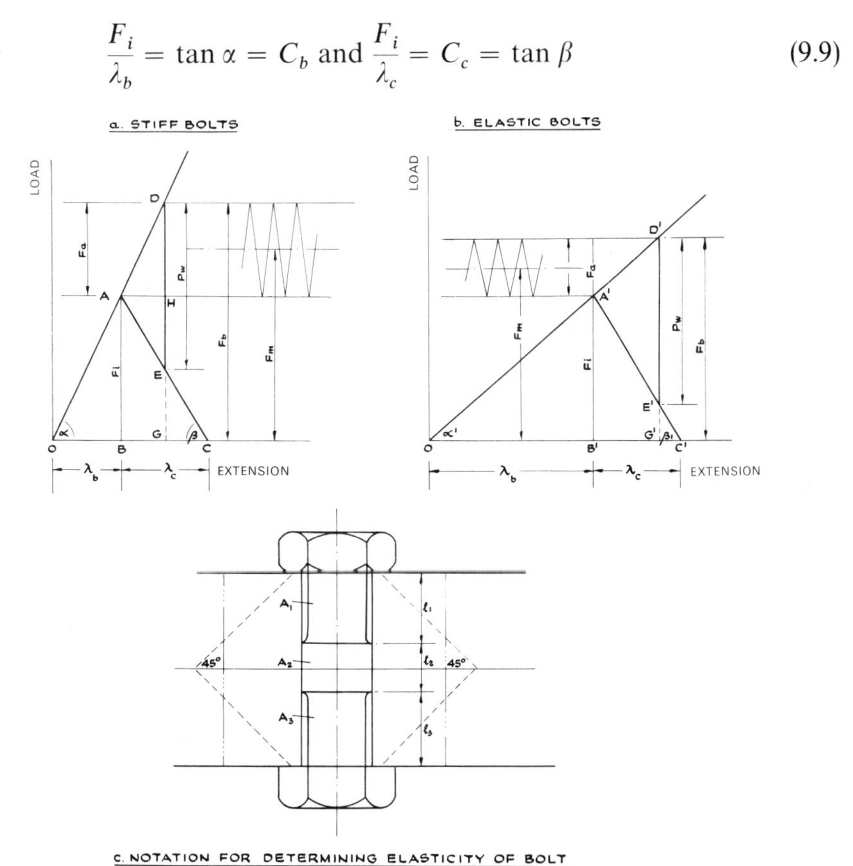

Figure 9.2 Load – deformation diagrams for bolted joint connections

where the suffixes b and c refer to the bolts and connecting rod. The symbols C_b and C_c may be defined as the elastic constants for the bolts and connecting rod. From the relation stress = (E . strain) it is easily seen that:

$$C_b = E_b A_b/l \text{ and } C_c = E_c A_c/l$$

where E_b, E_c are the moduli of elasticity, A_b, A_c the areas of cross-section of the bolts and connecting rod ends and l the effective length of the bolts. If the bolts are not of uniform section throughout the overall value is given by:

$$\frac{1}{C_b} = \frac{1}{E_b} \sum \frac{1_1}{A_1} \tag{9.10}$$

In the case of the connecting rod ends the difficulty is to assess how much of the metal in the two halves can be considered as being effectively compressed during the initial tightening of the bolts. One suggestion put forward by Rotscher assumes that this effective zone is that contained within the two conical frusta represented by the dotted line in Fig. 9.2c and that these can be replaced by a cylinder of the same cross-sectional area.

Corresponding to eq. (9.10) we have also for the connecting rod:

$$\frac{1}{C_c} = \frac{1}{E_c} \sum \frac{1_1}{A_1} \tag{9.11}$$

If now the assembly is subjected to an external load P_W the bolts will undergo an additional elongation represented in Fig. 9.2a by BG and the compression of the connecting rod will be relieved by the same amount. The additional load on the bolts is F_a and the total load $F_b = F_i + F_a$ which is less than $F_i + P_W$. As can be seen from the diagram, F_b is equivalent to a mean load F_m together with an alternating load $\pm F_a/2$.

The residual load between the two halves of the connecting rod is represented by EG.

Now

$$F_a = \text{AH} \tan \alpha = \text{AH}C_b$$

and

$$P_W = \text{AH} (\tan \alpha + \tan \beta) = \text{AH}(C_b + C_c)$$

from which

$$F_a = P_W \frac{C_b}{C_b + C_c} \tag{9.12}$$

The reduction in load between the two halves of the connecting rod is thus given by:

$$\text{EH} = F_i - \text{EG} = P_W - F_a = \frac{C_c}{C_b + C_c} P_W \tag{9.12a}$$

The load will vanish when EG = 0, that is when:

$$F_i = \frac{C_c}{C_b + C_c} P_W \tag{9.13}$$

which gives the minimum tightening load for the connecting rod bolts.

Under these conditions the bolt load becomes

$$F_b = F_i + F_a = P_W$$

which shows that the bolts must then carry the whole of the external load.

In Fig. 9.2b the elastic constant C_c for the connecting rod, that is $\tan\beta$, is the same as in Fig. 9.2a but the bolts themselves are more elastic, $\tan\alpha' < \tan\alpha$. For the same initial load F_i and external load P_W, Fig. 9.2b shows clearly that the additional load in the bolts, due to the external load, is reduced – as can also be seen from eq. (9.12). Consequently the alternating component of the load, viz. $\pm F_a/2$, is likewise reduced. The elastic characteristics of the bolts and connecting rod as defined by C_b and C_c are thus of paramount importance in determining the amplitude of the alternating component of the bolt load.

The above analysis is also applicable to – and leads to the same conclusions for – the attachment of the piston rod to the crosshead.

From what has been said above, it is unnecessary to emphasise that connecting rod bolts must be tightened equally and provision made for ensuring that the required pre-stress is achieved, for example by using a torque spanner, specifying the angular rotation of the nut (which should be suitably marked to facilitate this measurement), or the elongation of the bolt.

Because of the disastrous consequences which may result from a slack bolt – the whole load must then be carried by the remaining bolt or bolts – too much attention cannot be given to them in service, especially when a new machine is started up or after a general overhaul. Thus it is advisable after a few hundred hours running to check that the initial tightening load has been maintained, and periodically to remove a single bolt at random and examine it for signs of corrosion or cracks. At intervals of three to four years (during a general overhaul), all the bolts should be given a thorough inspection and metallurgical examination. If signs of corrosion are found the substitution of an alloy steel for carbon steel will not as a rule provide an answer, since although the latter may be significantly better under normal fatigue conditions the difference in fatigue limit when corrosion is present may be negligible. In such cases, some form of protection of the bolt, for example by a metallic or plastic coating, offers the best, though not a perfect, solution.

In compressor cylinders it is the cylinders for the high pressure stages – which are invariably of forged steel – which are the cause of most concern. At medium pressures for which cast steel can be used, it is usually possible to provide generous radii at all corners and there is more freedom in the arrangement of flow passages. Moreover, defects in the casting are frequently shown up by the hydraulic test, though sometimes they are not revealed until the final stages of machining. The use of cast steel for compressor cylinders has certain advantages, a saving both in material and in machining time, compared with forged steel, but their manufacture calls for considerable skill and experience. It is becoming increasingly difficult to ensure continuity in this skill, and with the great advances in welding techniques there is a growing tendency to adopt fabrication as a method of producing cylinders for which casting was formerly the only choice.

Since in a compressor cylinder the stress fluctuates cyclically between a maximum and minimum value (corresponding to the delivery and suction pressure respectively), design must be based on fatigue criteria. The allowable stress to which the material can be subjected is accordingly greatly reduced and, whilst there is now considerable data on the behaviour of thick cylinders with a plain bore (admittedly of small dimensions), there is as yet very little information on the effect on the fatigue behaviour of transverse openings, or on the scale effect. Tests on small bore cylinders have not in general extended much beyond 10^7 cycles; a modern compressor running at 300 rev/min will complete this number in about 23 days, but it is rare for a cylinder to fail within this period: for machines operating in the range of 350–450 bars failure (if it occurs at all) even of cylinders of identical design is more likely to take place within the period covered by $10^8 - 10^9$ cycles, and can invariably be traced to a notch or stress concentration effect. It is probable, however, that in many instances the crack which ultimately leads to failure may have been initiated much earlier.

In the manufacture of a high pressure cylinder, the first requirement is that the steel should be "clean" and of the highest quality. Where the pressure permits, a straight carbon steel is to be preferred to a low alloy steel, because of its lower notch sensitivity. The ingot must be sufficiently large to ensure adequate size reduction by forging and the attainment of uniform mechanical properties. This question of size reduction is important because of the need to ensure removal, by machining later, of the original centre zone of segregation in the ingot or—where complete removal is not possible—of minimising the effect of any which still remain and could act as a stress raiser.

Some of the methods of construction which have been devised to simplify the design of high pressure cylinders will be illustrated later. The most vulnerable areas are those immediately adjacent to the points of breakthrough of cross-bores such as valve pockets, lubrication holes, etc., and the facings in the valve pockets where the valves 'seat'. It is of vital importance that all sharp edges be smoothly rounded with the largest radius which is practicable, and that all internal corners, especially those on the facings referred to above where the intensity of stress is high, should be given as generous a radius as possible. Too often, the radius specified for internal corners is inadequate and insufficient care is taken to ensure that it is perfectly smooth. Lubrication holes present the most difficult problem from this point of view, because they are in general not easily accessible and because also of their small bore it is not easy to round off the edges to give a *smooth* radius.

It is not only transverse openings which give rise to stress concentration. Notch effects also occur wherever there is an abrupt change of section, such as at the junction of the bore of the cylinder proper and the larger bore in which the packing is housed. The cages or housing for the packing rings themselves afford a further example.

Stress concentrations cannot be entirely eliminated from the design of compressor cylinders for high pressure service and it is – or should be – the aim of the designer to avoid as many of the causes as possible, or at least to minimise the effect of a failure. In modern plants, where the tendency is to operate without any spare machine, the consequences of fracture of a compressor cylinder can be very serious, since replacement must be measured in months rather than days or even weeks; because it is the simplest – and cheapest – form of construction, many manufacturers are reluctant to depart from the monobloc cylinder. Whilst it is true that failures of the type discussed here are infrequent and represent only a very small percentage of the total number of cylinders of the same type in service, they are – like valve failures – unpredictable and have been known to be initiated at apparently well radiused edges or corners.

The most obvious way of minimising the results of a failure due to stress concentration is to divide the cylinder into a number of separate forged items, in which, for example, the end (or ends, in the case of a double-acting cylinder) containing the valve ports, suction and delivery passages, etc. are bolted to the cylinder proper. Apart from the fact that this method makes it easier to produce sounder forgings with more uniform properties, it permits a greater degree of flexibility in the arrangement of the valve pockets, etc., and improves the accessibility for removing sharp edges. There is, however, a significant increase in the cost of manufacture as compared with that of a monobloc cylinder. Also the need for a joint (subject to fluctuating pressure and temperature) between the cylinder head, cylinder and liner, is an additional problem.

As has been shown in Chapter 3, an effective way of improving the static strength of a thick cylinder is by compound construction – shrinking one cylinder over another. Unfortunately this is much less effective for compressor cylinders where the problem is essentially one of fatigue and, as Fig. 9.3 shows, for the hoop stresses, compounding does not alter the range of stress although it does reduce the mean stress; this only gives a small improvement to the fatigue endurance unless the direct stresses can be made wholly compressive. It is however used by some manufacturers for very high pressure machines, and in some cases the inner component is made of tungsten carbide (see Chapter 3, Section 3) which, by reason of its high elastic modulus, is able to reduce the stress range of the material behind it, although its own weakness in tension demands that it be so heavily shrunk as to be always in compression.

For the highest pressures in industrial use at the present time – 2,000 to about 4,000 bars – it is necessary to adopt the principle of breaking down the design into simple cylindrical or ring-shaped elements to the maximum possible degree. Such sections are relatively easy to calculate, and are free from notch effects, an important consideration at extremely high pressures where the use of alloy steels, with their greater notch sensitivity than ordinary carbon steels, cannot be avoided. Where the packing is stationary, the housings for the rings must also be built up by shrinking together two or more rings and the same principle followed for the valves.

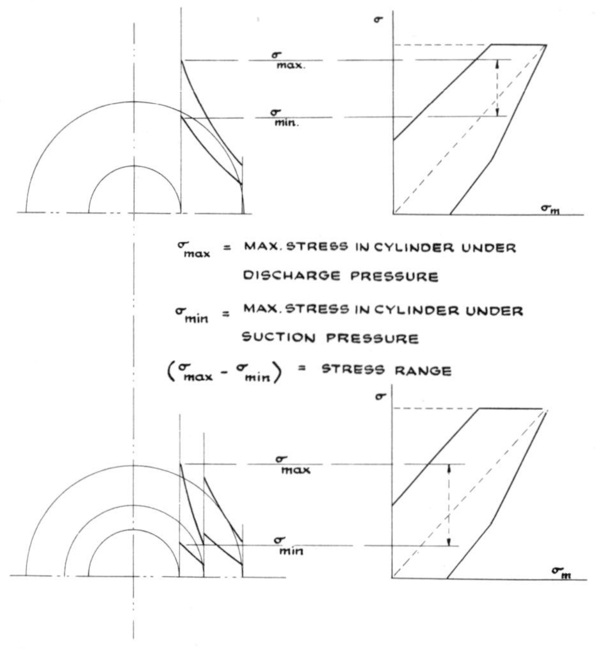

Figure 9.3 Comparison of stresses in simple and compound cylinders

One of the disadvantages of the compound form of construction is the difficulty of replacing the inner cylinder. A more elegant solution of the whole problem of the design of compressor cylinders for the highest pressures is for the inner cylinder to be mounted loosely in the outer cylinder, and the clearance between the two maintained at constant (static) compressor delivery pressure. With this "cartridge" design, not only is the risk of failure reduced due to the more favourable stress conditions, but replacement of the inner cylinder is simpler, quicker and less costly.

One consequence of the need to base cylinder design on the range of stress, is that whereas, normally, in a multi-stage high pressure compressor the ratio of compression in each stage is of the same order of magnitude (usually with a slight reduction in the higher stages), in hypercompressors it is the difference between the suction and delivery pressures which becomes the criterion.

9.5 Compressor Valves

The valves are the most crucial element in a compressor and the one, more than any other, responsible for the growing mistrust of this type of machine, as compared with the centrifugal compressor, in the Process Industry. Whilst it is possible to operate for a period with a hot gland or a leaking valve, an actual valve breakage requires that the machine be shut down within a reasonably short time. Because of the number of valves in a large multi-stage compressor such shut downs are inevitably frequent, even with a reasonably average valve life, but it is the fact of their unpredictability which is one of the main reasons for the unpopularity of the reciprocating machine even when it is the only choice.

The most widely used type of compressor valve is the one which consists, in some form or other, of a plate or ring-shaped disc. The former which was invented by Hoerbiger (and is still generally referred to by that name) towards the close of the last century is, basically, a plate with a series of slots at different radii. The plate is guided by flexible spring arms, integral with the plate, in which case the latter is clamped at the centre to the bolt holding the whole assembly together, and merely deflects, or is free to slide on a special guide ring. The plate may be replaced by one or more annular rings, an arrangement which has a number of advantages from the standpoint of manufacture and soundness. Thus plates of the Hoerbiger type frequently have the slots formed by stamping, with the risk of local incipient cracks at the highly stressed edges which are not always adequately rounded. A much superior method – which relatively few manufacturers of such valves adopt – is to mill the slots, but this increases the cost considerably, with the result that it is hard to justify the use of the milled plate on the basis of cost/average expected life. Ring or annular plates can obviously be manufactured with more certainty of soundness and freedom from cracks. It is good policy to check all new or reground plates by some form of crack detection technique before fitting them to a valve assembly.

Apart from the plate itself, the other elements of the complete valve are the seat, stop, cushion or damper plate to absorb some of the kinetic energy and reduce the noise, the lift washer which controls the lift and the valve springs to close the valve rapidly after pressure equalisation. In some designs the damper plate is replaced by damper springs.

The design of the closing springs is important because it is the failure of a spring which often leads to a broken plate. They must provide sufficient closing force combined with minimum pressure loss. If they are too weak and thus delay the closing of the valve, the plate may bounce off its seat and a state of hunting may be set up, leading to breakage in a relatively short time. Care must also be taken to ensure that when compressed under working conditions in the compressor adjacent coils do not touch and that the natural frequency of the spring is not too close to the speed of the compressor or a multiple of it. An obvious point, though one which is sometimes overlooked, is that the stress in the coils, when the spring is compressed solid, should be below the fatigue strength of the material.

The materials most commonly used for valve plates are carbon steel containing about 0·75 %C, high chromium (12 %–14 %) or low-alloy steels containing various proportions of Cr, Ni, Mo or Va. Great care must be taken to see that the steel is free from slag inclusions, and has a uniform structure. The closing springs are generally of spring steel, and they are often cadmium-coated to provide protection against corrosion. Coating with PTFE (polytetrafluorethylene) has also been tried with some success although there is a tendency for this to be rubbed off on the flat bearing surfaces of the end coils. Titanium springs have given excellent results in certain cases where corrosion was a serious problem.

Most compressor makers prefer to obtain their (Hoerbiger type) valves from specialist firms, relying on the empirical data and experience which the latter possess. In recent years many attempts have been made to investigate the actual movement of a valve plate, using oscillograph techniques, but the problem is not an easy one and it is not possible to predict with any degree of certainty exactly how a valve will behave in a compressor under specified conditions. If this were so it would be easier to analyse—and perhaps avoid—valve failures, which are often the most troublesome and baffling of the problems with which the compressor user has to deal. Sometimes the cause of failure is easy to decide, for example when it is due to corrosion or carry-over of liquid droplets from the previous stage, but at other times the reason is more obscure, particularly if a similar valve in a similar machine has given satisfactory service under identical conditions—a by no means unusual phenomenon. The indicator diagram is of considerable help in certain types of failure, e.g. valve bounce, but the solution even so is not always easy to find. Generally, the first step in difficult cases is to reduce the lift of the valve, and in a high pressure compressor of, say, several thousand kW and where trouble is being experienced with the valves in one stage only, the lift can be reduced very considerably without any measurable effect on the power consumption of the machine. Frequent spring breakage, which may be due to a variety of causes, is more difficult to overcome and more serious because all too often it is accompanied by a breakage of the valve plate.

There should be a difference in diameter between the seats—and corresponding pockets—of the suction and delivery valves in order to avoid the danger of inserting a suction valve assembly in a delivery valve pocket. Unfortunately, mainly for reasons of standardisation, valve manufacturers are reluctant to ensure non-interchangeability unless it is insisted upon by the purchaser. Valve assemblies of this type are easily confused and the consequences of making an error of the kind mentioned are disastrous.

At high, and especially at very high, pressures, "plate" valves are reduced to a single disc and the two valves—suction and delivery—are often arranged coaxially along the axis of the cylinder. This arrangement has the advantage of reducing the number of cross-bores in the cylinder head and it also facilitates the adoption of designs whereby the suction passage can be entirely relieved of fluctuating pressure. Its disadvantage, however, is that the plunger cannot simply be removed through the cylinder head, which has itself first to be removed, and this involves breaking the suction and delivery connection.

An entirely different design of suction valve is the mushroom type shown in Fig. 9.4, which is generally used in conjunction with the "poppet" type delivery valve, Fig. 9.5. The relatively large clearance volumes of these valves make them unsuitable for low pressures, but they are satisfactory at medium, and excellent at high, pressures. They can operate with lifts of two to three times that of plate valves, but it is evident from the illustrations that the gas flow area through them is, even allowing for the greater lift, less than that of a plate valve, so that the pressure drop is also greater. The effect on the overall power consumption of the machine is, however, small and is more than compensated for by the greater reliability due to the robustness of the design, which means that breakage of components is much less likely to occur and even if it does it is virtually impossible for any broken pieces to fall into the cylinder. Provided careful attention is paid to the heat treatment—hardening—of the valves to avoid cooling cracks,

Figure 9.4 'Mushroom' type suction valve *(P. Brotherhood Ltd)*

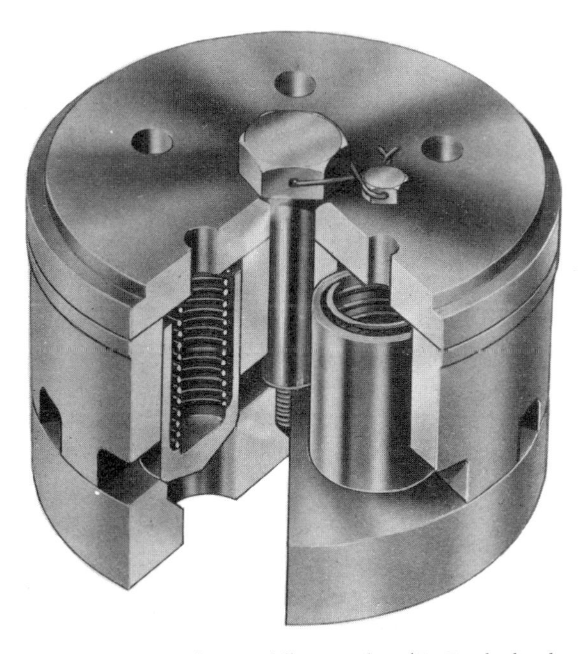

Figure 9.5 'Poppet' type delivery valve *(P. Brotherhood Ltd)*

failure of the valves themselves can virtually be ruled out. A spring failure, whilst it affects the operation of the particular valve, does not prevent the assembly as a whole from continuing to function and often enables the compressor to be kept on line until a more favourable opportunity for a shut-down arises.

A life of several years is not uncommon for the valves of the type under discussion and this raises the controversial question of whether—as some compressor manufacturers recommend—all valves should be changed periodically even if they are still in sound condition. Since—at least in the authors' experience—valve failures do not follow any logical pattern, there is no justification for changing a valve which is functioning correctly. Indeed the less the assembly is disturbed (even for cleaning) the better. A significant percentage of failures of new or reconditioned valves occurs early in their service life and if they survive this period there is every likelihood of their continuing to run for a year or longer.

9.6 Glands and Gland Packing

Next to the valves the method of preventing leakage of gas from a cylinder at a given pressure to a region of lower pressure is perhaps the most important detail in the design of a compressor. Where the cylinder is fitted with a piston, as is invariably the case at low and medium pressures, a conventional stuffing box or gland provides the solution, but at high pressures, where the "piston" often takes the form of a plunger, sealing against leakage is effected either by a stationary gland or by means of an assembly of rings on the plunger itself.

The stationary glands of modern (lubricated) compressors consist of self-adjusting metallic packing. A typical design for 350 bars is shown in Fig. 9.6, from which it will be seen that the packing itself consists of segmental rings grouped in pairs, each pair mounted in an L-shaped housing with sufficient clearance between faces to allow the packing to move freely in the space provided. Each ring is in three segments held together by a garter spring on the outside. In one ring of each pair the segments are cut tangentially, in the other radially, and it is of the utmost importance that they are correctly assembled in the housing with the former type nearest the cylinder, that is "facing" the pressure. Provision is made for supplying oil at one or more intermediate points depending on the number of groups of rings, which in turn is decided by the pressure. In some cases, particularly if the low pressure side of the gland is directly open to atmosphere, a separate connection is often provided to lead away any gas which leaks past the sealing rings. The latter are usually of cast iron or bronze running on a steel rod nitrided, induction or flame hardened where it passes through the packing. For service at very high pressures a compound construction is adopted for the L-rings to reduce the effect of stress concentration at the change of section.

A design of packing which is widely used by Continental compressor makers is the "Kranz" packing illustrated in Fig. 9.7. The packing shown is suitable for sealing against a pressure of 450 bars. As in the type described above, it consists of several housings—the number of which depends on the pressure—in which the packing elements are contained. The internal parts consisting of the three-piece packing rings mounted in the taper sleeves and counter cones are permitted

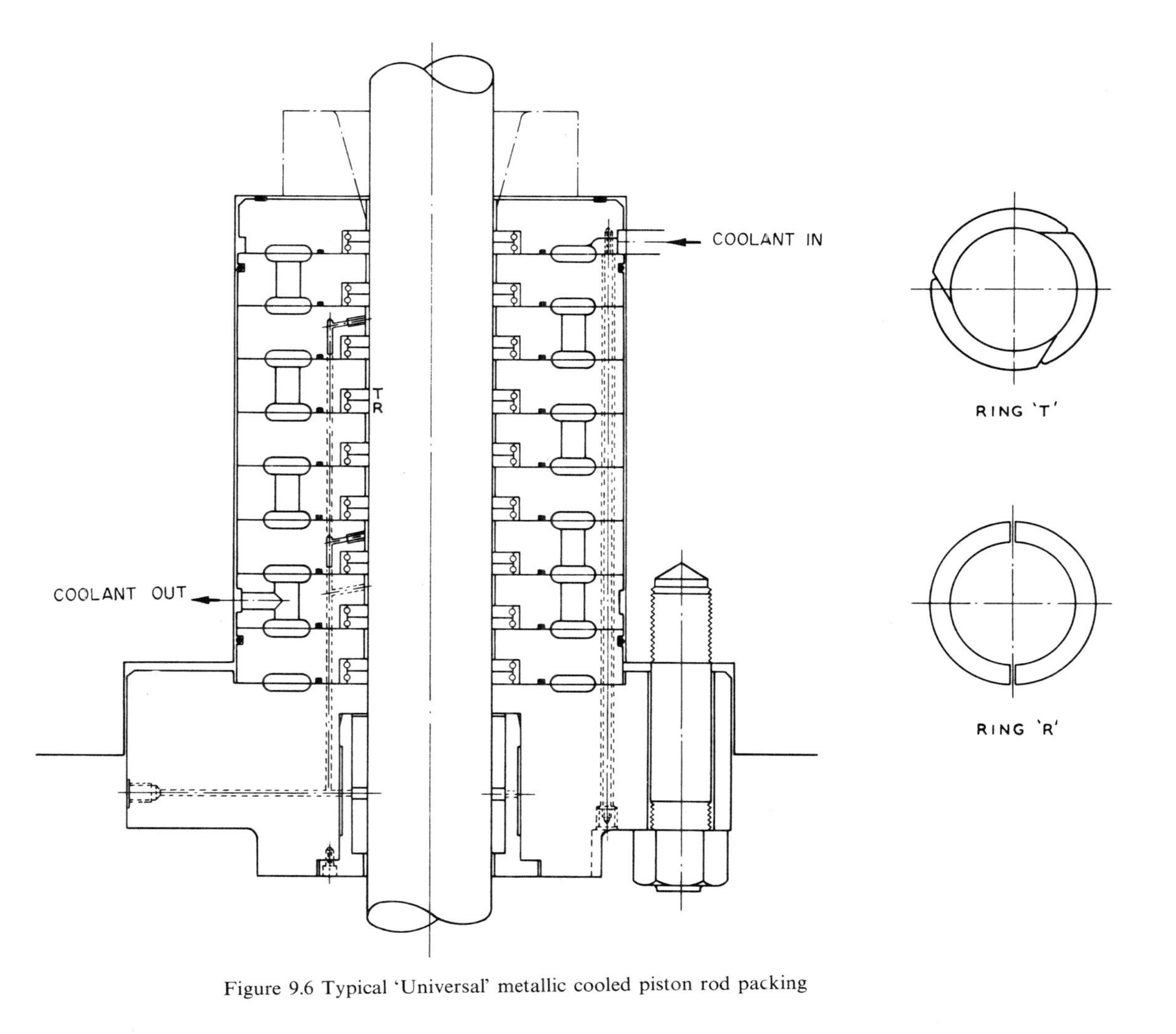

Figure 9.6 Typical 'Universal' metallic cooled piston rod packing

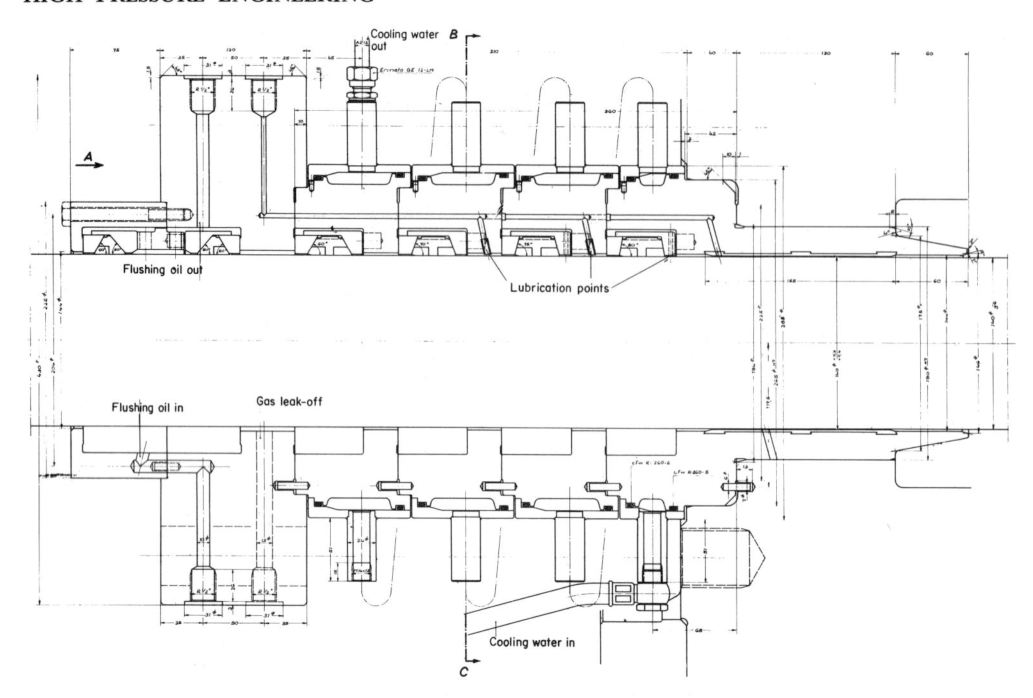

Figure 9.7 Flexible 'Kranz' type metallic packing *(Elementenwerk 'Kranz')*

a slight axial and radial movement. Sealing is effected by the working pressure pressing the segmental packing rings against the rod, and the cone angles are chosen so as (in theory) to maintain the radial pressure on the rod approximately constant over the length of the packing. A number of springs are incorporated around the housing to prevent the packing elements from moving when the compressor is running without load. As in the packing described above, the rings themselves are made of a variety of alloys—lead or tin bronze, bronze, etc.—to suit the gas being compressed and the pressure.

For pressure up to about 350 bars it is not necessary, with metallic packing, to cool the latter, but for higher pressures cooling is desirable, and in some cases essential. This can be achieved by circulating the cooling medium, which may be water or oil, round the outside of the packing cups or housings, as in the Kranz packing shown in the illustration, or through passages in the housings as in the first example. For non-lubricated packings, where the sealing rings are now usually of "filled" PTFE, it is essential to provide adequate cooling because of the poor conductivity of this material.

Manufacture of the packing calls for precision machining of the highest order. The joint faces between each pair of housings must be ground and lapped to ensure absolute tightness, not only to prevent leakage of gas round the outside of the housings, but also of lubricating oil, since the passages for the latter are drilled through them. The most difficult joint to seal is that between the cylinder and the packing housing nearest to it: the face on the cylinder must be perfectly square with the cylinder axis, since this is critical for the alignment of the whole packing assembly and if not satisfactory will result in "hot spots" on the rod, accompanied in extreme cases by spalling of the surface and, perhaps, bending of the latter. Accordingly, too much care cannot be devoted to the preparation of the individual packing elements and the assembly of the gland. The rings themselves must be bedded to the rod, but in spite of all these precautions there may be some leakage during the initial period and the packing may tend to run warm. This running-in period is also critical, especially in a new compressor, since however carefully the procedure for eliminating dirt, weld beads etc., has been followed there is always a danger that some remain and will find their way into the gland, scoring the packing and the rod. Accordingly, the gland should be over-lubricated at first and the amount of oil gradually reduced, keeping a close check on its temperature.

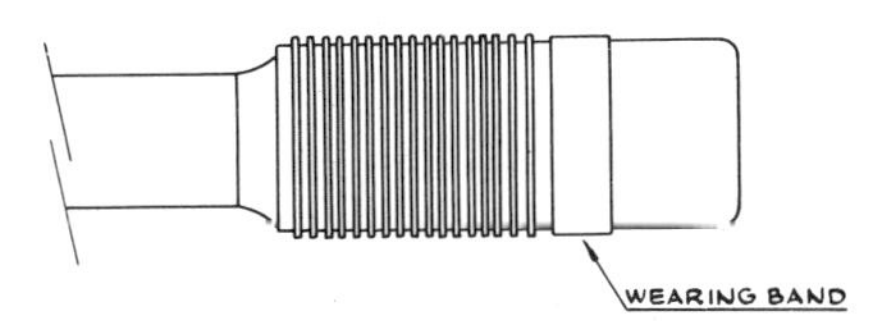

Figure 9.8 Solid type plunger with piston ring sealing

Where sealing is effected by packing on the plunger a number of variations are possible. The rings themselves are, in general, of cast iron of normal "piston ring" design although bronze rings are occasionally used. They may be assembled on the plunger itself if there is sufficient spring in them, as in Fig. 9.8, or they may be similarly assembled on a sleeve shrunk on the plunger. A third method is to build up the packing by fitting each ring in a separate L-shaped housing in much the same way as for the stationary gland packing described earlier, the housings then being assembled directly on the plunger or on a separate sleeve, Fig. 9.9. Which-

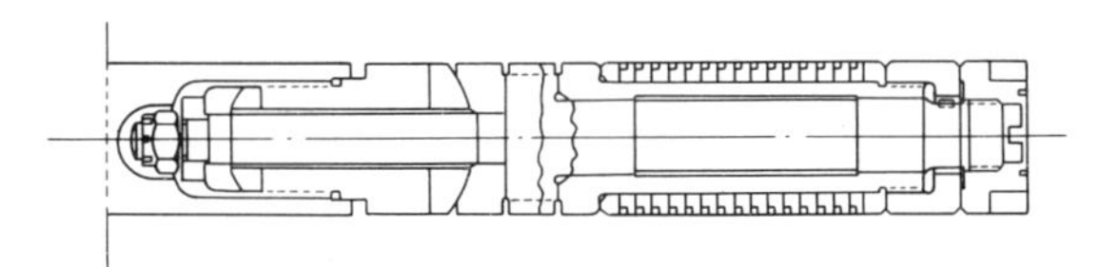

Figure 9.9 Self-aligning plunger with sleeve and carrier mounted piston rings

ever method is adopted, the diameter of the plunger is reduced over the length of the packing and great care is necessary to ensure a perfect pressure-tight joint between the plunger at the point where the reduction takes place and the adjacent housing or sleeve; otherwise if the initial tightening load is inadequate the whole assembly may be forced along the plunger until the nut on the end of the latter hits the cylinder end or valve head. Extreme care must also be taken to avoid stress raisers at this reduced section.

The piston rings may be of the simple form with the "cut" at 45° or of the type with overlap at the free ends, giving—when the ring is fitted—a gas-tight seal, whereas the former always permits a slight leakage through the gap. For running-in purposes the working face of the ring is often given a special surface treatment such as phosphate coating or, as in one special type, may be provided with a thin, very slightly projecting bronze strip which is quickly rubbed off on to the cylinder or liner wall forming a protective film on the latter. The metallurgical structure of the ring is most important for good life; it should be basically perlitic with medium-sized crystals of graphite and phosphide embedded in it. Preferably, the rings should be a little harder than the cylinder or liner.

Experience has shown that the rings in the assembly do not share the pressure difference across the plunger equally. This is taken, in the main, by one or two rings, not necessarily those nearest to the upstream pressure. When these, owing to wear, are no longer adequate to hold the pressure, the sealing is taken over by other rings. This behaviour can be verified by examining a set of rings which has been in service and which generally shows a quite irregular pattern of wear. For this reason, there is no simple rule for specifying the number of rings to seal a given pressure. The fact that fewer than 10 rings have been found adequate to seal pressures as high as 2,500 bars shows that at lower pressures, of the order of 350/450 bars where 12 to 15 rings are frequently fitted, this number is appreciably in excess of that required. The larger number may provide an additional safeguard and in some cases a longer life, but for obvious reasons compressor manufacturers are reluctant to depart from proven practice by making a drastic reduction in the number of rings.

For non-lubricated conditions piston rings of PTFE with various fillers, such as synthetic carbon and glass, are at present the most widely used. They are sufficiently flexible to be passed over a piston and slipped into the grooves. There is, however, a great deal still to be learned about the ideal composition, structure and process of manufacture of such rings. Even at medium pressures of a few hundred bars, performance is variable and unpredictable, and although it is claimed that the range of application extends to pressure differences up to 250 bars there is, as yet, insufficient data on the behaviour and service life over reasonable periods. But it has already been established that it is essential that the rings should be manufactured from an absolutely homogeneous mixture, and that the condition of the surface against which the rings run is also of vital importance, with some evidence of a critical degree of roughness for a reasonable, though in the long term not necessarily acceptable, performance.

In the preceding pages an attempt has been made to discuss in general terms the design of some of the principal parts of the reciprocating compressor with particular reference to the high pressure stages. However, the question of the detailed design of these components has not been considered as it is outside the scope of the present work which, as explained in the Introduction, is intended to deal with principles. In the next chapter, these principles are further illustrated by a number of examples of modern high pressure compressors.

10 Some Modern High Pressure Reciprocating Compressors

10.1 Introduction

In the preceding pages the principles underlying the design of some of the more important parts of reciprocating compressors have been explained with the emphasis on the high pressure stages. In what follows a few examples are given of high pressure compressors from laboratory machines to the largest industrial compressors at present in service and covering the whole range from 350 to about 3,000 bars.

10.2 High Pressure Laboratory Compressors

The principles which apply to the design of laboratory compressors differ in several important respects from those which must be followed in larger machines intended for process applications. In the first place the former are seldom required to operate continuously for extended periods: the runs are mostly of short duration, a few hours to a few days. This means that valves, glands, piston rings, etc. are not subjected to the same severe conditions as are these components in modern large high pressure machines which are expected to run for several months or longer with only an occasional shut-down, perhaps, for a valve change. Again, the laboratory compressor must, in most cases, be a general purpose machine, capable of compressing a wide variety of gases of different adiabatic indices and compressibilities. Thirdly, it must occupy small space since laboratory accommodation is usually restricted. Some concessions can also be made with regard to materials of construction for the high pressure stages where for certain gases, for example carbon monoxide, it would be necessary in an industrial machine to use special materials or to limit the maximum temperature at the end of compression to below about 130°C.

10.2.1 *Reciprocating compressors*

The firm of Andreas Hofer of Mülheim/Ruhr, Germany, have specialised in small compressors and manufacture a wide range of reciprocating machines of which, in the present context, the series of four, five and six stage compressors is of most interest. There are four basic types compressing gas from atmospheric pressure to 350 bars in four stages and to 1,000 bars in five stages. Certain of the machines within the range can also be manufactured with six stages for the same (1,000 bars) delivery pressure. All the machines are of the single crank type with all cylinders in line: the smallest compressor of the series occupies a floor space (excluding drive) of less than 2 m × 0·5 m and the largest 3 m × 0·6 m. Outputs range from 3·6 m^3/h at 200 rev/min to 200 m^3/h at 350 rev/min, the corresponding power requirements – for 1,000 bars final pressure – being 4·0 kW and 85 kW respectively. The pressure ratios are such that the maximum compression temperature does not exceed 140°C in any stage.

A sectional view through one of the largest compressors in the range 120 m^3/h, 1,000 bars pressure is shown in Fig. 10.1. Starting at the end nearest the crosshead the stages are arranged as follows, 3–2–1 (double acting)–4–5–6. There is thus only one gland through which leakage to atmosphere can take place and this (on the third stage) is at the moderate pressure of about 40 bars. The packing is of a special white metal or polyamide plastic and is provided with a liquid seal and sight glass which gives a visual indication of the leakage rate so that the gland can be tightened or new packing fitted as necessary. With the exception of the final (6th) stage which has a plain plunger, all the other pistons or plungers are provided with rings. In the first three stages these are mounted directly on the respective pistons, in the 4th and 5th stages in housings. For the normal lubricated machine the rings are of special cast iron with overlapping joints.

The valves are of the automatic disc or ring type and because of the range of gases of widely differing densities with which the compressor may have to deal they are conservatively designed with respect to pressure loading and gas velocities. After each stage the gas is cooled in a coil-type cooler located in a container housed in the base frame of the machine.

Hofer are now also able to offer compressors designed to operate without lubrication. These machines are of the vertical type, otherwise they are of the same basic design as the standard lubricated machine, with the exception that piston and gland rings are of graphite impregnated "Fluon" (Teflon) with special arrangements for cooling because of the poor thermal conductivity of the plastic material.

Fig. 10.2 is a section through the cylinder of a booster compressor also of Hofer design, for a suction pressure of 1,000 bars and a delivery pressure of 2,500 bars. Because of the intermittent nature of the duty which such a booster compressor is likely to be called upon to perform it is possible to dispense with certain features in the design which would be considered desirable – if not absolutely necessary – in an industrial machine for the same final pressure, for example compound construction for the cylinder, dispensing with a liner. Less importance can also be attached to stress raisers such as screw threads, cross-bores, etc., though the finishing of these, including the rounding of corners and sharp edges, still demands particular care.

These 1,000/2,500 bar compressors are made in three model sizes and for each model a range of speeds permits a range of outputs. The smallest actual

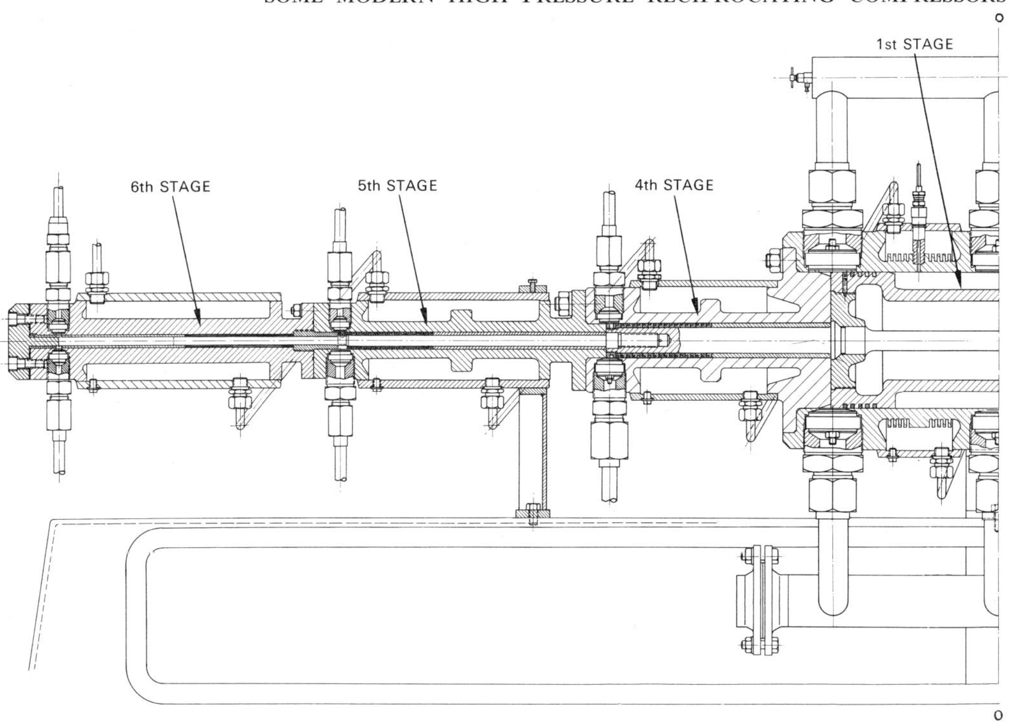

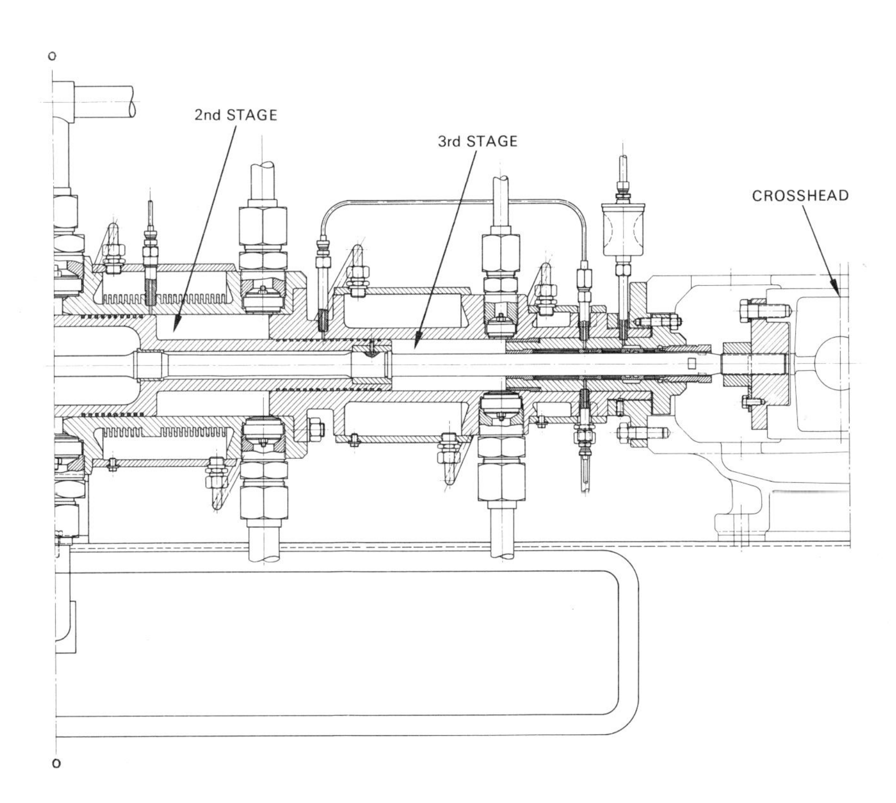

Figure 10.1 6-stage compressor for 1,000 bars *(Andreas Hofer GmbH)*

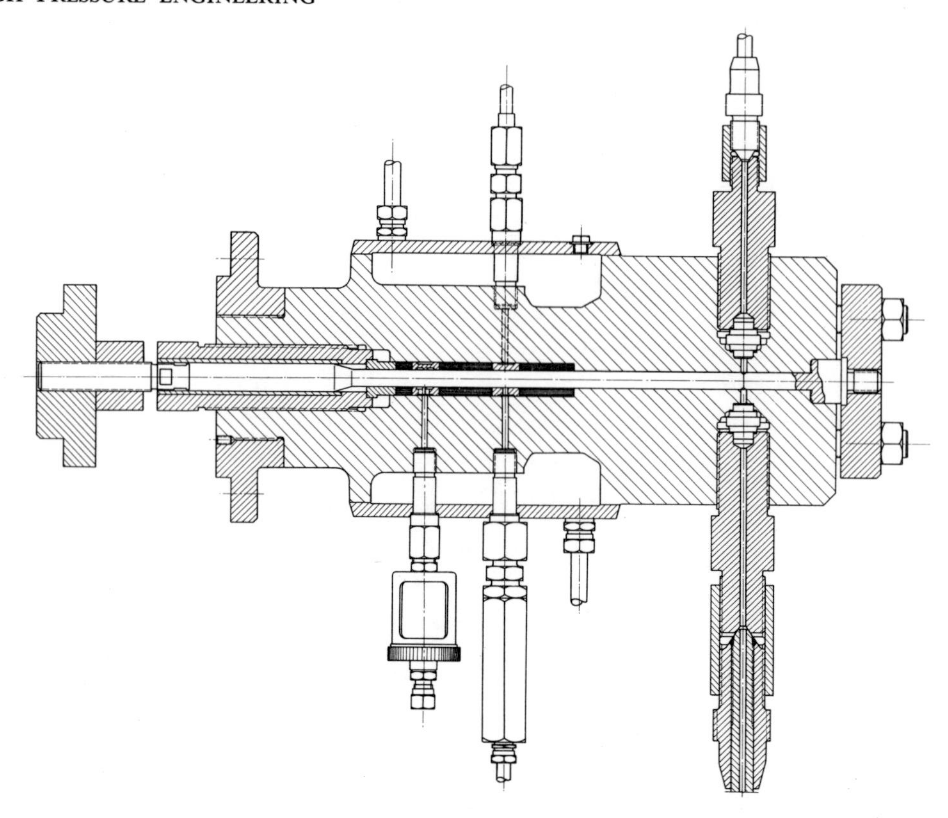

Figure 10.2 Cylinder of booster compressor to 2,500 bars *(Andreas Hofer GmbH)*

suction capacity at 1,000 bars is 9 litres/hour and the largest 420 litres/hour, the corresponding speeds being 78 and 215 rev/min respectively.

It will be appreciated that the major problem in these booster compressors is the gland, a problem which is all the more difficult because of the smallness of the plunger diameter. Exceptional accuracy is therefore called for in manufacture, alignment in the guiding of the plunger, in the surface finish of the latter and the packing rings, since the ultimate seal is that provided by the oil film. Extreme care is needed in assembly to ensure perfect cleanliness – the slightest scratch or score mark on the plunger is sufficient to lead to rapid failure of the packing.

Booster compressors of basically similar design to that described have been made for pressures as high as 5,000 bars.

10.2.2 *Diaphragm compressors*

An alternative to the conventional reciprocating compressor for laboratory service is the *diaphragm* or *membrane* compressor. In this type the piston or plunger of the former is replaced by a thin diaphragm clamped between two plates (diaphragm plate and perforated plate). The plates are slightly coned on the face towards the diaphragm forming a double-coned chamber in which the flexible metallic membrane oscillates. The oscillatory motion is produced by the reciprocating motion of a piston which displaces a pressurised fluid – usually oil. On the forward stoke of the piston the oil passes through the perforated plate and forces the diaphragm against the cover, discharging the gas in the space bounded by the two; on the return stroke gas is drawn into this space as the suction action causes the diaphragm to flex in the opposite direction towards the perforated plate. There is thus no contact between the hydraulic fluid and the gas to be compressed, so that the latter is not contaminated in any way. Very high compression ratios – of the order of 15 to 18 : 1 are possible in a single stage but the large area for cooling presented by the diaphragm and cover ensures that the temperature of compression is approximately the same as that in a reciprocating machine for the same output with, however, the much lower stage compression ratio usual in this type. Another valuable feature of the diaphragm compressor is the absence of a gland and therefore complete freedom from leakage.

The weakest element in the diaphragm compressor is, as might be expected, the diaphragm or membrane itself, which is subjected to high stresses and sensitive to small solid particles which may enter with the gas. It is usually made of nickel-chrome-molybdenum stainless steel but a wide choice of materials is possible depending on the gas, pressure level and other factors. Unfortunately in spite of rigid manufacturing and inspection specifications diaphragm life remains unpredictable. Figures of up to 5,000 hours have been quoted with an average life of 1,000 hours but the manufacturers of such machines recommend that the diaphragm should be changed at regular intervals established by the operating conditions.

The valves are of the disc or ring type of stainless steel. Since they operate completely dry they do not have the advantage of the cushioning effect of the oil film which is present in lubricated compressors, a factor which must be borne in mind in determining the pressure loading on, and gas velocity through, them.

As already mentioned, oil is mainly used in the hydraulic system and the choice of oil is important. It must, of course, have good lubricating properties and, in addition, must be free from foaming, have a flat viscosity curve and low compressibility factor. Consideration must also be given to the consequences of the failure of a diaphragm, which would result in the oil coming into contact with the gas. This is especially important when compressing oxygen; in this case water is used as the hydraulic fluid and owing to its lack of lubricating properties steps must be taken concerning the lubrication of the motion work and sealing of the piston. Any leakage of hydraulic fluid past the piston rings is taken care of by a compensating pump, excess fluid being returned via a pressure relief and overflow valve which also protects the system against excessive pressure.

Among the best known compressors of this type are those manufactured by Corblin in France and Hofer in Germany. The machines built by the former cover vertical single-stage compressors up to 15 bars, two-stage in V-form for pressures up to 250 bars and three-stage machines, also of V-form for higher pressures. The Hofer compressors, on the other hand, are horizontal. They use the same motion work and embody many other features to be found in the reciprocating machines made by this firm. The standard range comprises single stage compressors up to 15 bars, two stage up to 350 bars. For higher pressures a separate booster compressor is normally provided and Fig. 10.3 is a section view through such a machine designed to compress to a final pressure of 3,000 bars from a suction pressure of 1,000 bars. The piston adopted at lower pressures has become a plain plunger well guided as in the reciprocating booster compressor, and the cylinder is of compound construction. Of particular interest is the attempt to minimise the effect of stress raisers by designing many of the parts subject to pulsating pressure as simple cylindrical or ring-shaped components, and the well-rounding of all corners and sharp edges.

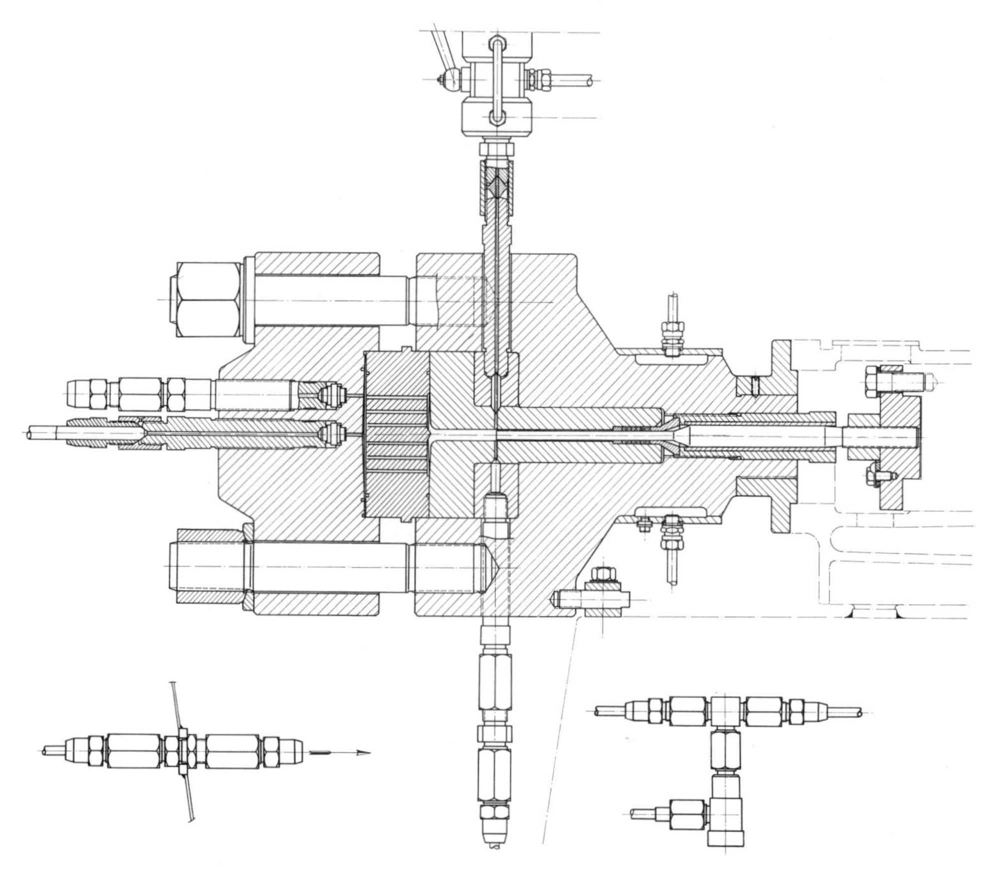

Figure 10.3 Section through diaphragm compressor for 3,000 bars *(Andreas Hofer GmbH)*

Corblin compressors are also made under licence by Maschinenfabrik Burckhardt in Switzerland and by the American Instrument Co. in the U.S.A. Another U.S.A. manufacturer of this type is Pressure Products Industries and a section through one of their machines for a delivery pressure of about 2,500 bars is shown in Fig. 10.4. As in the compressors of other manufacturers the swept volume of the oil plunger is appreciably greater than that of the diaphragm. A non-return valve on the suction side of the hydraulic – usually oil – system enables any leakage to be made good. An interesting feature of the Pressure Products Industries design is the method of sealing the diaphragm. The upper face of the valve

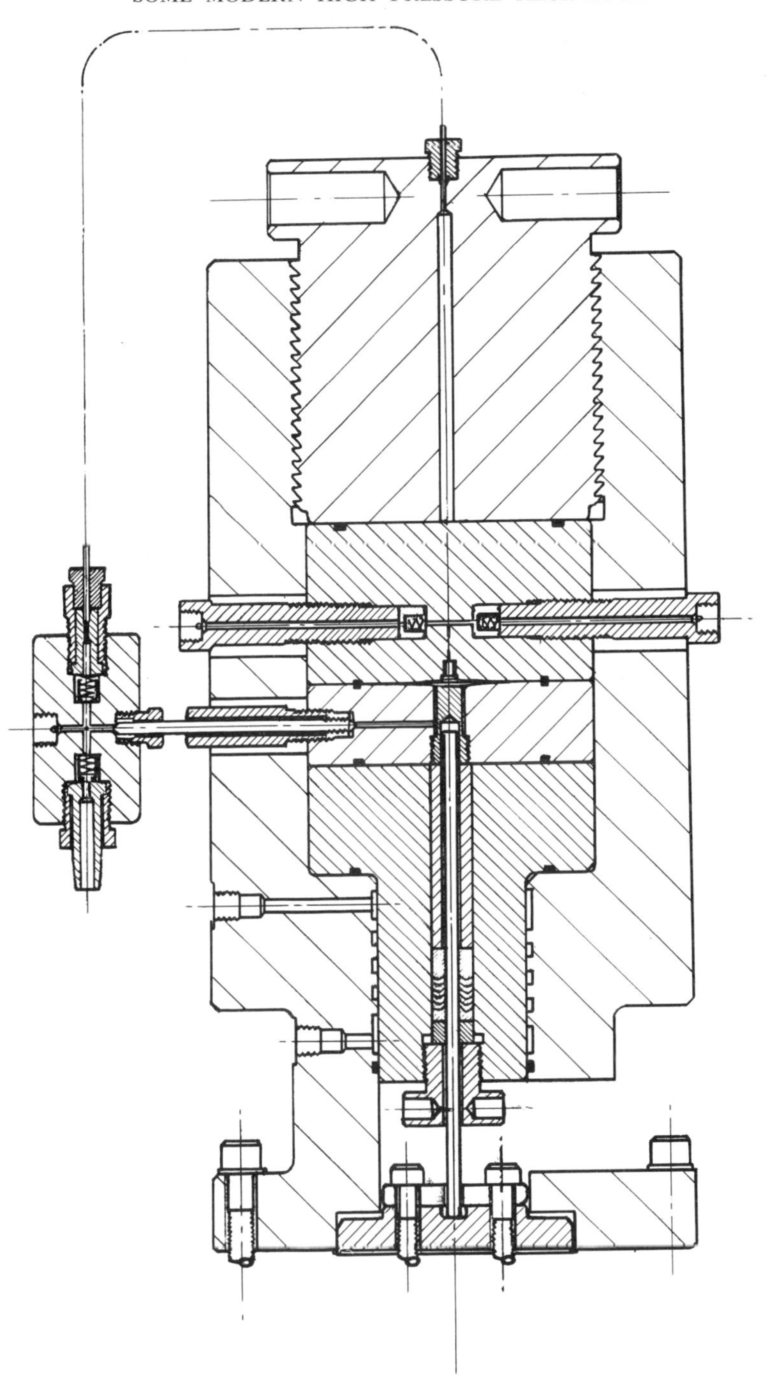

Figure 10.4 Section through diaphragm compressor for 2,500 bars *(Pressure Products Industries Co.)*

block is maintained at pump delivery pressure and the diameter of the O-ring joint between this face and the screwed plug is greater than that of the similar joint on the opposite face. There is thus a net positive load to provide a seal for the diaphragm.

In spite of the limitation imposed by the uncertainty of the life of the diaphragm the membrane compressor also finds a useful field of application in the compression of rare and highly toxic gases for industrial use where relatively small volumes at medium or moderately high pressures are involved.

10.3 High Pressure Compressors for Industrial Purposes

10.3.1 *Historical development*

For the beginnings of development of high pressure compressors we must go back to the close of the last century. None of the processes for the manufacture of the basic chemicals – soda ash, sulphuric acid, chlorine, dyestuffs – in use at that period required pressures of more than a few bars and the same was true of the "blowing engines" in use in blast furnace plants and steelworks. Some of these machines were of quite large size, driven by steam, and later by gas, engines and in spite of their low operating pressure the crank loading was relatively high. They were to set the pattern for the design of the large compressors for the first chemical process – the synthesis of ammonia – operating at high pressures.

It was the work of Linde in the field of refrigeration which may be said to have provided the impetus for the development of compressors for high pressures. He realised the potential industrial implications of air (and gas) liquefaction and his apparatus, built in 1895, was the first continuously operating plant for this purpose. This process was later to play an important part in the development of another industry, the synthesising of ammonia from its elements. The Linde process required air at a pressure of 200 bars – a pressure far beyond any pressure previously used; and this, in the early plants which had an output of only a few litres of liquid air per hour, was developed in a two-stage compressor with horizontally opposed cylinders, immersed in water with their respective coolers.

This same pressure of 200 bars was chosen by Haber and Bosch as the operating pressure for the first pilot – and later full-scale – plant for the production of ammonia from its elements. The compressors for this plant (designed originally for an output of 30 tonnes/day) which was commissioned in 1913 were of a quite different order of magnitude, with regard to size, from the compressors used in air liquefaction plants. Although from the work of Amagat and others thermodynamic data on the behaviour of the more common gases at high pressures was available, virtually nothing was known of the behaviour of materials when subjected to cycles of fluctuating pressure nor of the effect of stress raisers such as sharp corners and transverse openings in cylinders under internal pressure. It is not surprising therefore that there were many failures of high pressure cylinders and serious problems with valves and gland packing. It must be remembered that, at that time, the choice of materials of construction was limited –

cast iron, mild steel and one or two non-ferrous metals with alloy steels still in their infancy.

In spite of these difficulties the demand for ammonia in Germany during the period of World War I led to the building of a much larger plant designed for a slightly higher pressure (250 bars) and requiring bigger compressors. Like the earlier machines they were of the duplex type with six stages—two sets of three in tandem, an arrangement which was retained in Germany right up to World War II and adopted for the numerous hydrogenation plants built in that country during the 1930's. These compressors operated at a pressure of 325 bars and were driven by 4,000 kW synchronous motors at a speed of 125 rev/min—though the mean piston speed was nearly 4·2 m/s. Some of the later plants operated at 700 bars, achieved by a "booster" compressor, although single units, of seven stages, for pressures as high as 1,000 bars had been manufactured as early as 1919 for the Claude ammonia process developed in France. The duplex design was adopted by most manufacturers in Europe (except the U.K.) and in the U.S.A. In the U.K. the vertical design was preferred; several high pressure compressors of the two-crank six-stage vertical design were also manufactured by Sulzer Bros. in Switzerland.

The slow speed duplex horizontal arrangement is expensive; it occupies considerable floor space and requires large costly foundations. In the post-war period it has been entirely superseded by the multi-crank short-stroke faster running machines, predominantly horizontal, with the cylinders arranged vis-à-vis on opposite sides of the crank case though the vertical design still has its protagonists. In Europe the former is usually referred to as the "boxer" type although it originated in the United States during the 1930's.

In the past 10–12 years there has been a spectacular increase in the size of high pressure reciprocating compressors. Whereas in the early 1950's a machine of 5,000 hp (3,750 kW) was considered large (though few exceeding about 3,000 hp (2,250 kW) had actually been built at that date), today compressors of over 12,000 hp (9,000 kW) are in service and all the leading manufacturers in Europe and the U.S. are prepared to offer machines of up to 15,000 hp (11,250 kW). Thus after a period of more than a quarter of a century in which there had been hardly any demand for an increase in the power of compressors for ammonia synthesis plants, manufacturers were called upon to produce, in little more than five or six years, machines of more than twice the maximum power previously available, and it is scarcely surprising that each step was accompanied by a crop of teething troubles, both minor and major, from which no manufacturer was exempt.

This rapid increase in the size of reciprocating compressors was due entirely to a corresponding increase in the size of ammonia plants, for which the preferred pressure is still regarded by many as within the range of 300–450 bars. From the reciprocating compressor maker's point of view it is perhaps unfortunate that the supremacy which, up to now, he has enjoyed in this field should have been challenged—and indeed virtually destroyed—by centrifugal machines where an output of more than 600 tonnes of ammonia per day is required, a figure which is likely to be further reduced in the not too distant future. This output by present-day standards is a small plant so that there would appear to be little future for these exceptionally large reciprocating machines.

Until the discovery of polyethylene during the early 1930's the only high pressure processes were those for the manufacture of ammonia, methyl alcohol and urea. The production of low density polyethylene required pressures of quite a different order besides posing other problems which will be referred to later. This is the one field in which the reciprocating compressor is likely to maintain its supremacy. Initially the type of compressor used for this duty was one in which the ethylene was compressed from 250 bars to about 1,500 bars by an oscillating mercury "piston", the movement being produced by high pressure oil acting on top of the mercury. In principle therefore the machine resembled a U-tube and accomplished its main purpose of ensuring steady production of this vital material during the war years (see Chapter 6). In the meantime some work had been proceeding on the development of a compressor of the more conventional type but it was not until after the war that any real progress was made towards a solution of the very formidable problems, affecting almost every feature of the design. Some measure of the success which has been achieved can be gauged from the fact that in the intervening years the maximum operating pressure has risen steadily as has also the size – output – of the compressor unit. Today a pressure of 2,500 bars is regarded as normal and industrial machines are also operating at 3,000–3,500 bars.

In an industry which is liable to rapid – and almost revolutionary – changes within a very short space of time it is dangerous to attempt to prophesy. Even in the mid-50's few could have foreseen the sudden escalation in the size and power of high pressure reciprocating compressors and already, in spite of the fact that all the leading compressor makers advertise that they are capable of manufacturing compressors up to a maximum of 11,250 kW, it seems likely that few machines of this type above about 4,500 kW will be installed in the future. This is certainly true so far as plants for the manufacture of ammonia are concerned – ironically the process which initiated the demand for exceptionally large units. But for some other processes, for example the production of urea and carbonylation products, the reciprocating machine will continue to hold the field – at least in the foreseeable future.

10.4 Horizontal v. Vertical Compressors

A question which has exercised the mind of all designers of high pressure compressors is that of the relative merits of horizontal and vertical machines. In examining this question the installation as a whole, and not simply the machine, must be considered. In the past the auxiliary equipment such as coolers, oil separators, pipework, etc. has received all too little attention, with the result that not only the appearance, and accessibility to important parts, of the machine suffered, but serious vibration problems due both to mechanical causes and to gas pulsations have arisen.

So far as the horizontal high pressure compressor is concerned we need only consider the type with opposed cylinders, usually (as already mentioned) referred to in Europe as the "boxer" design, since this is nowadays the arrangement almost universally adopted. One of the outstanding characteristics of this type is the excellent balance of the reciprocating parts. If the pistons of each pair of adjacent opposed cylinders were of the same weight, both primary and secondary forces would be perfectly balanced, but owing to the offset of the cylinders there would be small residual primary and secondary unbalanced moments. Usually it is not possible even in a two-crank machine (of this type) to obtain perfect balance of the inertia forces but in a compressor with several cranks, by suitably arranging the cylinders and crank angles, the degree of unbalance can be reduced to an insignificant amount. It must be remembered that the crank angles cannot be decided on inertia balance alone; they must also be selected to minimise peak torques and torsional forces and also to provide the smoothest gas flow with minimum pulsation in the delivery pipes. Indeed the complete design of a modern large multi-crank compressor with its associated pipework, coolers, etc., and the driving unit requires the aid of a computer.

A degree of balance comparable with that of a boxer compressor can only be achieved in a vertical machine with a minimum of four cranks. Practically complete inertia balance can be obtained in five- and six-crank vertical compressors. Whereas boxer machines have been manufactured with as many as ten cranks in a single frame, so far as the authors are aware no vertical compressor with more than six cranks has so far been built.

A specially attractive feature of the horizontal compressor is the excellent accessibility to all parts of the motion work as well as to the valves and glands. By contrast, the vertical does not, on the whole, offer such convenient accessibility. So far as cylinders and valves are concerned this disadvantage can be overcome by arranging the operating floor at cylinder base level, thus dispensing with the necessity for stairways and working platforms. Even this, however, does not solve the problem if some cranks are fitted with cylinders in tandem. But access to the crosshead still leaves much to be desired, particularly if—as in most compressors of this vertical type—the guides are cylindrical.

Undoubtedly the part of a compressor installation which poses the most problems is the layout of the pipework and auxiliary plant such as coolers, pulsation dampers, separators, etc. There can be no question that the boxer compressor offers greater freedom and flexibility in this respect than the vertical machine. Pulsation dampers, for example, to be most effective should be mounted direct on the cylinder or as close to it as possible, a requirement which can be met in the horizontal but not in the vertical machine because of the need to avoid interference with access to valves. Nor is it easy in this latter type to lead the piping to and from the various stages without its creating an obstruction at some point or other. It is true that some elegant solutions have been proposed to overcome these difficulties, such as increasing the size of the suction and delivery pipes adjacent to the machine, but there is a limit to what can be achieved in this way so that it can only be regarded as a compromise.

The boxer design takes up more floor area than the vertical compressor but the latter requires a higher building and the smaller floor area may not compensate for the extra cost of adding height, particularly if some stages are arranged in tandem to reduce the number of cranks in a five- or six-stage machine.

In recent years more and more importance has been attached to freedom from oil in the compressed gas and continuous efforts are being made to raise the pressure at which operation for long periods can continue without lubrication of cylinders and gland packings. Even at medium pressures – 100/150 bars – results are still too variable to justify the conclusion that the problem has been satisfactorily solved at such pressures and, although dry running machines have been manufactured for pressures as high as 350 bars, work in this range is still only in the experimental stage. In this field of non-lubricated compressors the vertical type appears to offer certain advantages over its rival, from the point of view of reducing cylinder and ring wear; so far, however, no clear preference for either type has emerged. It is natural that a compressor manufacturer should seek to adapt, for oil-free service, the designs he has developed for normal lubricated operation. American compressor makers have on the whole tended to retain the horizontal type, but European manufacturers have adopted a more flexible approach, and, especially for high pressures, have shown a preference for the vertical arrangement.

It is convenient in the present context to consider high pressure compressors as those machines designed for a final delivery pressure up to 450 bars – a pressure at which many modern ammonia plants are operating – and hypercompressors as those with delivery pressures in the range of 1,500 to about 3,500 bars for which the only industrial application at the present time is the production of low density polyethylene. Between the upper limit of the first and the lower limit of the second group are to be found a small number of compressors for pressures of the order of 700 bars – a range in which other processes (Casale, Claude) for the manufacture of ammonia operate and which was also chosen as the one for several of the plants built in Germany during the 1930's to produce petrol by the hydrogenation of coal.

In developing a range of high pressure compressors the criterion usually adopted is the maximum piston rod loading which serves as a basis for the design of the frame and motion work etc., and later defines the limitations of a particular machine in the range with respect to the duty for which it is suitable. Largely because, as mentioned briefly on p. 209, the horizontal opposed type has established itself as the preferred arrangement, most European manufacturers have rationalised their programme to give a series of standard frames in which the piston rod loadings are based on geometric progressions with the full range usually a combination of the two progressions with the ratio of successive terms $\sqrt[5]{10} \approx 1{\cdot}6$ and $\sqrt[10]{10} \approx 1{\cdot}2$. Thus a typical range of one manufacturer which, with an appropriate number of cranks, covers power inputs from 150 to 9,000 kW consists of seven frame sizes with rod loadings of 6·5, 10, 16, 22, 26, 40 and 60 tonnesf whilst another manufacturer has, in addition, four intermediate sizes with rod loadings of 8, 12·5, 32 and 50 tonnesf. There is some overlap between successive frame sizes and this provides a degree of flexibility in selecting the optimum frame and arrangement of cylinders for a specified duty. For example, in the first mentioned series of frame size a duty requiring a power input of 4,500 kW can be met by a six-crank 40 tonnesf frame, a four-crank 60 tonnesf or a six-crank 60 tonnesf frame. In high pressure compressors the most economical solution is usually that with the minimum number of cranks, provided these are also economically loaded, since the cost of adding an additional pair of cranks is high. In deciding on the best layout for the various stages, the designer must endeavour to equalise as far as possible the rod loading on the inward and outward strokes and to make the resultant loading the minimum possible. The devices which are available to achieve this result are too numerous to mention here, but include the use of stepped pistons, balance spaces at some suitable intermediate stage (or atmospheric) pressure and the "splitting" of a stage between two cranks. There is the further advantage in choosing the minimum number of cranks of a slight reduction in mechanical (friction) losses with a corresponding improvement in efficiency and reduction in power costs, which, in a compressor of several thousand kW is by no means negligible. In Europe the maximum size of frame available in general limits the number of cranks to six, though at least one manufacturer can offer eight; in the U.S.A. all the leading manufacturers of high pressure compressors have frames capable of taking ten cranks. With a separate bearing between each pair of cranks any number of the latter – odd or even – up to the maximum can be provided but where there is no such bearing ("bicycle crank" arrangement) only even numbers are possible.

American firms, on the whole, tend to adopt a "one crank, one cylinder" arrangement and to make less use of balance spaces and tandem cylinders. With the lower cost of power in the U.S. there is not the same incentive to aim for the minimum number of cranks nor is the question of space in general so important. There is, however, as the illustrations which follow show, little difference between U.S. and European practice in detail design of the principal items. Mean piston speeds and valve velocities too are of the same order.

A practice which has been widely adopted in recent years, especially for ammonia plants based on the modern process for manufacturing the synthesis gas by the steam reforming of naphtha or natural gas under pressure, is to carry out some or all of the compression duties on the same frame. Such multiple service machines may therefore be designed to compress air for the secondary reformer, synthesis gas, ammonia for refrigeration as well as recycle of synthesis gas. With this arrangement there is considerable variation in the loading of individual cranks. Moreover, due to the fact that different sections of the plant must be progressively brought into commission, it is necessary to analyse a whole range of torque diagrams for the various conditions which might arise. All firms now have access to – or their own – analogue computers on which can be analysed not only the torsional oscillation phenomena but also the often highly complex piping system in order to determine the optimum layout and to ensure freedom from vibrations due to mechanical causes or gas pulsations. These latter have frequently been a source of considerable trouble in the past and among the most difficult problems to solve.

10.5 Some examples of industrial high pressure compressors

Fig. 10.5 shows the general layout of a four-crank synthesis gas compressor

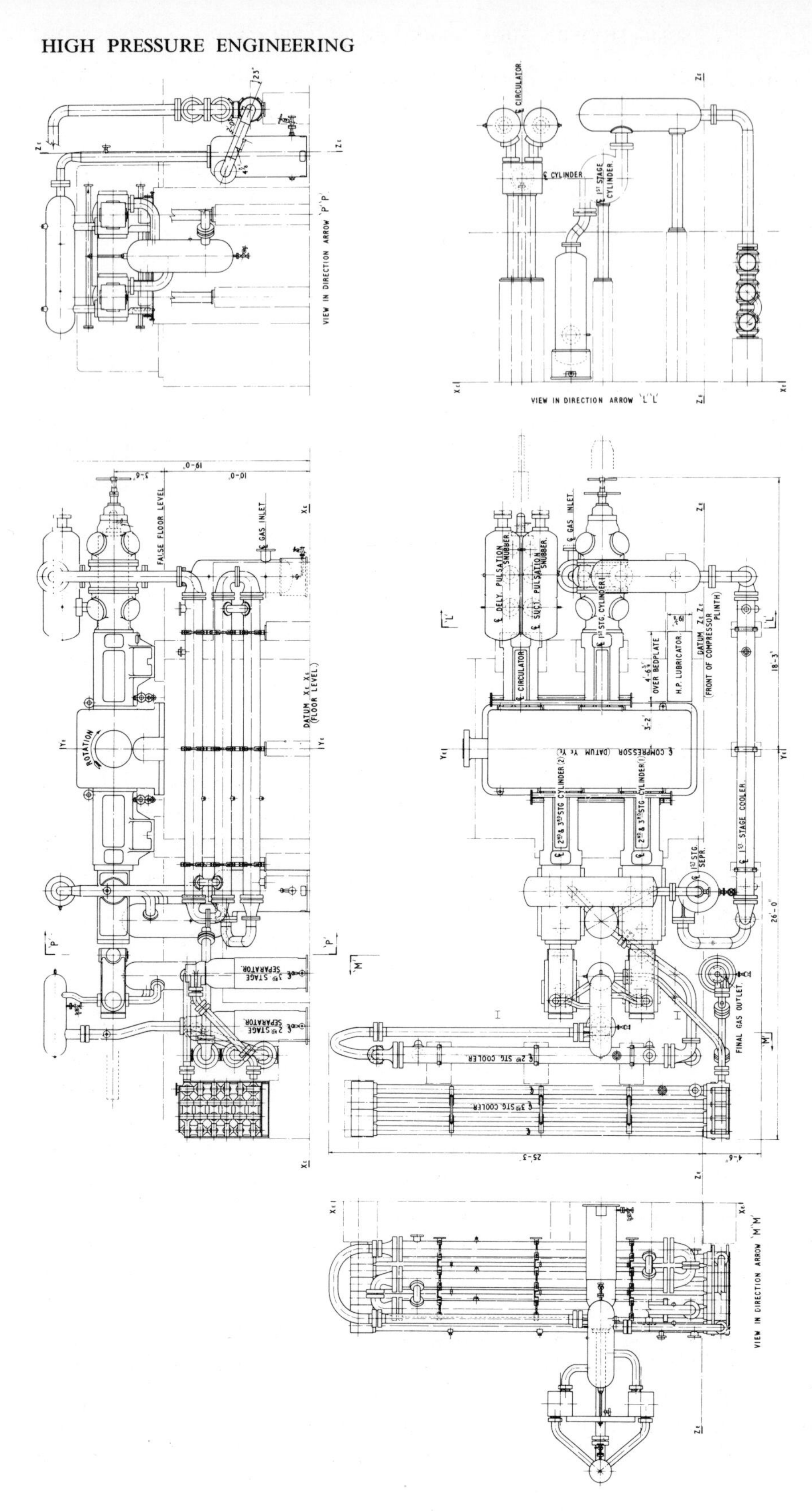

Figure 10.5 General arrangement of 3-stage plus recycle gas compressor: 25/345 bars (*P. Brotherhood Ltd*)

designed by Messrs. Peter Brotherhood – incidentally the only firm in the U.K. manufacturing large high pressure compressors of wholly British design. This machine – for a modern ammonia plant – compresses 31,350 m^3/h of synthesis gas from naphtha feed stock from 25 bars to 345 bars in three stages, one double-acting first stage and two single-acting second/third stages in tandem. The double-acting cylinder, with tail rod, for recycle gas is mounted on the fourth crank. The stroke is 15 in (381 mm) giving at 300 rev/min the very conservative piston speed of 3·8 m/s. The power absorbed at the compressor coupling is 4,560 kW.

A section through one of the second/third-stage cylinders is shown in Fig. 10.6. The third-stage plunger is bolted to the end of the second-stage piston by a split collar. It is of cast iron with lead bronze pads to provide a low friction running surface between the piston and cast iron liner. The second-stage piston has a cast iron sleeve which is a light interference fit on the enlarged portion of the rod and which is also provided with lead bronze pads on the rubbing face. Leakage past the third-stage plunger and second-stage piston is led back to the first-stage suction from the balance space formed in the third-stage cylinder. The packing for the second-stage gland (which is at a pressure of about 170 bars) is of the metallic type discussed earlier; it is cooled by an independent closed circulating system, the coolant being lubricating oil. In general even with a mean piston speed of 4·2 m/s it is not considered necessary to cool glands operating at such a moderate pressure, so that at the low piston speed of 3·8 m/s this design must be considered conservative. Similarly, the number of rings – 18 on the second-stage piston and 30 on the third-stage plunger – is generous for the pressure difference across them of approximately 145 bars and 325 bars respectively. Both cylinders are cooled by circulating water through transverse holes drilled in the cylinder blocks. In addition they are surrounded by water jackets which are provided with baffles to increase the circulation rate.

The suction valves are of the mushroom and the delivery valves of the poppet type described earlier, with a lift in each case of 5·6 mm.

Lubrication of cylinders and glands is from a lubricator of Messrs. Brotherhood's own design. It was developed originally to meet the very stringent requirements of minimum lubrication of compressors supplying air to oxygen plants. Most high pressure lubricators incorporate a variable stroke or lost motion principle in the design and their performance at low settings of the regulator (corresponding to low delivery rates) tends to be erratic; the output from two similar – and allegedly interchangeable – pump units may differ considerably at the same regulator setting. In the Brotherhood lubricator, of which a sectional view is shown in Fig. 10.7, there are separate suction and delivery pumps each of which at all times operates at full stroke, the output being regulated by a by-pass between the two pumps which by-pass returns the excess oil to the oil reservoir. Linear control of the rate of feed to each point can be obtained over a wide range – 0·1 ml to 4 ml – and, for a given setting of the regulator, the rate is consistent and reproducible.

The lubricator has been fully tested at pressures up to the order of 700 bars and a typical calibration curve is also illustrated in Fig. 10.7 and shows the wide

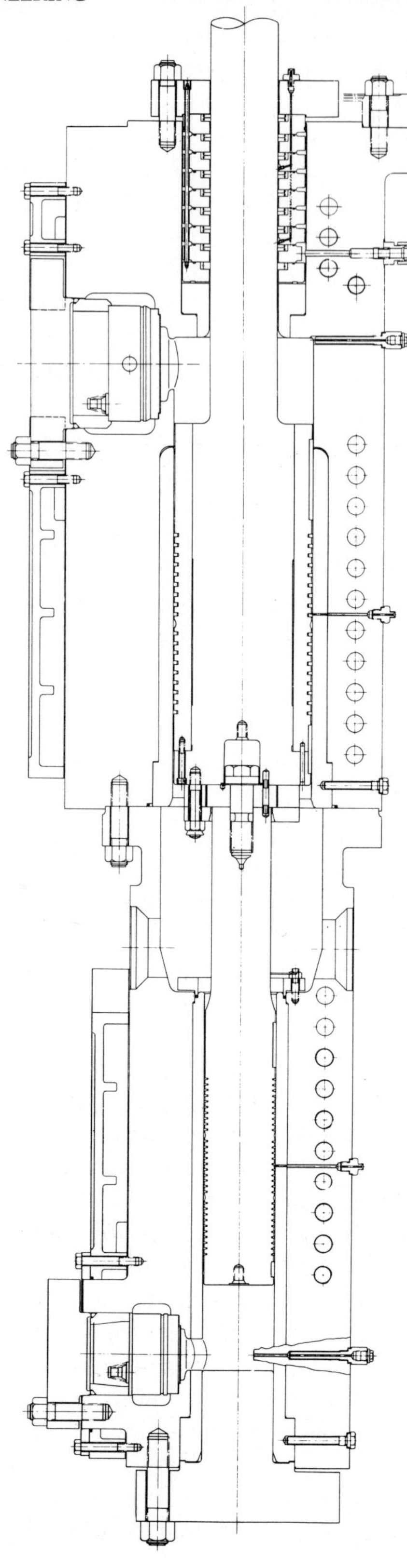

Figure 10.6 Section through 2nd/3rd stage cylinder of compressor shown in Fig. 10.5 *(P. Brotherhood Ltd)*

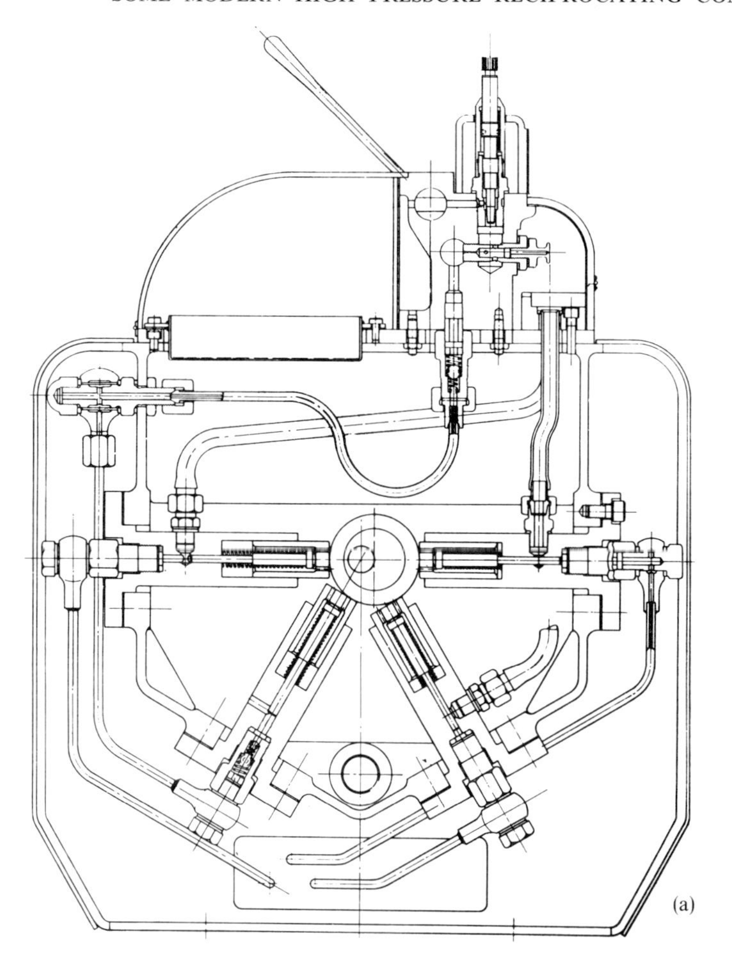

(a)

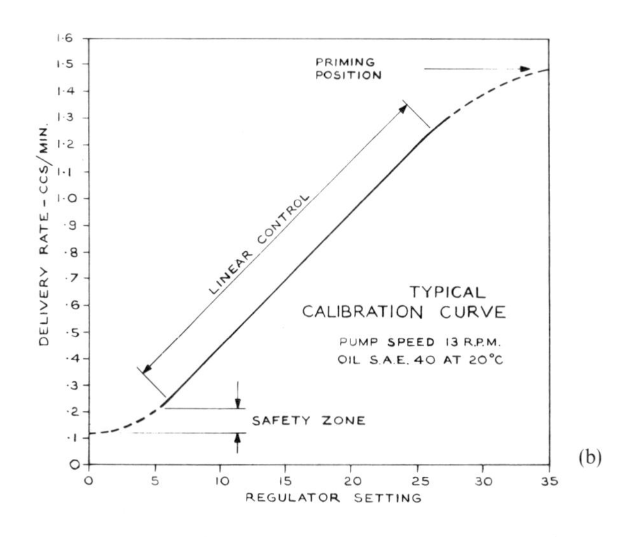

(b)

Figure 10.7 (a) Section through high pressure lubricator *(P. Brotherhood Ltd)* (b) Calibration curve for high pressure lubricator *(P. Brotherhood Ltd)*

range over which the output is linear. Even with the by-pass fully opened the feed to a particular point cannot be stopped completely. A unique feature of the lubricator is the priming device, which enables full output to be obtained from all points without disturbing the individual settings; this is most useful when starting up a machine initially or after a shut-down.

The third stage cooler is of the jacketed type and is of interest because of its compactness. The normal design of such coolers occupies considerable space because of the necessity to provide clearance for the large diameter high pressure flanged joints between the gas tubes and return headers. In the design illustrated in Fig. 10.8 these external bends and joints have been eliminated and replaced by special cast steel housings which incorporate separate passages for the gas and water flow from one element to the next. Silicone O-rings in the housings seal the end of each water jacket whilst at the same time allowing relative movement of the gas tubes. The use of O-rings at this point is another example of the simplification of design which can thereby be introduced and of their value as a static sealing element even at high pressures. Further examples of their versatility will be found in other illustrations of compressor details in this Section. In the case of the type of cooler under discussion, before the adoption of O-rings the only solution was to fit soft packed glands to seal the ends of the water jackets—always a source of trouble due to leakage caused by vibration of the cooler tubes. This wide use of O-rings has, of course, only become possible since the development of synthetic materials—Neoprene, Viton, silicones—capable of withstanding temperatures higher than ordinary rubber and also resistant to attack from many media.

A four-stage six-crank compressor for synthesis gas made by Halberg Maschinenbau, Ludwigshafen, Germany, is shown in Fig. 10.9. As can be seen, there are two first stages and one second stage, all double-acting and three third/fourth stages. The machine is designed to compress 27,500 m^3/h from 10 bars to 160 bars in the first three stages after which the carbon monoxide is removed and the remaining gas (21,500 Nm^3/h) compressed to the final pressure of 355 bars in the fourth stage. The compressor runs at 250 rev/min and the mean piston speed is 4·17 m/s. The power at the compressor coupling is 3,875 kW.

Of general interest in this design is the extremely robust crank case which is of double wall construction and providing sufficient rigidity to make it unnecessary to fit tie rods or tension bolts across the top as is essential with the normal single wall design. In the boxer arrangement the crosshead bears on the upper face of the guide on one side and on the lower face on the other. This can be clearly seen from the sectional view through one line of the first and third/fourth stages (Fig. 10.10), which also shows the method adopted by this firm for the attachment of the piston rod to the crosshead. This is by means of a coupling, in halves, with conical bearing faces mating with corresponding cone faces on the crosshead and piston rod nut. This results in a more favourable stress distribution when the rod is in tension than when the latter is screwed directly into the crosshead. This illustration also brings out clearly one of the main problems in compressors with cylinders in tandem. This is the ensuring of satisfactory alignment. In the present example there are eight faces which must be accurately machined

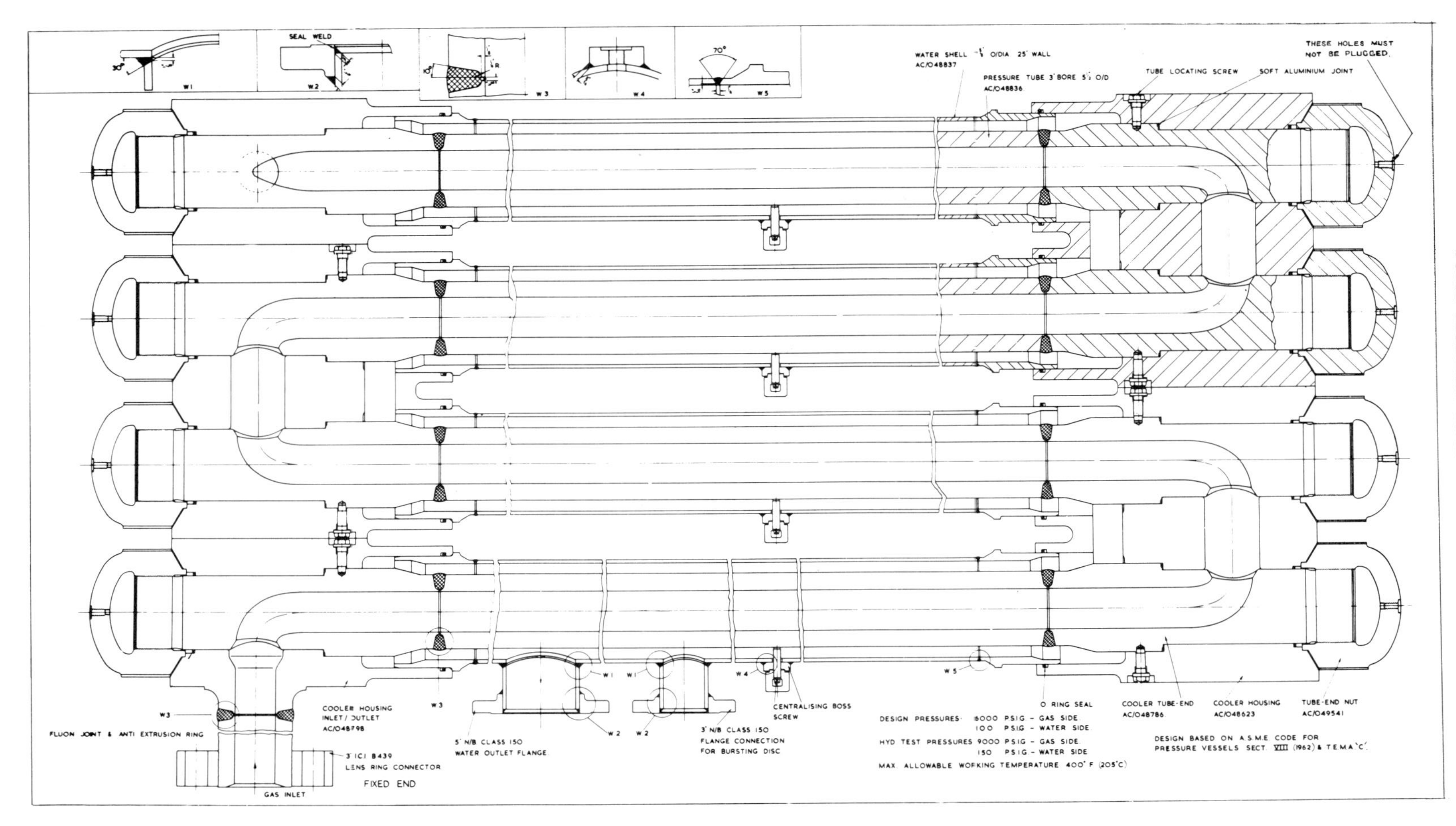

Figure 10.8 3rd stage gas cooler of compressor shown in Fig. 10.4 (*P. Brotherhood Ltd*)

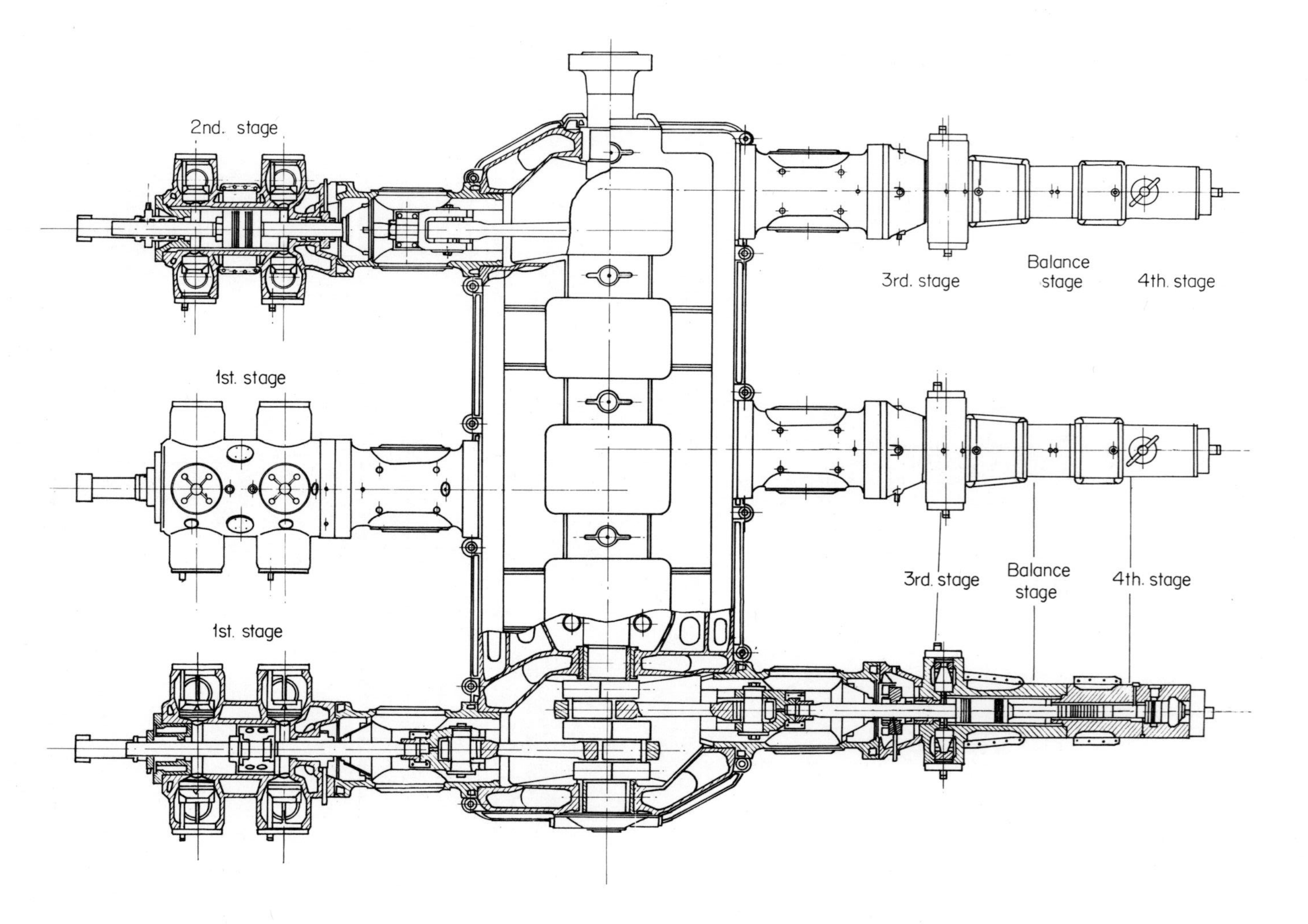

Figure 10.9 4-stage synthesis gas compressor: 10/355 bars *(Halberg Maschinenbau GmbH)*

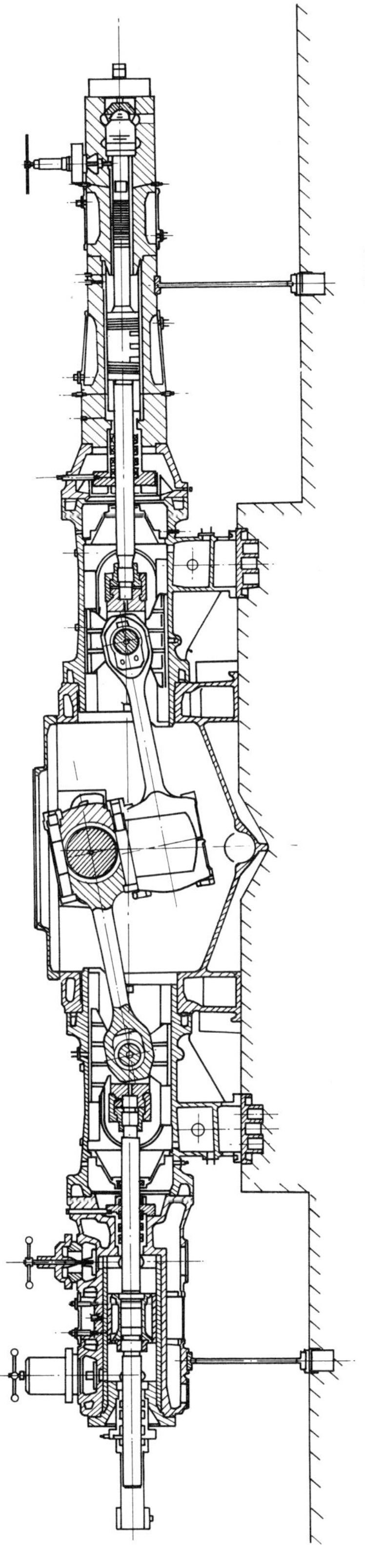

Figure 10.10 Section through 1st, 3rd/4th stages of compressor shown in Fig. 10.9 (*Halberg Maschinenbau GmbH*)

square with the horizontal axis of the compressor and a similar number of spigot surfaces. A lack of squareness in any of the joint faces is sufficient to throw the assembly out of line and much time and labour may be involved in finding and correcting it. Undoubtedly the ideal design from the technical point of view-though rarely from the cost angle – is a single cylinder per crank. The arrangement of the stages on the third/fourth-stage leg follows the same pattern as that of the stages on the second/third-stage leg of the Brotherhood compressor described earlier. In the present instance, however, the balance space is at first stage–suction pressure so that the pressure differences across the rings of the third-stage piston and the fourth-stage plunger are higher – 150 bars and 350 bars respectively – and both are fitted with fewer rings.

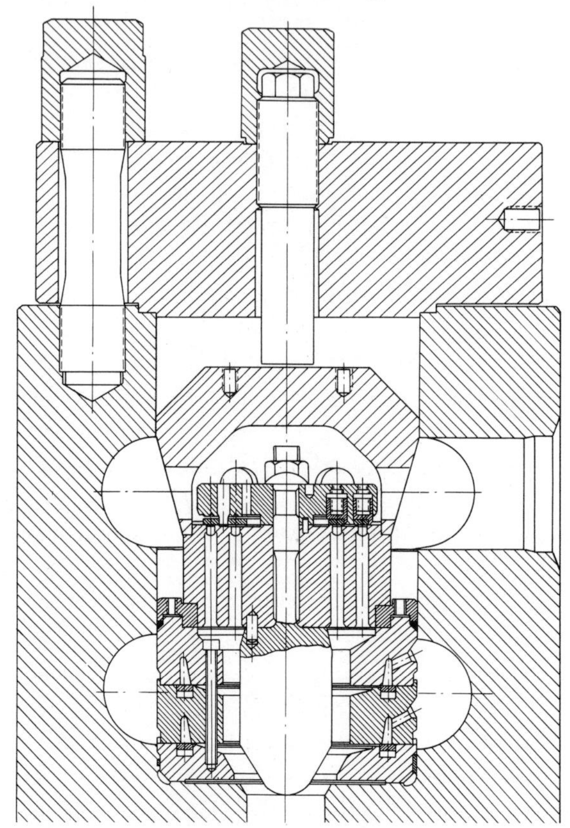

Figure 10.11 Co-axial suction/delivery valve: 210/445 bars *(Halberg Maschinenbau GmbH)*

The suction/delivery valve assembly in the fourth stage consists of a combined unit mounted axially in the end of the cylinder. An assembly of this type is shown in Fig. 10.11. Whilst the combined valve has simplified the design (shape) of the forging for the fourth-stage cylinder, it is doubtful whether it has contributed much to the overall reliability, and failure of a cylinder at any of the points of stress concentration still means that the whole cylinder must be replaced. It is surprising that compressor manufacturers are content to perpetuate designs in which the risk of failure will always exist, with little attempt to minimise the consequences to the user – in loss of output from the plant.

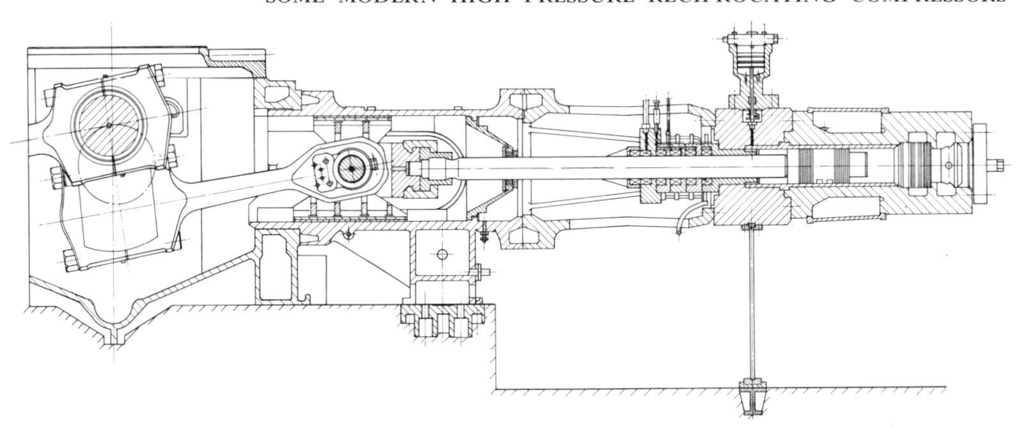

Figure 10.12 Section through 6th/5th stage of 6-crank 6-stage synthesis gas compressor: 1/445 bars *(Halberg Maschinenbau GmbH)*

An arrangement which avoids the introduction of a balance space is that of the fifth/sixth stages of the six-crank compressor of which Fig. 10.12 is a section through the fifth and sixth stages. This machine compresses synthesis gas from atmospheric pressure to 441 bars with removal of CO_2 and CO at intermediate pressures. Referring to the fifth/sixth stage, it will be seen that the fifth- and sixth-stage cylinders are separate forgings and that the latter stage is formed by the annulus between the piston rod and the diameter of the fifth stage. Many designers in the past have tended to avoid this construction because it means that the gland is subjected to the final delivery pressure of the machine – in this case 441 bars – in spite of the fact that the omission of a balance space results in a significant shortening in the length of the tandem cylinders. The details of the gland have already been illustrated in Fig. 9.7. Both fifth and sixth stages are fitted with co-axial valves and the respective cylinder forgings are largely relieved of pulsating load since the suction and delivery passages are under constant corresponding pressure.

It has been remarked earlier that whilst, for lubricated compressors, the horizontal type predominates, many European manufactuers prefer to adopt the vertical design for conditions where oil-free gas is required or minimum lubrication desirable. A three-crank four-stage compressor, of Halberg design, for compressing 4,600 m^3/h CO_2 from 4 bars to 130 bars is shown in Fig. 10.13. The stroke is 0·315 m giving the mean piston speed of 3·07 m/s – a very moderate figure – and the power absorbed is 680 kW.

The piston rings and gland packing rings are graphite-filled Teflon. The final stage is fitted with a wet liner of nitrided steel to assist in carrying away the heat from the poorly conducting Teflon rings. This cylinder also has a separate valve head with co-axial valves.

This firm manufactures a range of vertical high pressure compressors with from one to four cranks for both lubricated and non-lubricated service for low to medium powers and contend that in this range – up to about 4,000 hp (3,000 kW) – the vertical design results in a cheaper machine than the boxer type.

It has already been mentioned that the early high pressure compressors were

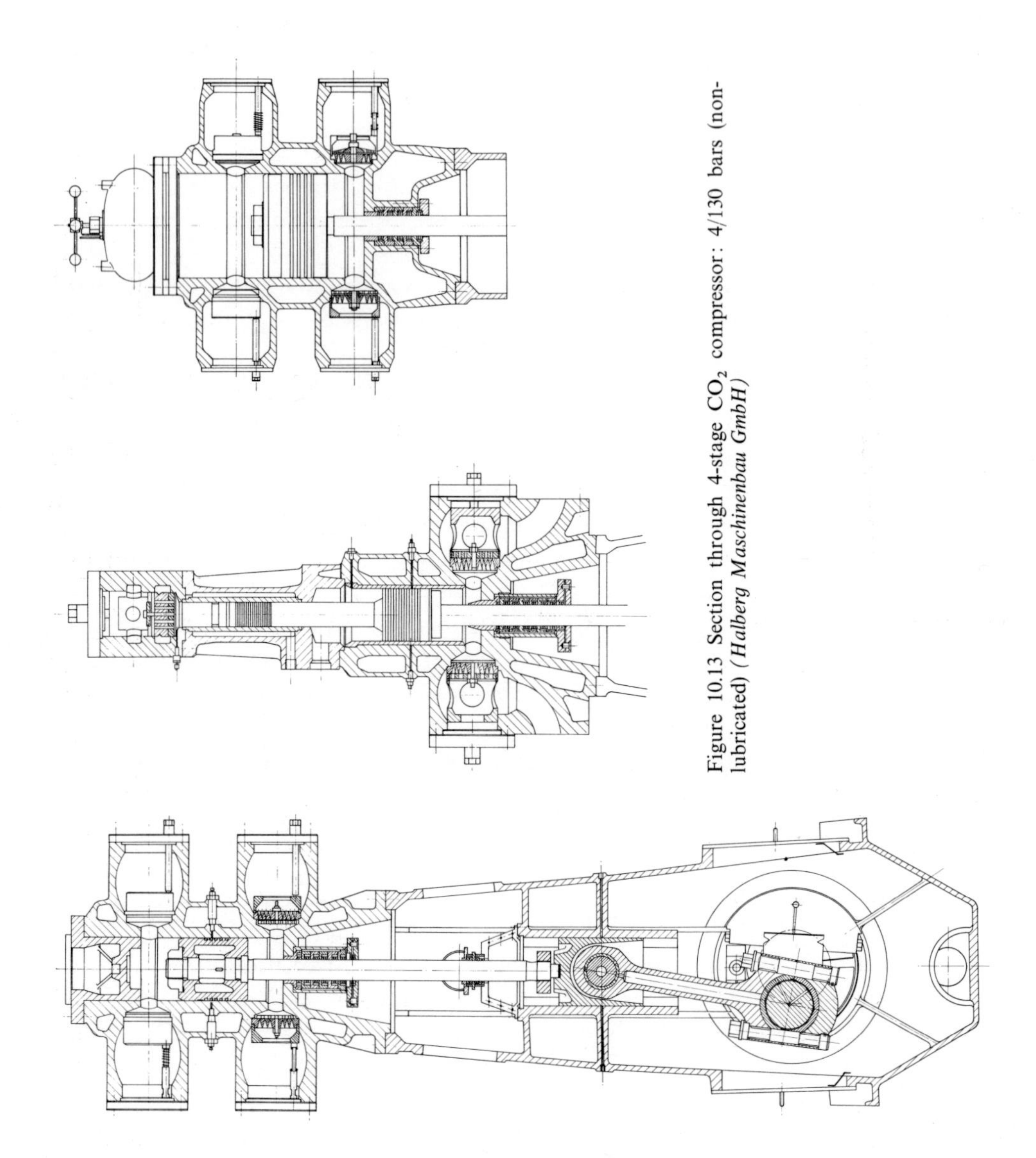

Figure 10.13 Section through 4-stage CO_2 compressor: 4/130 bars (non-lubricated) (*Halberg Maschinenbau GmbH*)

slow-running machines, of necessity over-designed. They owed much to the experience already available from the large steam – and particularly the gas-engine driven – "blowing engines" in use in blast furnace and steel plant, the general design of which was retained for the motion work and low pressure cylinders of the compressor. The large demand for high pressure compressors which arose in Germany for ammonia, and later, hydrogenation plants in the decade or so preceding the outbreak of World War II led to a high degree of standardisation and indeed to interchangeability of parts of machines made by different manufacturers.

It was largely to meet this challenge from German compressor firms that the Swiss firm of Buckhardt, at that time themselves a leading manufacturer of the slow-speed horizontal type, turned their attention to the vertical design and began the development of high pressure compressors running at about twice the then current speed and designed to give a smaller and lighter machine. In this they drew largely on Diesel engine practice; thus they adopted a similar form of construction for the frame, with the crosshead running on a flat guide on one face only, forked connecting rod and so on. This arrangement provides a degree of accessibility to the motion work comparable with that offered by the horizontal arrangement and which is not possible in a vertical machine with three or more cranks and cylindrical crossheads. By mounting only one cylinder on each crank and bolting all the cylinders together to form a rigid block assembly, Burckhardt produced a compressor of relatively low height which offers many of the advantages of the alternative type. The extent to which they have succeeded can be gauged from Fig. 10.14 of a five-stage machine for ammonia synthesis, suction

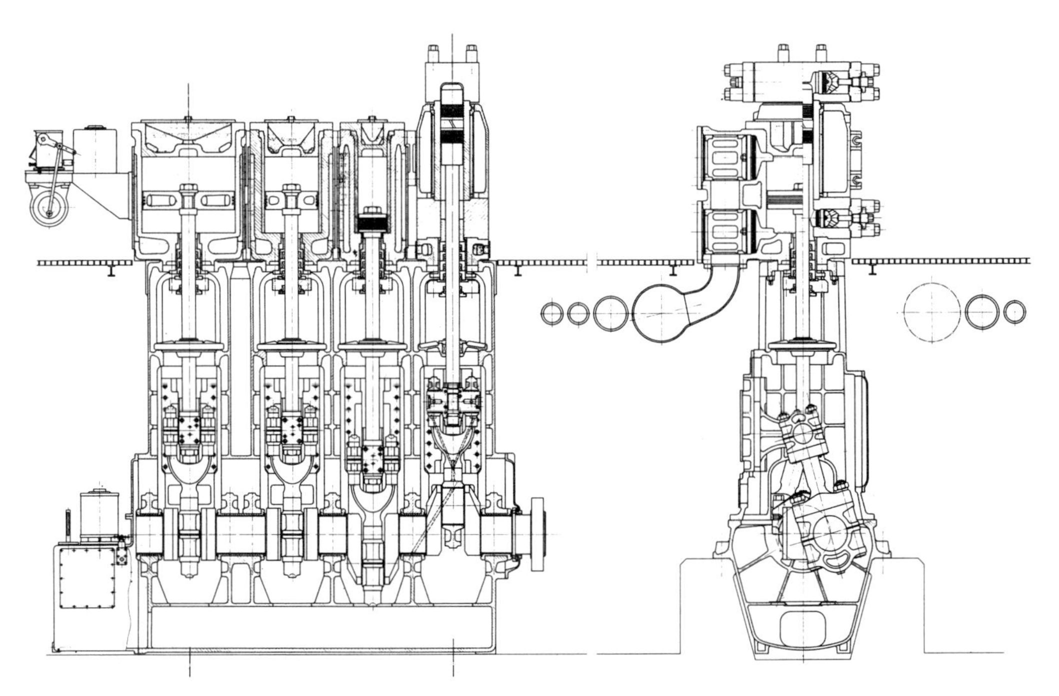

Figure 10.14 Section through 4-crank 5-stage synthesis gas compressor: 3/295 bars *(Maschinenfabrik Burckhardt)*

volume 13,280 m^3/h, suction and delivery pressures, 3 bars and 295 bars respectively, power absorbed 2,660 kW. The fourth/fifth stages are arranged in one cylinder, the final stage being formed by the annulus between the bore of the cylinder and the outside diameter of the piston rod – as in the compressor shown in Fig. 10.12.

A similar arrangement has been adopted in the three-crank two-stage compressor illustrated in Fig. 10.15. There are three identical cylinders to compress 40,000 m^3/h of synthesis gas from 53 bars to 260 bars. The actual cylinder is held between the upper and lower valve heads which are connected by long through bolts tightened hydraulically. Each cylinder is hard chrome-plated and mounted in a C.I. frame, the three frames being bolted together to give the rigid "monobloc" construction mentioned above. The nitrided piston rod is of

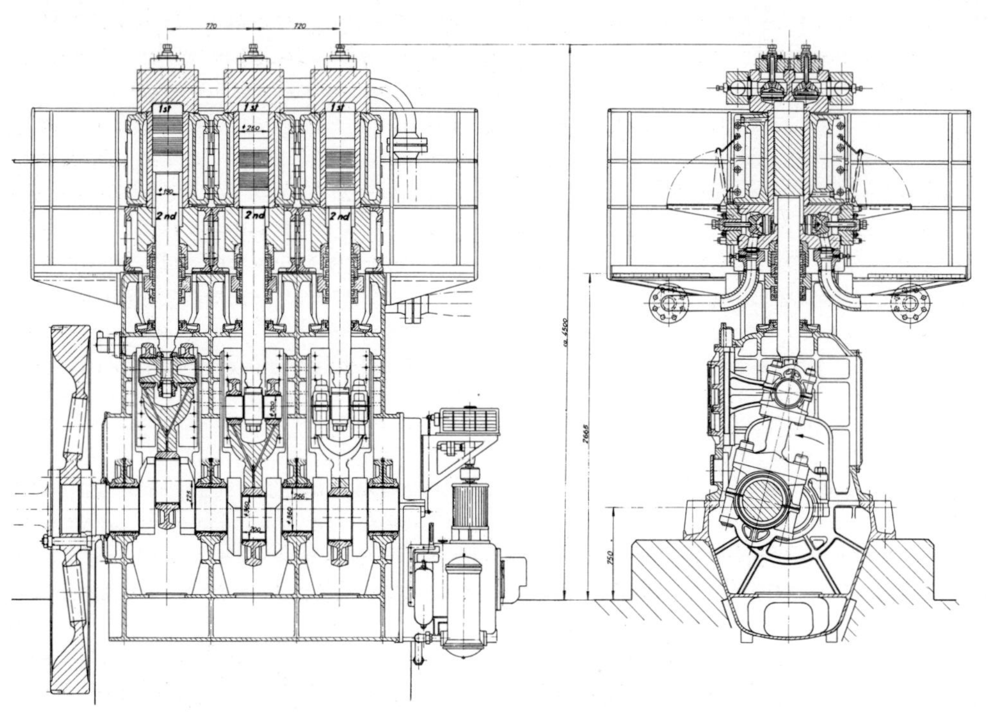

Figure 10.15 Section through 3-crank 2-stage synthesis gas compressor: 53/260 bars *(Maschinenfabrik Burckhardt)*

unusually large diameter – 190 mm. The compressor is driven by a steam turbine through double reduction gearing, the power required at the compressor coupling being 2,800 kW. At the maximum speed of 230 rev/min the mean piston speed is 3·45 m/s. The installation presented some interesting problems in torsional analysis as there are three critical speeds in the operating range and these must, of course, be avoided as running speeds.

The examples already given are sufficient to show that in the medium high pressure range – say up to pressures of the order of 450/500 bars – between the various manufacturers a measure of standardisation has been reached both in respect of the design of certain details and in the general arrangement and layout of the machines as a whole. We shall accordingly conclude with brief descriptions

Figure 10.16 4-crank 5-stage synthesis gas compressor: 11/835 bars *(Demag AG)*

of a few compressors of special interest by firms other than those mentioned above.

A five-stage compressor made by Demag is shown in Fig. 10.16. The particular feature of interest in this compressor is the introduction on each leg of the machine of an additional crosshead to facilitate accurate alignment of the plain plungers in each of the fourth- and fifth-stage cylinders. This compressor compresses 15,000 Nm^3/h of synthesis gas from 11 bars to 835 bars. It has a stroke of 450 mm and runs at 210 rev/min giving the very moderate mean piston speed of 3·15 m/s. The power at the coupling is 3,350 kW. The interposition of the additional crosshead means that there are three glands on each leg, a total of twelve for the whole machine. This is no doubt considered to be justified by the easier alignment referred to above and the improved access to the packing in the fourth and fifth stages, which in the latter case must be able to seal against the full discharge pressure of 835 bars. The details of the final stage can be seen in Fig. 10.17 and the arrangement of the fourth stage is similar. Great care has been taken in the design of the cylinder to avoid stress raisers and the adoption of co-axial suction and delivery valves results in simple forgings for both the cylinder and separate head. The plunger is mounted in the crosshead in a spherical seating providing a degree of self-alignment. It is also internally oil-cooled as are the packing boxes. A separate primary gland with a flushing oil system is fitted to provide an oil film on the rod and also help to wash off any small particles of foreign matter. This illustration shows clearly how many of the principles for improving the

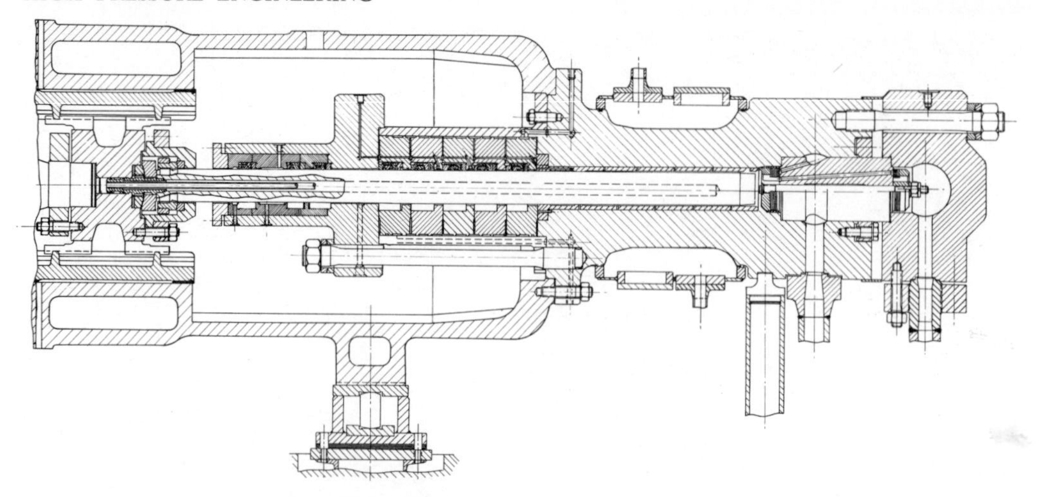

Figure 10.17 Section through 5th stage of compressor shown in Fig. 10.16 (*Demag AG*)

behaviour under pulsating pressures of the high pressure cylinders of compressors referred to earlier have been applied to produce a sound design, as free as possible from sources of weakness.

Fig. 10.18 shows the final stage piston of a six-crank four-stage boxer compressor for a delivery pressure of 320 bars and the arrangement of the co-axial suction and delivery valves. The floating piston is loosely mounted on the end of the rod and seats on a spherical face, permitting it to align itself initially in the cylinder. This feature – of which there are many variations in design – is widely used by Continental manufacturers but is not so common in U.K. or American built compressors.

Demag do not include vertical machines in their programme but for medium output manufacture a series of angle or L-type compressors. This design is equivalent to a 90°V and thus possesses the same excellent characteristics with regard to inertia balance. For high pressures it can be built with up to six stages, usually as a two-crank machine with the two low pressure cylinders vertical and the four remaining cylinders horizontal. The coolers for the first two stages are built into the frame, avoiding external piping and so giving a compact arrange-

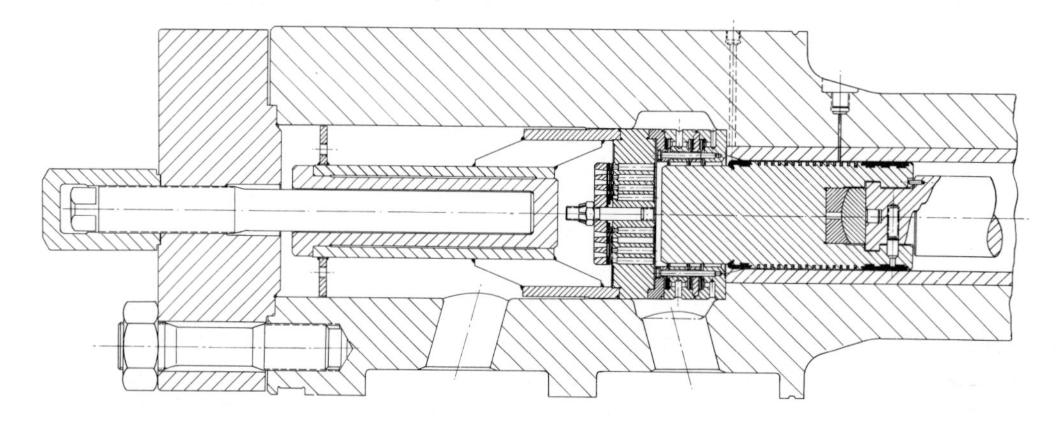

Figure 10.18 Section through 4th stage of 6-crank 4-stage synthesis gas compressor: 10/320 bars *(Demag AG)*

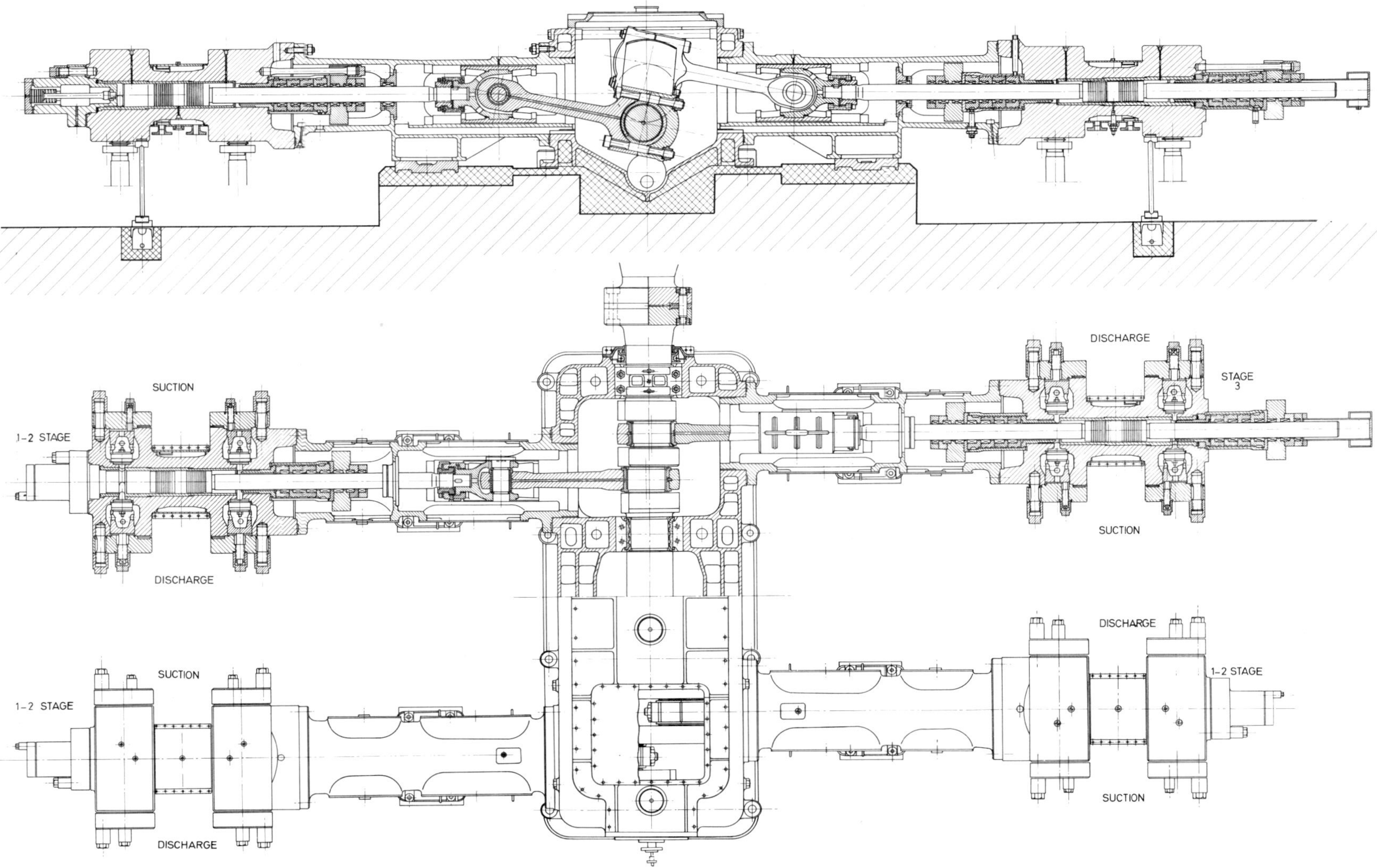

Figure 10.19 Sectional arrangement of special 4-crank 4-stage synthesis gas compressor: 70/460 bars (*Ehrhardt & Sehmer*)

ment. Even quite large compressors of this design can be shipped complete except for the coolers and piping associated with the higher pressure stages.

One of three special ammonia synthesis gas compressors designed by Erhardt & Sehmer, Saarbrücken, is shown in Fig. 10.19. It compresses 22,500 m^3/h from 70 bars to 460 bars in three stages and because of the high suction pressure it has been possible to make all four cylinders interchangeable, the only modification required to alter the duty being to change the cover side studs. Thus when a cylinder is used as a first/second stage the studs are short and serve to hold the first-stage clearance pocket; when used as a third-stage cylinder a tail rod is fitted to equalise the loading on the in and out strokes and in this case long elastic studs which hold the packing box for the tail rod gland are required. The first-stage clearance pockets are maintained at second-stage delivery pressure. When the machine is started up the pockets are opened by means of the springs and close automatically as the pressure increases.

The interchangeability extends to the valves and packing which accordingly must be designed to seal against 460 bars. The packing is oil-cooled and between the primary (high pressure) and secondary packings flushing oil is fed into a chamber to cool the rod. In this way the gland is maintained at a temperature of about 60°C.

The compressor runs at 250 rev/min and has a stroke of 450 mm, giving the fairly high piston speed of 4·5 m/s. It absorbs approximately 2,080 kW.

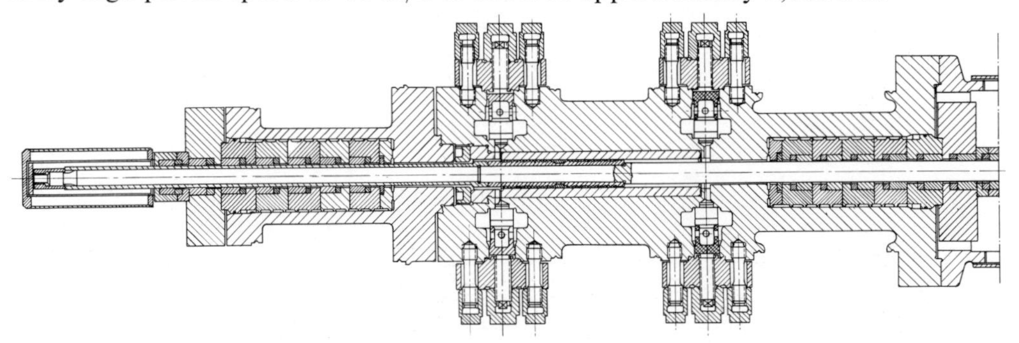

Figure 10.20 Section through cylinder of CO_2 booster compressor: 300/685 bars *(Erhardt & Sehmer)*

The cylinder of a booster compressor for carbon monoxide is shown in Fig. 10.20. Because of the high pressure difference across the packing – 685 bars – and extreme toxicity of the gas the greatest care must be taken to centre the packing element and ensure uniform tightening without the risk of any tilting of an individual ring. A satisfactory solution was found by mounting the cover side packing – of Kranz design – in a housing, which forms an extension of the cylinder, and the tail rod packing in a separate substantial cylindrical housing. Two gas leak-off connections and a flushing-out connection are provided in the packing.

The booster compressor operates between pressure limits of 300 bars and 685 bars with the very low piston speed of 1·0 m/s (stroke 200 mm, speed 150 rev/min). The power absorbed is 44 kW.

For medium powers Erhardt & Sehmer also manufacture a range of single- to four-crank vertical compressors, the maximum power being 3,000 kW.

Fig. 10.21(a)

Fig. 10.21(b)

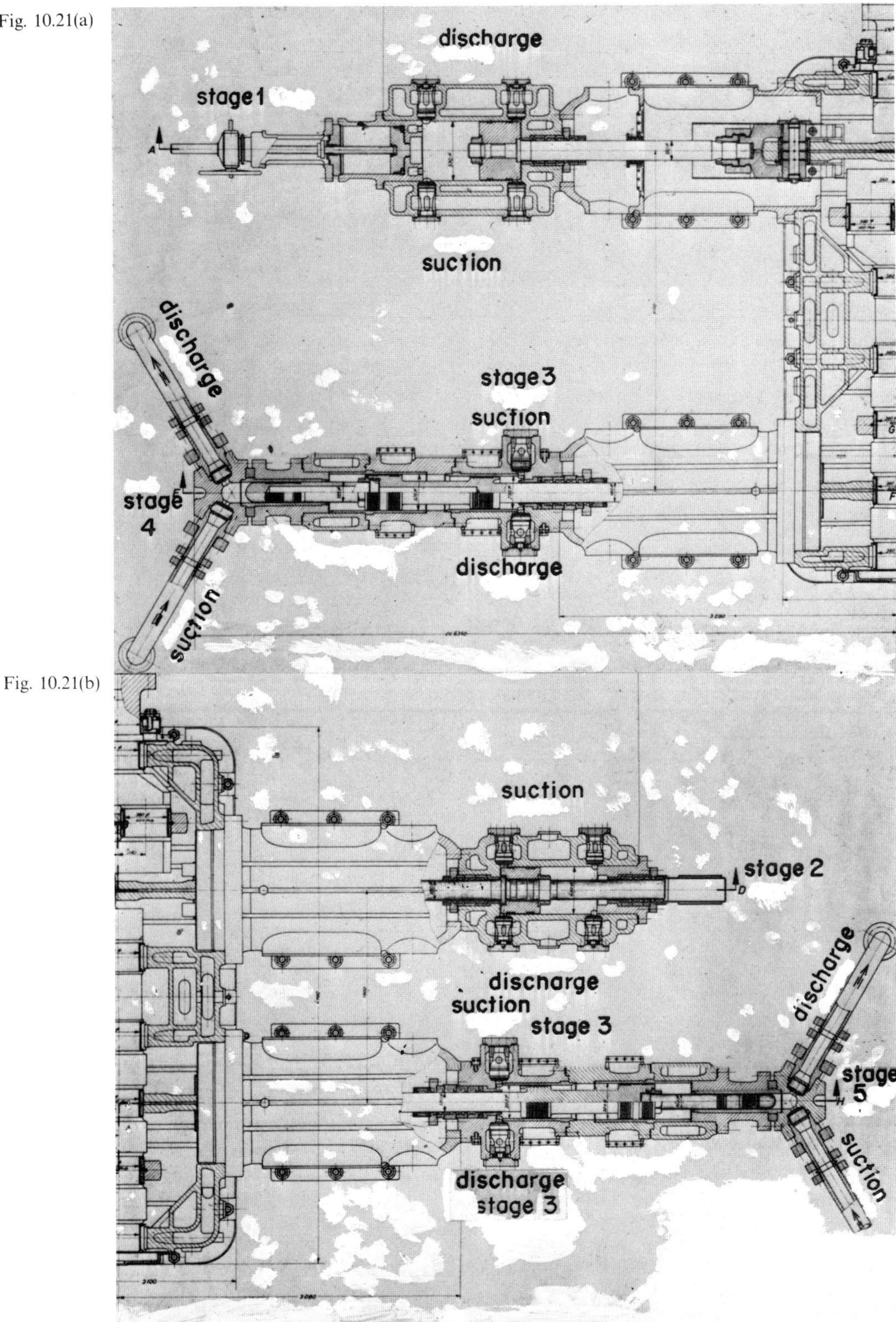

Figure 10.21 4-crank 5-stage synthesis gas compressor: 9/490 bars *(Maschinenfabrik Esslingen)*

A recent four-crank five-stage synthesis gas compressor made by Maschinenfabrik Esslingen of Esslingen, Germany, is illustrated in Fig. 10.21. Fig. 10.21(a) is a section through the first, third and fourth stages, whilst Fig. 10.21(b) is a section through the second, third and fifth stages. There are thus two single-acting third stages, each associated with a balance space, the purpose of which, as explained below, is different in the two cases. The fourth and fifth stages are also single-acting. In the first stage 19,170 m^3/h of the raw gas are compressed from 9 bars to 25 bars and in the remaining stages 24,000 m^3/h from 21 bars to 490 bars with intermediate pressures 72, 220 and 400 bars. The first and second stages are both double-acting, the latter with a tail rod to help equalise the crank loading and a stroke of 500 mm giving, at a speed of 249 rev/min, a mean piston velocity of 4·16 m/s. The third/fourth stage has a balance piston between the two for the same purpose; the balance piston between the third and fifth stages on the other hand is for the purpose of giving a more uniform turning effort diagram. The fourth and fifth stage pistons are sealed by piston rings and are free to adjust themselves laterally. The separate cylinder heads each have a hemispherical compression space and radially arranged valve passages and pockets to obtain as favourable a stress distribution as possible. The stroke of the third- to fifth-stage pistons and plungers is only 400 mm giving a mean piston speed of 3·33 m/s.

Although the method adopted for holding in the valves in the two last stages is ideal from the standpoint of producing a simple symmetrical design of cylinder head, it has the disadvantage that a short length of pipe must be removed to take out a suction or delivery valve.

Maschinenfabrik Esslingen have used the same method of sealing the high pressure plungers, viz. piston rings, at pressures up to approximately 700 bars.

As a final example of a compressor in the "medium" high pressure range we select a multi-duty machine. This compressor manufactured by Ingersoll Rand and a photograph of which is shown in Fig. 10.22 is one of the largest machines of this type, on a single frame. It compresses 39,000 m^3/h of NH_3 synthesis gas from 16·5 to 333 bars in four stages, and in addition handles refrigeration and recycle services. There are six cranks and thus, in line with the practice favoured by American manufacturers, one cylinder per crank. All the cylinders are double-acting and fitted with cast iron liners. On the synthesis gas service the power absorbed per crank is fairly evenly distributed – 1,445, 1,265, 1,190 and 1,375 kW – but the refrigeration and recycle services absorb only 490 and 835 kW respectively; the total power absorbed is accordingly 6,600 kW. Multi-duty compressors have been made in which the air for the reformer is also compressed – usually in three stages – and on this duty the corresponding cranks are lightly loaded. Thus in multi-duty machines – and this applies even more so if air compression is also carried out in one compressor – the full potentiality of the single frame cannot be utilised. Nevertheless the arguments in favour of the multi-service, as opposed to a separate machine for each duty, are overwhelming.

Reverting to the Ingersoll Rand compressor, it has a stroke of 394 mm and runs at 300 rev/min, giving a mean piston speed of 3·94 m/s. This is also the mean speed of the recycle piston, but some manufacturers prefer to operate with a lower mean speed in recycle or circulator cylinders and, in the type of machine

Figure 10.22 'Multi-service' compressor: synthesis gas/refrigeration/recycle (*Ingersoll Rand Co.*)

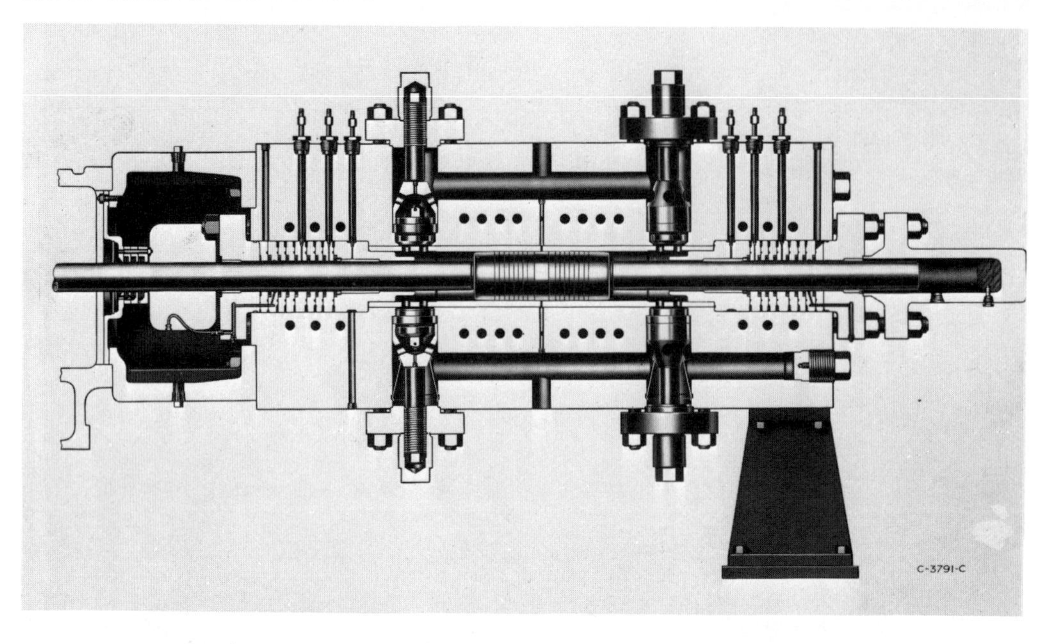

Figure 10.23 4th-stage cylinder of compressor illustrated in Fig. 10.22 *(Ingersoll Rand Co.)*

under discussion, to achieve this by having a shorter throw on this particular crank.

The fourth-stage cylinder itself is shown – in section – in Fig. 10.23 and some idea of its size can be obtained from the photograph. The stage compression ratio is 2 : 1 and the adiabatic discharge temperature 104°C – conditions which may be considered conservative. The cylinder is machined from a single forging of rectangular section and is not so expensive initially as one with separate valve heads though from the user's point of view the latter is preferable since the cost of replacement in the event of a fatigue failure is less. The gland packing, which is of the type illustrated in Fig. 9.6 is oil-cooled – as incidentally is the packing in the third stage which operates at a discharge pressure of only 159 bars, but the modern tendency is to cool such packings at pressures well below that at which this was previously considered necessary.

The valves are of the annular ring plate type, made by Ingersoll Rand themselves – one of the few compressor manufacturers to do so. Suction and discharge valves cannot be interchanged and care has been taken, in the design of the liner, to see that in the event of the bolt, holding the valve assembly together, failing no part of the valve can fall into the cylinder, with of course disastrous consequences. Such failures in the small highly stressed bolts in the valves of high pressure cylinders are not uncommon.

10.6 Compressors for Very High Pressures (Hypercompressors)

10.6.1 *Historical development*

During the past decade the supremacy of the reciprocating compressor in the field of high pressures has been challenged by the centrifugal machine. It is ironic that the same reason which led to the spectacular rate of increase in the size –

capacity and power requirements – of high pressure compressors, namely the need for larger fertilizer (ammonia) plants to give more economic units by lowering production costs, should provide the impetus for the entry of the centrifugal machine into a province hitherto regarded as the exclusive prerogative of its rival.

There is, however, one field in which the reciprocator is likely to hold its own, namely for very high pressures which in this context may be taken to mean pressures in excess of 1,500 bars up to the present limit – for industrial machines – of about 4,000 bars.

Compressors to operate in this range are generally referred to as "hyper" compressors. They were developed specifically for the compression of ethylene used in the manufacture of low density polyethylene and this, indeed, has remained so far the only application for them. As mentioned in the historical introduction (p. 4) this material was discovered by Imperial Chemical Industries in the early 1930's and the first machines (of the conventional type) were designed by this Company – for a pressure of 1,500 bars. With the licensing of the process to other concerns and the rapid growth of the market for polyethylene, both I.C.I. and its licensees sought to collaborate with certain selected compressor makers in the development, first, of larger machines and later of compressors for higher and higher pressures. With the feedback of operating experience from the users these firms acquired a technical "know-how" which has enabled them to establish a virtual monopoly in the manufacture of these hypercompressors. In Europe there are three such firms, Maschinenfabrik Burckhardt, Maschinenfabrik Esslingen and Nuovo Pignone; in the U.S.A., Ingersoll Rand and Clark Bros.

From the thermodynamic point of view the design of a compressor to handle ethylene up to pressures of the order quoted above does not present any problem, since the properties of this gas are known up to, and even beyond, the highest pressure at which any plant for the manufacture of polyethylene is today operating. The problems arise mainly from two sources – lubrication and mechanical design for fatigue conditions. Mineral oils are highly soluble in ethylene and so would contaminate the product; special "lubricants" with less satisfactory lubricating properties must therefore be used and, moreover, for the aforementioned reason lubrication must be kept to a minimum. Design problems are considered in what follows.

Although it is more than twenty years since the development of machines for pressures of 1,500 bars and over began, no single type has yet emerged as the preferred design. Each firm specialising in this field has tended to develop along its own lines and, understandably, users too have been somewhat reluctant to introduce designs differing too much from the one of which they have had experience. The veil of secrecy which surrounded the early machines has now to some extent been lifted, or perhaps it would be more accurate to say has been shifted to the region of higher pressures, so that, as will be seen, one finds great similarity in certain features and design details in the compressors from the different manufacturers.

In referring to compressors for service in polyethylene plants it has become customary to refer to the size – output – in terms of the weight of ethylene (tonnes) per hour. This is convenient because the machines usually operate with a suction pressure in the region of 250 to 300 bars. However the conversion to m^3/h is easily made since the density of ethylene at 0°C and 1·02 bar is 1·25 kg/m^3. An indication of the enormous variation in the compressibility of ethylene in a hypercompressor working from 300 to 3,000 bars can be seen from the fact that the values of Z at these pressures and 25°C are 0·75 and 5·3 respectively.

The first generation of hypercompressors for the above duty had an output of 4 to 5 tonnes/h and were designed for 1,500 bars, though originally they operated at a slightly lower pressure. This was in the early fifties and a measure of the progress since that date is indicated in Fig. 10.24 which shows the present manu-

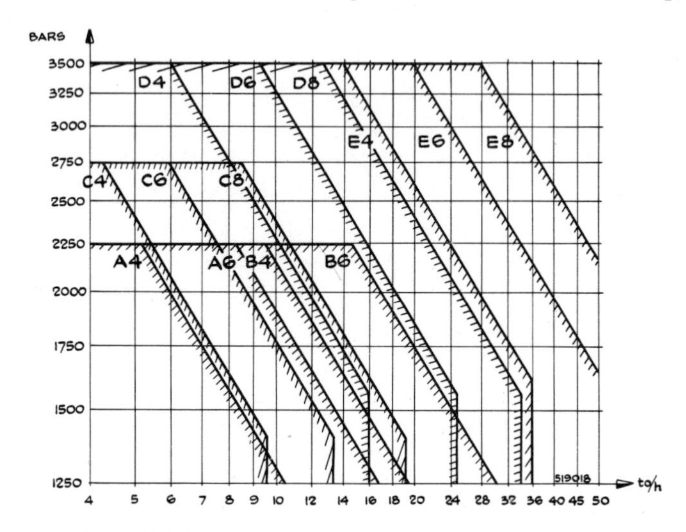

Figure 10.24 Hypercompressor manufacturing programme *(Maschinenfabrik Burckhardt)*

facturing programme of Maschinenfabrik Burckhardt. Series D and E, about which more will be said later, have hydraulic transmission, series C are horizontal machines with a link motion work, and series A and B vertical compressors using the standard motion work developed for lower pressures. Thus capacities of about 32 tonnes/h are now possible for a pressure of 3,000 bars and 28 tonnes/h for 3,500 bars. Series A and B are four- and six-crank machines; the remaining series are built with four, six and eight cranks.

10.6.2 *Some examples of modern hypercompressors*

Even when operating over such a wide difference of pressure between suction and delivery as 300/3,000 bars only two stages are required but the intermediate pressure is selected with regard to the range of pressure in each stage rather than the pressure ratio, as explained in Sections 6.2–6.4. The high stresses associated with these extreme pressures, in some cases approaching the yield strength of the material, call for the development of new design techniques and a wide use of experimental methods of stress analysis to determine the stresses at critical points in certain of the components. Broadly speaking, the higher the pressure the greater the extent to which components subject to fluctuating pressure must be broken down into simple ring or cylindrical sections which are susceptible to fairly accurate calculation and free from notch effects. The latter, the possible source of which cannot always be eliminated, require particularly careful consideration; design precautions which would be satisfactory at lower pressures are quite inadequate at pressures of the order we are discussing here.

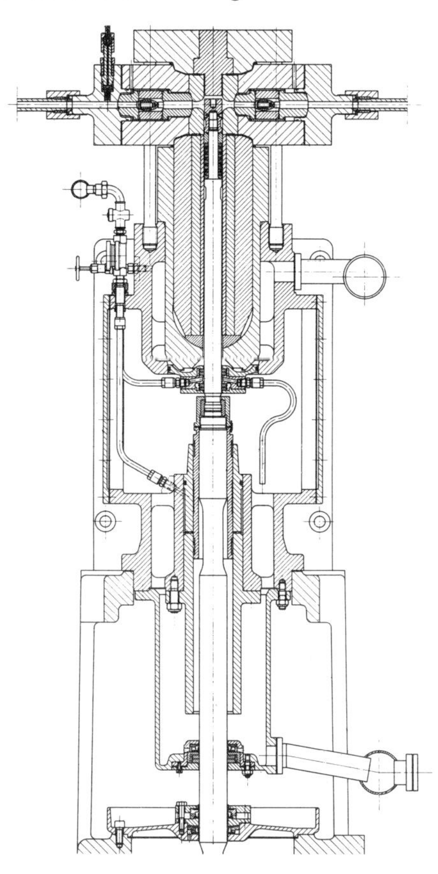

Figure 10.25 2nd stage cylinder of hyper-compressor for 2,000 bars *(Maschinenfabrik Burckhardt)*

The principles outlined above are well illustrated in the accompanying Fig. 10.25, which shows the cylinder of a two-stage hypercompressor for a final pressure of 2,000 bars. It consists of a two-component compound cylinder with a tungsten carbide liner. The assembly is completely free from stress raisers and is mounted in a steel housing which has only to transmit axial forces. Tungsten carbide has been found to be the most satisfactory material for the cylinder liners of ethylene hypercompressors in which sealing is effected by piston rings of conventional design. The cylinder or valve head, with separate suction and delivery valves as shown, with its two cross-bores, presents a particularly difficult problem and recourse must be had to photo-elastic measurements or strain gauge tests to determine the location of the points of critical stress concentration and an indication of the magnitude of the stresses and their distribution. Here again the solution is a shrink fit construction in which the shrinkage stress must be sufficiently high to ensure that the resultant stresses under operating conditions in these critical regions are always compressive (see Section 3.3).

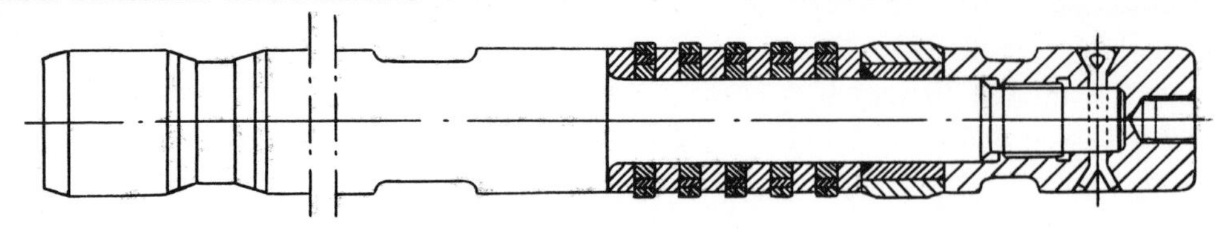

Figure 10.26 High pressure plunger with piston ring packing *(Maschinenfabrik Burckhardt)*

A detail of the piston is shown in Fig. 10.26. The grooves for the piston rings are formed by a succession of plain supporting and distance rings, the assembly being held together by a prestressed nut. The stress in the thread is relieved under operating conditions by the gas load on the end of the nut which causes additional compression of the piston ring assembly and, because of the difference in this gas load between suction and delivery, the thread is subjected to a fluctuating load which can cause a fatigue failure. Accordingly, special attention must be paid to the transition from the threaded portion to the plain section of the piston and similar care must be given to the junction between the reduced and full diameters.

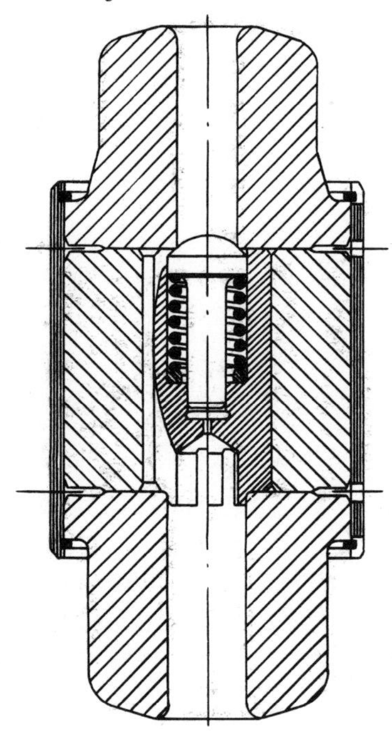

Figure 10.27 High pressure valve *(Maschinenfabrik Burckhardt)*

An enlarged view of a valve is shown in Fig. 10.27. At the pressures with which we are here dealing ethylene is already highly incompressible and tending to behave as a liquid, so that the problems can be regarded as associated more with pumps than compressors, and a type of valve frequently used in the former is also suitable for hypercompressors. The valve body is, of course, subjected to high fluctuating stresses so that its design follows the same principles as have already been fully discussed with reference to the cylinder.

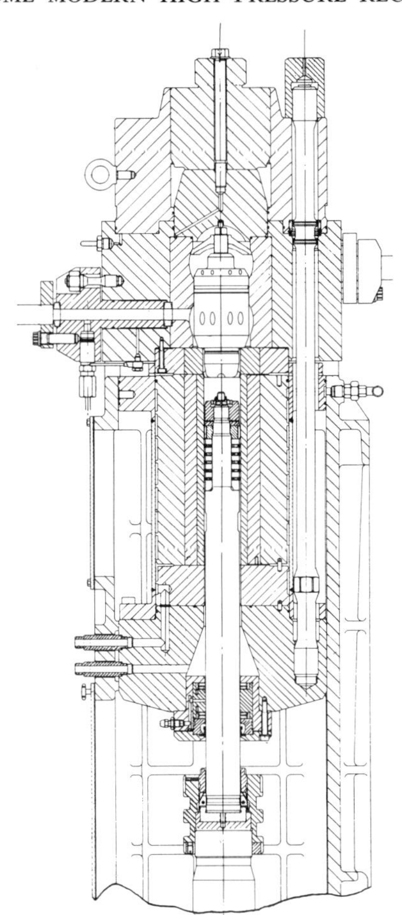

Figure 10.28 Hypercompressor cylinder for 2,500 bars *(Maschinenfabrik Burckhardt)*

With slight modifications the cylinder design described above is used for pressures up to 2,500 bars. In this case, as shown in Fig. 10.28, the outside is cooled. The design of the valve head, however, has been changed to accommodate a co-axial valve so that the head itself is subjected only to steady pressure loading.

For still higher pressures—up to 3,000 bars—an alternative design with a plain plunger and stationary packing is illustrated in Fig. 10.29. The same arrangement can, of course, also be used at lower pressures. The plunger is self-aligning and, as in some other Burckhardt designs, a secondary cylindrical cross-head is interposed between the main cross-head and the plunger to provide accurate guiding of the latter. It will be noted that the annular elements which form the housing for the packing rings are of compound construction and that the cylinder itself is very short—little more than the length of the stroke. Also of interest is the extensive use made of O-rings as static (joint) sealing elements.

Burckhardt compressors for pressures up to about 2,250 bars are built on this

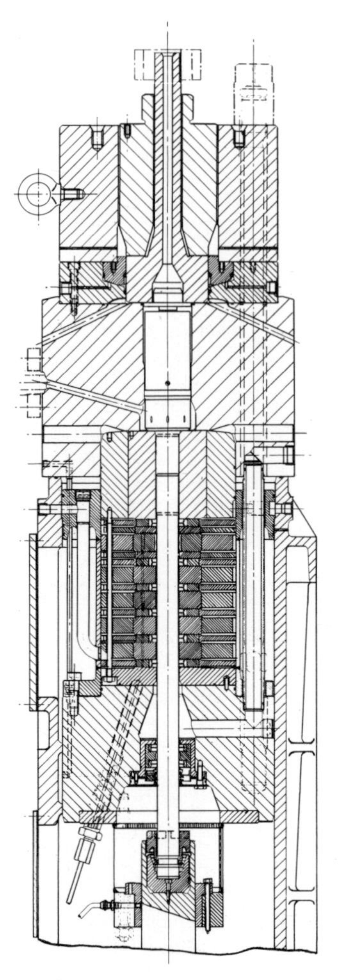

Figure 10.29 Hypercompressor cylinder for 3,000 bars *(Maschinenfabrik Burckhardt)*

firm's standard vertical frame and motion work. For higher pressures – up to 2,750 bars – Burckhardt have adopted, for three sizes only, a horizontal layout in which the main cross-head is driven by a link mechanism and the cylinders are horizontally opposed. The driving force, and thus the load on the motion work, is reduced to that resulting from the difference in the pressures in the two cylinders. This series is usually manufactured as a single-stage hyper-machine in which the suction pressure is of the order of 900/950 bars and which must be controlled fairly closely to avoid overloading the link mechanism. A section of a typical compressor of this series is shown in Fig. 10.30 and a photograph of a machine to compress 10 tonnes of ethylene per hour – the largest in the range – in Fig. 10.31. This particular compressor operates between 900 and 2,300 bars, absorbs about 1,500 kW, has a stroke of 200 mm and runs at 272 rev/min giving a mean piston speed of 1·82 m/s. This is significantly lower than for the vertical design where the figure is of the order of 2·5 m/s.

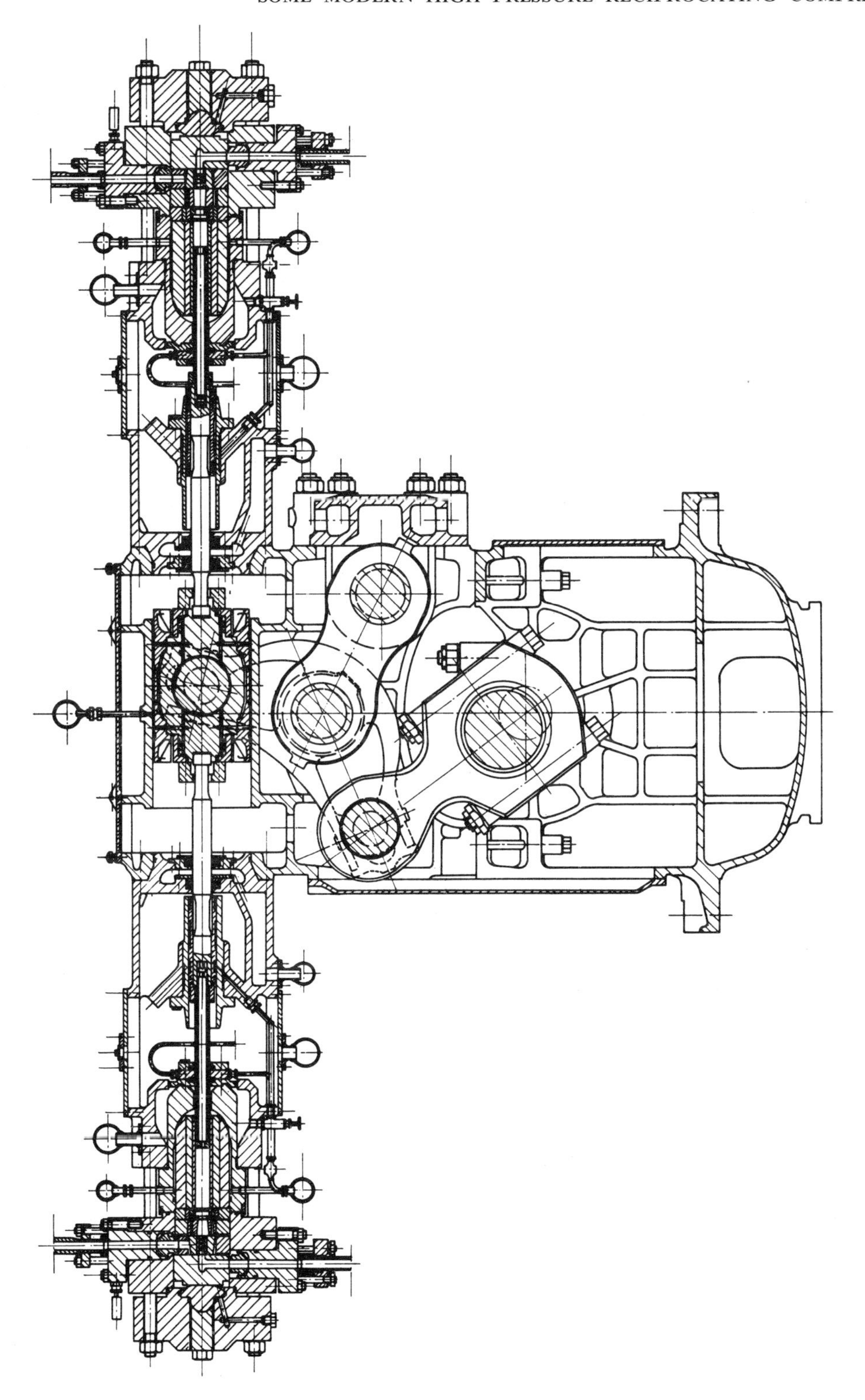

Figure 10.30 Sectional arrangement of hypercompressor with link motion and horizontally opposed cylinders. *(Maschinenfabrik Burckhardt)*

Figure 10.31 3-crank 6-cylinder single stage hypercompressor: 900/2,300 bars *(Maschinenfabrik Burckhardt)*

Without doubt the most interesting hypercompressor made by Maschinenfabrik Burckhardt is their latest development in which power is transmitted to the plungers hydraulically. One of the main considerations which led to this design was the desire on the part of the manufacturers of polyethylene for a machine capable of a variable output. Variable speed electric motors are costly, especially if suitable for flameproof operation, though apart from by-pass control the only practical solution. A further factor which is sometimes overlooked is the number of different grades of polyethylene which the manufacturer is required to produce—in widely varying tonnages. With compressors driven at constant speed and conventional mechanical drive this requirement can only be met by several machines with relatively low output capable of being run singly or in parallel according to the demand for a particular grade. A hydraulically driven machine, on the other hand, with its stepless variation can cover a wide range of outputs.

Other advantages are claimed for this design among which may be mentioned: minimum space requirements for a specified output, low piston speed giving long life of piston and packing rings, good accessibility to all parts resulting in easy maintenance.

This range of compressors is built in two basic sizes, each with four, six or eight cylinders, for pressures up to 3,500 bars.

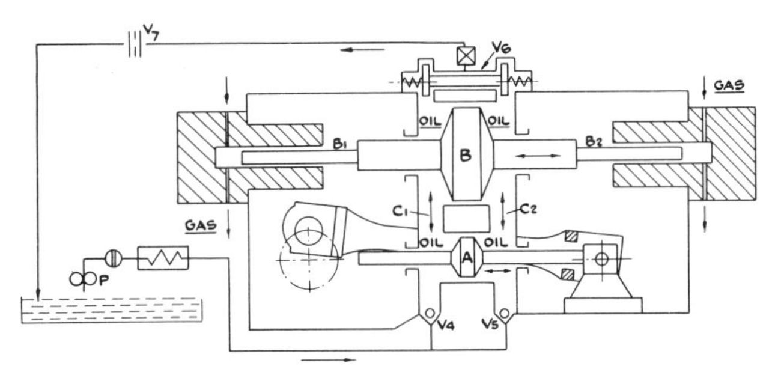

Figure 10.32 Principle of hydraulic drive for hypercompressor *(Maschinenfabrik Burckhardt)*

The principle of operation can be followed from Fig. 10.32. A conventional crank/connecting rod mechanism drives the primary piston A which, due to the high connecting rod/crank ratio, has a very nearly simple harmonic motion. The piston A transmits its reciprocating motion to the secondary piston B which is directly connected to the plungers B_1 and B_2 of the hypercompressor gas cylinders; thus the motion of B is also approximately simple harmonic and the strokes of A and B are inversely proportional to their respective areas. Accordingly the stroke and mean speed of B_1 and B_2 can be made quite small depending on the ratio of the above areas. A low mean piston speed is, of course, desirable from the packing life aspect, but the low speed in turn means that B_1 and B_2 are relatively large in diameter. An oil pump P supplies oil to the system continuously, pumping against the pressure at which the expansion or reducing valve V_7 is set, thus maintaining the whole system under a certain over-pressure. When the compressor is operating, the delivery pressure is superimposed on the feed oil pressure, the secondary piston B being operated by the difference between the two. The function of the two check valves V_4 and V_5 is, on the one hand to protect the pump when, for example, the primary piston A is moving towards the left and the pressure in the chamber C_1 is high and on the other hand, to permit oil to continue to flow through C_2. The double check valve V_6 fulfils a similar function: it closes on the left when the pressure in C_1 is high (so that this pressure must operate the secondary piston B), at the same time opening on the right to allow the oil in C_2 to return via V_7 to the oil tank.

The output can be controlled by inserting a throttle valve between C_1 and C_2 but this method of variation is wasteful since even at reduced capacity the primary piston A must displace the entire volume of oil against full final pressure so that there is no saving of power.

A more elegant method in which the power consumption is approximately proportional to the output is by by-passing oil from C_2 to C_1 and vice versa when the pressures in the two chambers are the same, which occurs once on the suction and once on the delivery stroke of each plunger. Fig. 10.33 explains the method by reference to the indicator diagrams of the two cylinders: 1 2 3 4 1 represents the diagram for one cylinder, 5 6 7 8 5 for the other. It will be seen that the compression line for the former and the expansion line for the latter intersect at point 10 and similarly the expansion line of the former and the compression line

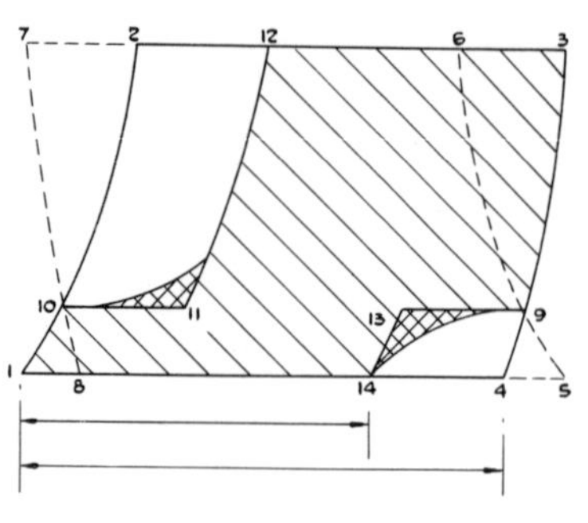

Figure 10.33 Principle of output control by by-passing operating fluid (*Maschinenfabrik Burckhardt*)

for the latter intersect at 9. At these two points the gas pressures in the two hyper-cylinders are equal so that the oil pressure in C_1 must be the same as that in C_2. If now a valve connecting C_1 and C_2 is suddenly opened oil which, as explained above, is being displaced in the chamber C_2 say (if the plunger B_1 is on its compression stroke) can pass into C_1 and the secondary piston will not move. After a certain time which can be chosen arbitrarily the valve is closed and the secondary piston can now continue its interrupted compression stroke—from 11 to 12. Delivery takes place from 12 to 3.

A similar state of affairs exists on the suction stroke: the expansion stroke is interrupted for a period (9–13) and subsequently continued (13–14). The suction volume has thus been reduced from (4–1) to (14–1).

Owing to the fact that the by-pass valve cannot be closed suddenly (at points 11 and 13), to avoid shock which would result from having to accelerate the secondary piston from rest to full speed, but must be closed slowly, the corners 10–11–12 and 9 – 13 – 14 are rounded; the enclosed (shaded) areas represent the losses which occur in this system. The actual operation of the by-pass valve will not be described here nor does space permit description of the various other refinements which are necessary to compensate for compressibility of the oil, to apply a correction for the position of the length of stroke and to ensure that at part load the reduced stroke always occurs in the centre of the cylinder (to maintain equal cylinder clearance in both cylinders).

A section through a hypercompressor with hydraulic power transmission is shown in Fig. 10.34 from which the various elements of the system referred to above can be seen. This type of compressor is manufactured by Burckhardt for pressures from 2,000 to 3,600 bars, outputs from 10 to 30 tonnes of ethylene per hour and powers from 1,500 to 4,500 kW. A 3,300 kW hypercompressor of this series for a delivery pressure of 2,600 bars and an output of 20 tonnes/hour is shown in Fig. 10.35.

The distinctive feature of the hypercompressors made by Ingersoll Rand Co., U.S.A., is the introduction of a yoke or secondary cross-head to which the plungers of an opposed pair of cylinders are attached. The arrangement can be seen from Fig. 10.36 of a two-crank two-stage compressor. This form of construction was first used over 25 years ago by Maschinenfabrik Esslingen for a vertical

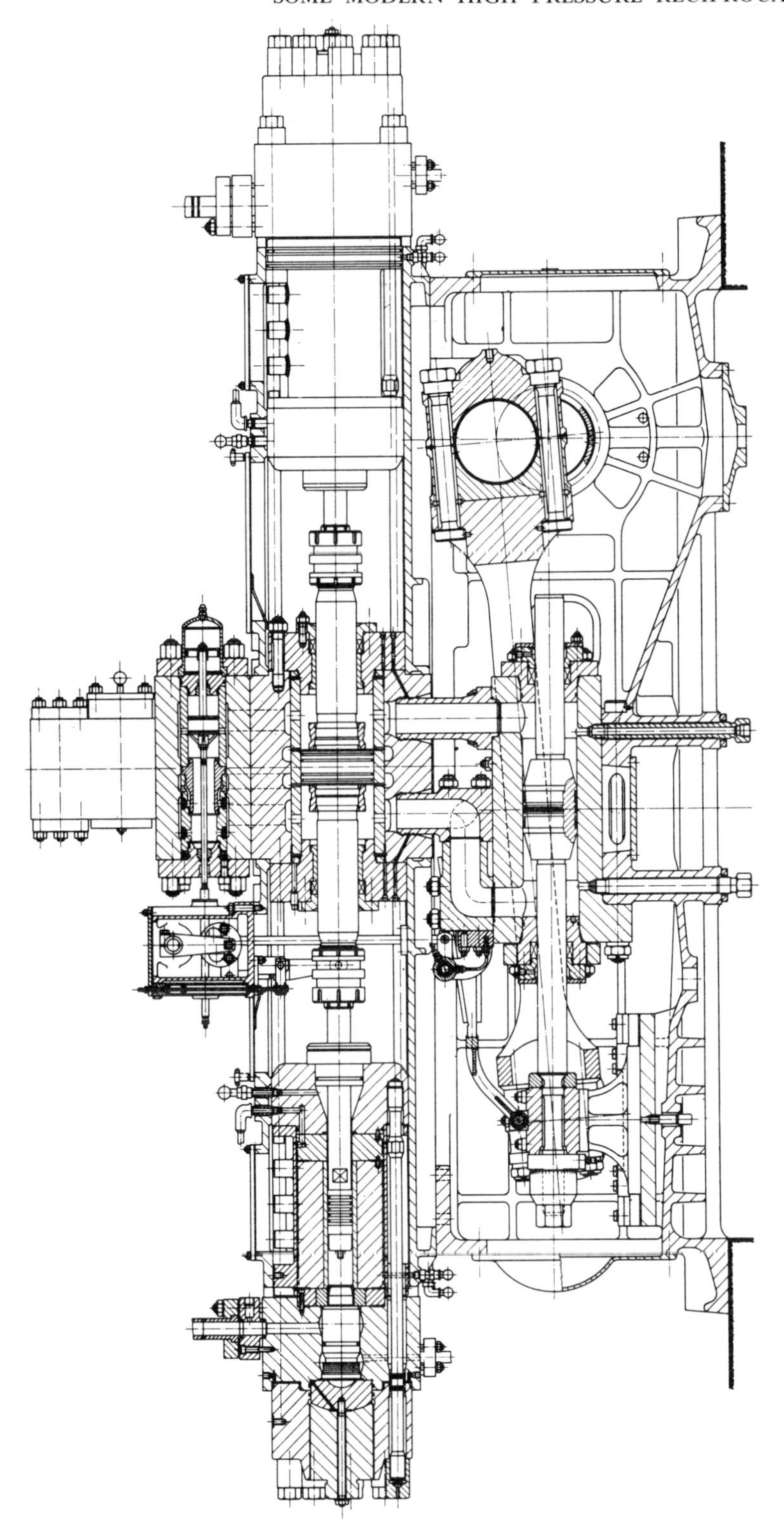

Figure 10.34 Sectional arrangement of hypercompressor with hydraulic drive *(Maschinenfabrik Burckhardt)*

Figure 10.35 Hypercompressor with hydraulic drive: Type D.8 *(Maschinenfabrik Burckhardt)*

Figure 10.36 Hypercompressor with secondary crosshead *(Ingersoll Rand Co.)*

hypercompressor designed for 4,000 bars, but the machine did not in fact reach this pressure and the construction was abandoned, mainly because, being a vertical machine, accessibility to the lower cylinder and gland was awkward.

The use of a secondary cross-head eliminates side thrust and permits very accurate lining up of the plungers, with beneficial results to packing and ring life. However, as the photograph shows, accessibility to the inner cylinder etc. is not ideal in spite of the fact that it can be removed without the necessity for dismantling the drive and tie rods. The chief disadvantage of the arrangement is its effect on the overall length of the compressor and the space needed for installation, particularly if it is arranged as a boxer design.

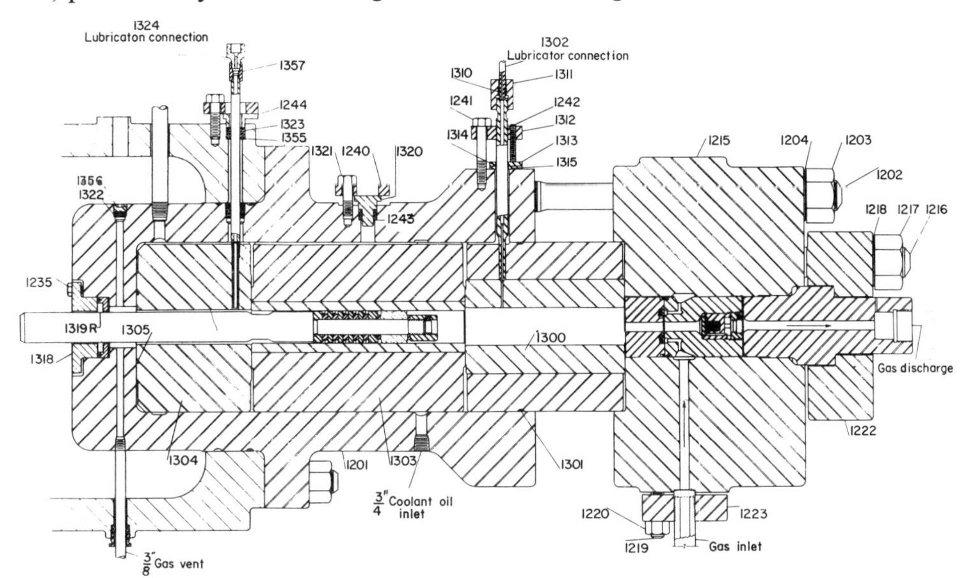

Figure 10.37 Section of hypercompressor cylinder, movable piston ring packing on plunger *(Ingersoll Rand Co.)*

Fig. 10.37 is a section through the hypercylinder of an Ingersoll Rand compressor with piston ring sealing, which shows admirably how the principle of breaking down the construction into simple cylindrical or ring-shaped elements has been followed. Only the two end sections have transverse bores for lubrication purposes and are easy to replace should a failure by fatigue occur. The centre 'working' part is a simple monobloc cylinder with shrunk-in tungsten carbide liner and is entirely free from cross-bores. The section adjacent to the head end of the piston is of compound construction but for the other end section, which is not subject to the same high pressure, a simple monobloc cylinder suffices. The valve is co-axial, suction and delivery valves being of the disc type. Only the valve assembly is subject to fluctuating stresses. The cylinder is oil-cooled.

A corresponding design with a plain plunger and conventional static packing is shown in Fig. 10.38 and, as can be seen, uses the same housing. There is a slight difference in the arrangements for lubrication but otherwise the two designs are interchangeable. The outside of the packing housings – which are of compound construction – is oil-cooled.

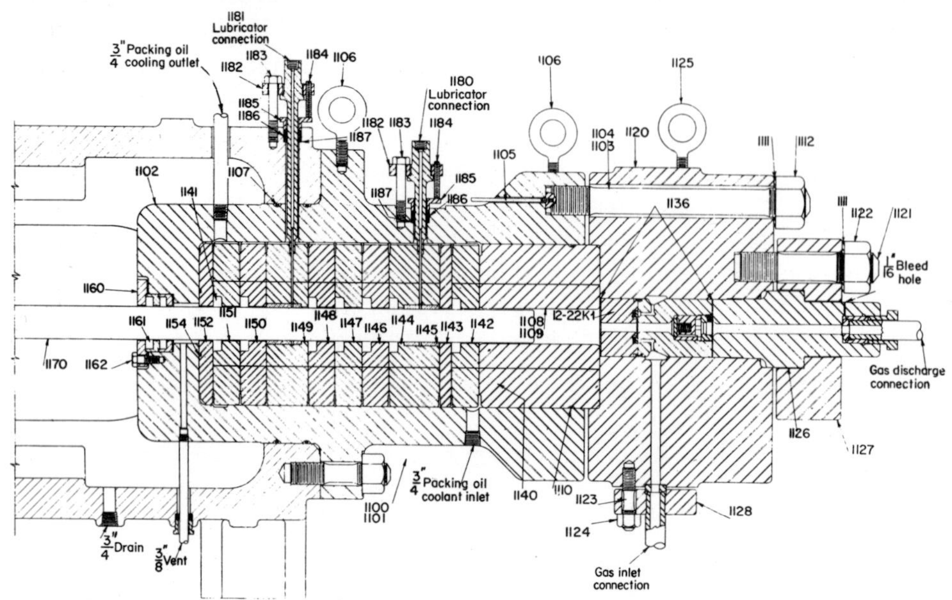

Figure 10.38 Hypercompressor cylinder with stationary packing and plain plunger *(Ingersoll Rand Co.)*

Of interest in both designs is the large use which is made of "O" rings to provide static sealing, with considerable saving in the time of assembly required for various components.

In the U.S.A. as in Europe the output from, and delivery pressure of, machines for compressing ethylene has increased rapidly in recent years. All the leading manufacturers of hypercompressors mentioned earlier can offer machines capable of an output of 40 tonnes/h and designs for pressures up to at least 3,500 bars. Though few installations are actually operating at this pressure at the present time there are no fundamental difficulties in designing for it merely by applying the principles already enunciated. Indeed the problems become those associated with the choice of suitable materials and their heat treatment—a not unfamiliar pattern in the whole history of the development of high pressures where metallurgical progress has sometimes tended for a time to lag behind design techniques.

An elegant solution of the yoke drive is to be found in the hypercompressors manufactured by Maschinenfabrik Esslingen. In this design the axis of the crankshaft lies between the crosshead and the yoke as can be seen from Fig. 10.39. The effect is to shorten very considerably the length of the machine compared with the Ingersoll Rand design. The Esslingen range of hypercompressors comprises single-stage machines for delivery pressures up to 1,800 bars (suction pressure about 250 bars) and two-stage machines for design pressures up to 3,500 bars. For the highest outputs and the corresponding pressures there are two second stage cylinders to handle the output from a single first-stage cylinder. The cylinders are thus in H or HH form with the motor between the two groups.

The motion work is standardised in four sizes for strokes of 250, 300, 320 and 400 mm with corresponding piston loads (including inertia loading) of 45, 45, 80 and 130 tonnes. The maximum speed for the first three sizes in the range is 214 rev/min and for the last 167 rev/min giving mean piston speeds of 1·78, 2·14, 2·28 and 2·26 m/s.

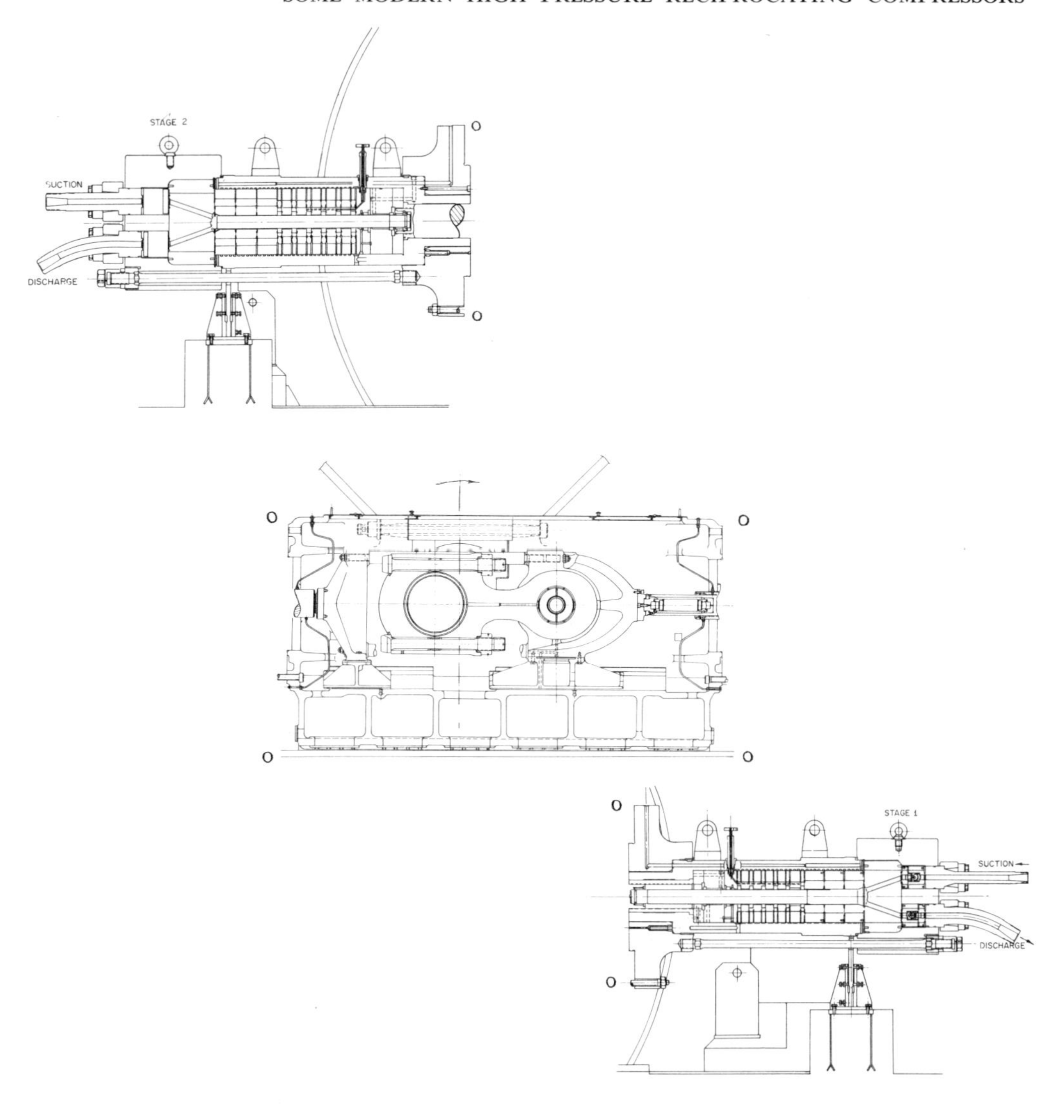

Figure 10.39 Section through 2-crank 2-stage hypercompressor in H form (with cartridge type cylinders) *(Maschinenfabrik Esslingen)*

Referring to Fig. 10.39, the sectional arrangement through one pair of second-stage cylinders of a relatively small three-crank two-stage hypercompressor, the main features are the additional plunger guide and the triple compound cylinder and packing ring housings where the pressure is highest, reducing to a two-piece shrink-fit construction and finally a plain cylindrical section at the lower pressure end of the gland. In this modern design the cylinder packing and plunger are enclosed in a cartridge-type housing which enables the whole assembly to be removed and replaced in a relatively short time by a pre-assembled unit, with considerable saving in time. Moreover this pre-assembly can be done under ideal conditions so that there is less likelihood of trouble on restarting.

The necessity for accurate control of the shrink-fit allowances in multi-layer construction, together with the problems associated with the actual shrinking operation, makes the manufacture of the various components for hypercompressors built up in this way a difficult and costly procedure which increases with the number of cylindrical elements. Moreover, as was pointed out in Chapter 3, with three or more cylinders unacceptably high residual stresses may be set up by the compounding operation if the overall diameter ratio K exceeds a certain value; for three cylinders – perhaps the maximum number likely to be used in hypercompressors – K is about 4·5.

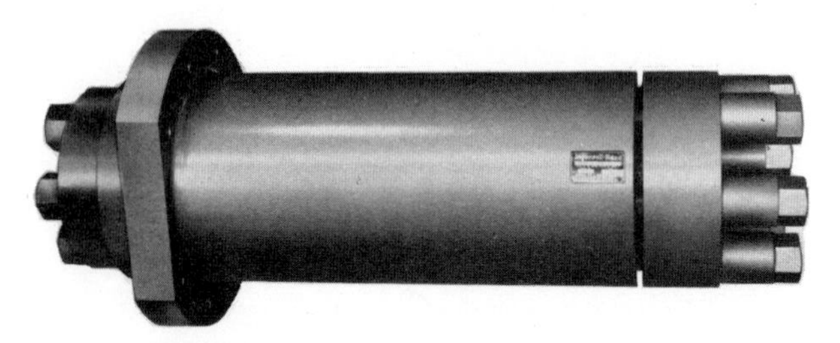

Figure 10.40 Cartridge type cylinder for hypercompressor *(Ingersoll Rand Co.)*

"Cartridge" cylinders are also used by other manufacturers and Fig. 10.40 is a photograph of an Ingersoll Rand cylinder of this type.

A method of obtaining the initial residual compressive stress in the innermost component of a compound cylinder which avoids the shrinkage problems referred to above is that known as the "fluid ring" construction. In this method the stress is produced by a fluid pressure in the annulus between each pair of adjacent cylinders – which pressure in the ideal case will vary for each pair. The application to the cylinder of a hypercompressor is relatively simple; a convenient source for the "jacket" fluid pressure is the delivery from the machine itself.

Adequate and accurate pre-stressing of studs and bolts subject to fatigue loading is vital, and increasing use is being made of hydraulic or electro-hydraulic equipment for simultaneously uniformly pre-tensioning such components. An important feature of such devices is that the stud or bolt is subjected only to direct – tensile – load; torsional loading is not involved. The simpler equipment for this purpose relies on the pre-calculated value of the hydraulic pressure to provide the required elongation – which can of course be checked by a clock gauge or strain gauge attached to one of the group of studs being pre-tightened. The more elaborate devices, as exemplified by the Heller system, incorporate automatic electronic control of the pre-set tensile force to within $\pm 1\%$ and residual pre-load within $\pm 2\%$.

11 Centrifugal Compressors

11.1 The Centrifugal Compressor for High Pressures

Manufacturers of ammonia have long been aware of the many advantages which the centrifugal compressor has to offer as a means of raising the synthesis gas to the operating pressure, but it is only within the past five years that it has become possible to achieve the pressures at which the majority of ammonia plants operate at the present time – 300 to 350 bars. Indeed in the first plants incorporating a centrifugal machine this operated at about 150 bars and was followed by a single-stage reciprocating compressor to raise the gas to the final pressure. Apart from the fact that the inherent weaknesses of the reciprocating type were still present – though admittedly to a less degree – this combination meant that one of the main points in favour of the centrifugal compressor, namely the delivery of oil-free gas, could not be fully realised. Accordingly the next step was the development of processes for producing ammonia operating at about the pressure referred to above (150 bars). However, many ammonia producers believe that, depending on local circumstances, the optimum synthesis pressure for this process is in the range of 300–450 bars and further development has resulted in centrifugal machines capable of compressing synthesis gas to 300 bars with designs available for pressures up to at least 400 bars.

The first successful centrifugal machines for handling synthesis gas were pioneered by Clark Brothers Company in the U.S.A. Most of the "single-stream" ammonia plants which were being introduced in the early 1960's were equipped with compressors made by this firm who have continued to play a major role in its developments.

The chief factor which determines the minimum size of a centrifugal compressor is the amount of gas which can be handled by the final impeller and until fairly recently a figure of around 500 m^3/h actual volume was regarded as the lower limit. However, the adoption of new techniques – which will be referred to later – for the manufacture of impellers has enabled this figure to be reduced to about half; even so it is generally accepted that for ammonia outputs below 550/600 tonnes per day the turbo-compressor is not yet able to compete economically with the reciprocating machine. Thus the larger the size of plant the more favourable is the case for the centrifugal compressor. It is only about fifteen years since 150 tonnes per day, and less than ten years that an output of 600 tonnes per day were regarded as large plants. To-day there are numerous plants in operation with outputs of 900 tonnes per day, and some with considerably larger.

The problem mentioned above of the width of the impellers in the final stages of compression is greatly eased by the need to recycle some four to five times the fresh or make-up gas volume. This recycle gas is mixed with the fresh gas either in the compressor itself or external to it and introduced at the inlet to the last wheel, the width of which is correspondingly increased.

The period of rapid escalation in the size of ammonia plants also coincided with the introduction of pressure steam reforming of naphtha for producing synthesis gas and this process—using naphtha or (where it is available) natural gas—has become almost universally adopted for modern plants, giving a gas at the suction of the compressor at a pressure of the order of 25/30 bars. Thus for the range of synthesis pressures mentioned above the overall compression ratio lies between 12 and 15 and this is accomplished in a train of three or four units in series.

The ammonia reaction is highly exothermic and the heat of the reaction is used to generate steam at high pressure, part of which provides the energy for the turbine driving the train of compressors—the ideal form of drive since the speeds of the compressor and turbine can be directly matched. Accordingly the fact that the full-load efficiency of the turbo-compressor is some 10% or so lower than that of an equivalent reciprocating compressor is not of great importance, but whereas the efficiency of the latter remains substantially constant at part load that of the former falls off sharply as the output is reduced, so that most, if not all, of the advantage of lower cost of ammonia from the centrifugal compressor plant may be lost if the plant has to operate for long periods below, say, 70% output.

The development of the high pressure centrifugal compressor for the purpose discussed above owes a great deal to the work which had previously been done on, and the experience gained in the running of, centrifugal machines in petrochemical plants. Though, in the latter, operating pressures are much lower, effective seals had been developed and many problems on details of design had been satisfactorily solved. There still remained, however, the question of the manufacture of impellers with a width of only a few millimetres. Techniques such as milling from a solid blank and welding, which in the high tensile alloy steels normally used for such impellers is fraught with many dangers—formation of hair cracks, precipitation of carbides, residual stresses, difficulty of ensuring complete weld penetration, etc.—cannot be used when the width of the impeller at outlet is, say, 4 mm or less. The answer has been found in "spark" or electro-erosion techniques which have been successfully used for impellers having a width at outlet as low as 2 mm. Moreover, this method produces an impeller with exceptionally smooth passages, a matter of great importance from the efficiency point of view, especially since such narrow impellers cannot be expected to have a high efficiency.

Centrifugal compressors for high pressures have heavy forged steel casings of cylindrical form and are therefore usually referred to as "barrel" type. The compressor proper, consisting of the rotor, diaphragms, etc., is a separate segmented, horizontally split, assembly which is inserted in the outer casing as a complete unit; the seals and bearings are located in the end covers. A balance piston takes care of most of the end thrust arising from the sum of the unbalanced pressure differentials across each impeller, and a double-thrust bearing is provided to absorb the residual end load. In a multi-cylinder compressor this question of end thrusts can occasionally give rise to some difficult problems. Usually in such arrangements each casing is designed to take care of its own "internal" end thrust, leaving the intermediate coupling between two adjacent casings to absorb any end float due, for example, to temperature effects. This has not always proved satisfactory because of the restraint on end float imposed by the torque load.

Control of the recycle gas rate is by variable guide vanes at the inlet of the impeller where this gas is admitted. This gives an independent control of the circulating gas and in some designs similar adjustable guide vanes are installed in the inlet of the machine to provide for part load running.

A section through the casing of a high pressure centrifugal compressor is shown in Fig. 11.1, from which most of the features referred to above can be identified. This particular design is due to Pignone of Italy who were the first

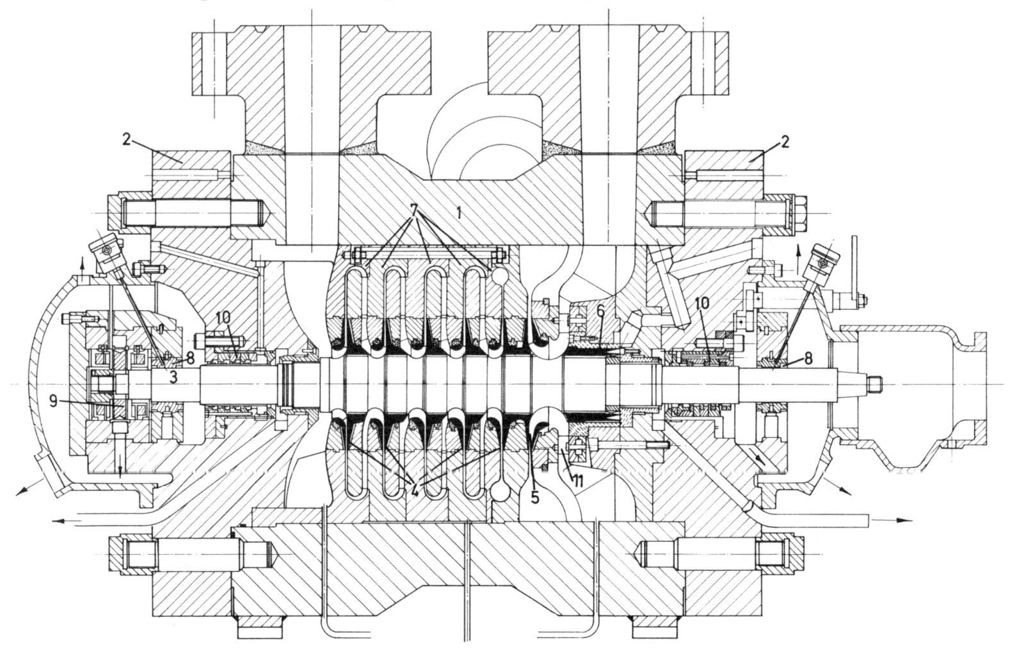

Figure 11.1 Sectional drawing of high pressure compressor with back-flow controlled by diffusers *(Nuovo Pignone)*

1 – Casing
2 – Casing heads
3 – Shaft
4 – Synthesis gas wheels
5 – Recycle gas wheel
6 – Thrust balancing cylinder
7 – Intermediate diaphragms
8 – Journal bearings
9 – Thrust bearing
10 – Oil seals
11 – Inlet guide vanes for capacity control of recycle gas

firm to design – and test – a machine to operate at a delivery pressure exceeding 320 bars. In the arrangement shown in Fig. 11.1 it will be observed that the recycle impeller is mounted back to back to the group of impellers handling the synthesis gas to facilitate the installation of the mechanism for operating the guide vanes.

The seal used in such compressors is of the oil film type in which oil at a pressure slightly above that of the pressure against which it has to seal is admitted into an annular chamber surrounding the shaft. A slight flow of pressurised seal oil takes place towards the gas zone opposing outward leakage of the gas and is drained off to a reservoir from which the gas mixed with the oil can be separated. Most of the seal oil, however, passes to the atmospheric drain, the pressure being broken down by a floating ring or rings according to the pressure. Oil leakage to the gas stream is further prevented by a series of labyrinths. Actual designs differ from manufacturer to manufacturer, but the principle remains the same. Thus in a typical seal used by Ingersoll Rand the seal oil is injected between two stationary floating rings (one single and one double) whilst Clark Bros. use L-shaped rings with vertically lapped faces between which the oil is introduced. Pignone employ a single babbit sleeve or ring for the high pressure seal and a series of rings for the low pressure or atmospheric seal.

As is usual in centrifugal compressors, the pressure at the delivery end of the machine is balanced off with that at the suction end so that the pressure against which the oil has to seal is the suction pressure of the particular casing. Accordingly the oil pressure must be automatically maintained slightly above the gas pressure. In a test on the final casing of a three-casing machine by Pignone, for which the suction and delivery pressures (of this casing) were 187 and 290 bars respectively, the oil/gas differential pressure referred to above was 0·7 bar; the flow of oil through the high and low (atmospheric) seals was 0·01 to 0·02 m^3/day and 30 to 40 m^3/day respectively.

A line diagram for the seal oil system of a three-casing compressor is shown in Fig. 11.2. Since the same oil can be – and frequently is – used for both the sealing and lubrication systems, the pumps for the former can, as shown in the diagram, take their suction from the discharge of the pumps for the lubrication circuit; alternatively they may have a separate suction from the common oil supply tank. The necessity for duplicating essential auxiliary equipment and the elaborate system of controls and alarms to ensure continuity of seal oil supply results in a complicated set-up, particularly in an installation consisting of three or four casings. Some simplification (compared with the arrangement in Fig. 11.2) can be obtained by using a single set of pumps to supply all the seal oil with appropriate reducing valves to reduce the pressure to that required in each casing.

11.2 Some Theoretical Considerations

In spite of the considerable attention which has been devoted to obtaining a better understanding – both qualitatively and quantitatively – of the flow in a centrifugal compressor the design of such a machine is still largely empirical, relying on test or performance data on single impellers for which manufacturers

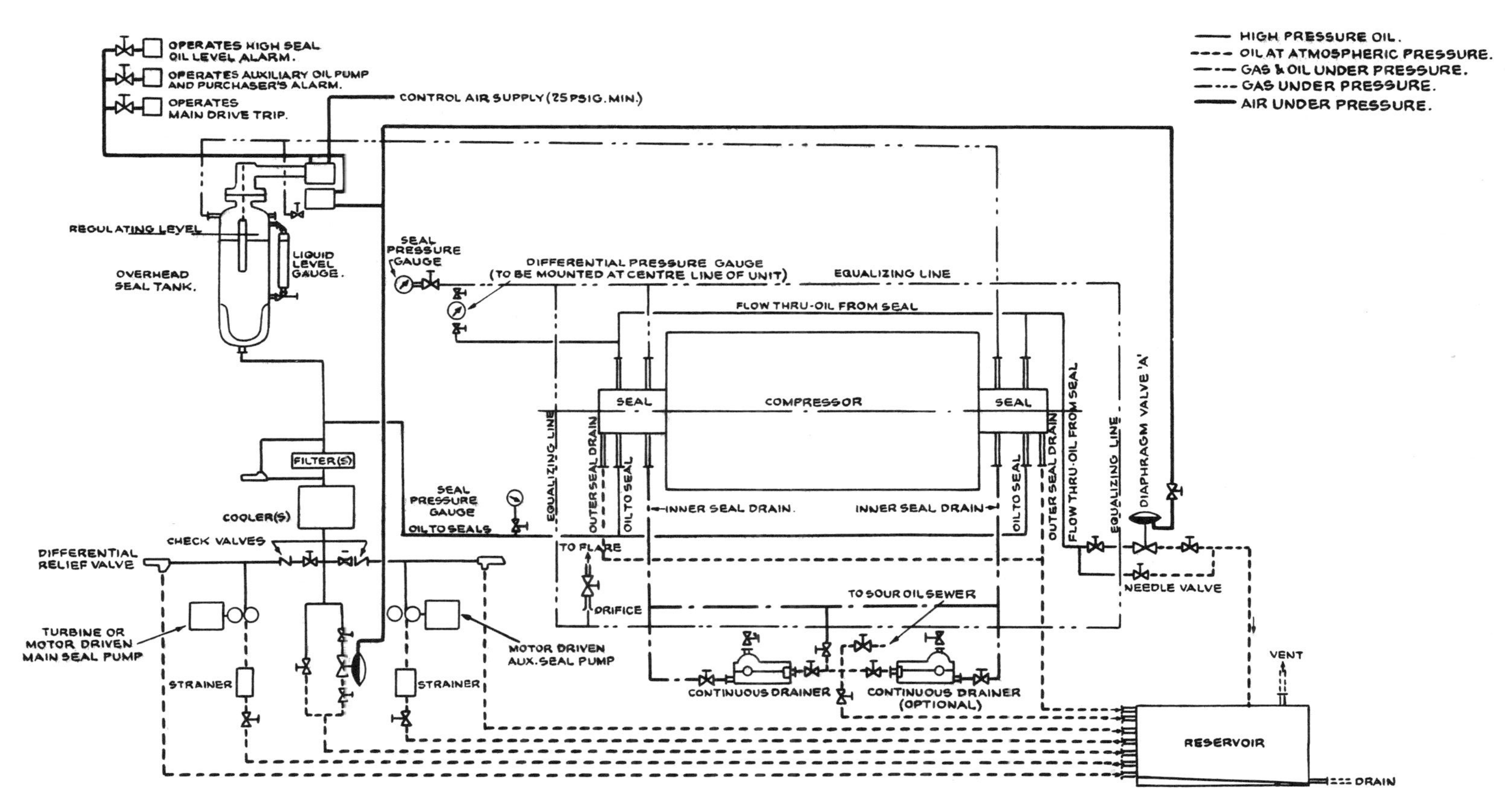

Figure 11.2 Schematic diagram of typical seal system

have standard sizes covering a wide range of diameters and for which the characteristics are known at various speeds.

No attempt will be made here to discuss in detail the theory of the radial type centrifugal compressor. This is treated in the works of Eckert and Schnell(11.1), Traupel(11.2) and Stepanoff(11.3), but a few general observations may be of interest.

From thermodynamic considerations and with the notation of Chapter 8, the total adiabatic head developed by a centrifugal compressor is given by:

$$H_{ad} = (\sqrt{(Z_1 Z_2)}) \frac{\gamma}{\gamma - 1} RT_1 \left[\left(\frac{P_2}{P_1} \right)^{(\gamma-1)/\gamma} - 1 \right] m \tag{11.1}$$

where the suffixes 1 and 2 refer to the inlet and outlet.

Eq. (11.1) also gives the adiabatic work in m kgf/kg.

The pressure ratio of a particular stage is given by:

$$\frac{P'_2}{P'_1} = \left[\frac{H'_{ad}}{(\sqrt{(Z'_1 Z'_2)}) RT'_1} \frac{\gamma - 1}{\gamma} + 1 \right]^{\gamma/(\gamma-1)} \tag{11.2}$$

where the accented notation refers to the particular stage to distinguish it from eq. (11.1) which is for the whole compressor. In these expressions the pressures and temperatures are total pressures, that is the sum of the static and dynamic pressures.

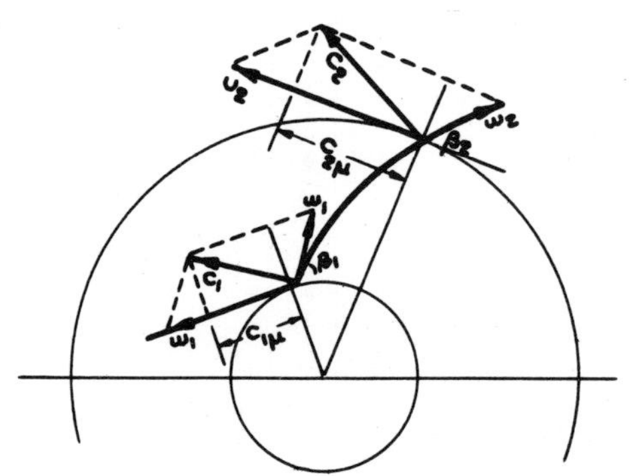

Figure 11.3 Velocity diagrams for impeller with backward curved vanes

Elementary theory for the flow of the medium through the impeller is due to Euler and assumes an infinite number of vanes, which thus correspond to the flow filaments. The total pressure rise in the impeller—which we denote by ΔP—is given by the difference in the moments of momentum at inlet and outlet. From Fig. 11.3:

$$\Delta P = \frac{\rho}{g} (u_2 c_{2\mu} - u_1 c_{1\mu}) \tag{11.3}$$

which can be transformed thus:

$$\Delta P = \frac{\rho}{2g} [(c_2^2 - c_1^2) + (u_2^2 - u_1^2) + (w_1^2 - w_2^2)] \tag{11.4}$$

or in terms of the total head H_{tot}:

$$H_{tot} = \frac{1}{2g} [(c_2^2 - c_1^2) + (u_2^2 - u_1^2) + (w_1^2 - w_2^2)] \tag{11.5}$$

which is independent of the density of the fluid and which is accordingly valid also for compressible flow. Moreover, since the expressions for ΔP and H_{tot} consist only of the sum of the squares of the velocities at inlet and outlet they hold also for a finite number of vanes.

The first term represents the increase in kinetic energy of the fluid in the impeller which can be converted to pressure energy in a diffuser or volute following the impeller. With frictionless conversion the pressure rise in the diffuser is $\rho/2g$ $(c_2^2 - c_d^2)$ where c_d is the velocity at outlet from the diffuser.

The second term gives the static pressure rise due to the centrifugal forces whilst the third term expresses the effect of the decrease in relative velocity in the impeller arising from the increase in area from inlet to outlet. Converted to a static pressure rise this becomes $\rho(w_1^2 - w_2^2)/2g$.

It can be shown that the optimum inlet angle β_1 – which is determined by the flow conditions at inlet – is approximately 30°. The outlet angle β_2, however, can be chosen at will depending on the form of the vane and Fig. 11.4 shows the effect of the form on the velocity triangle at outlet.

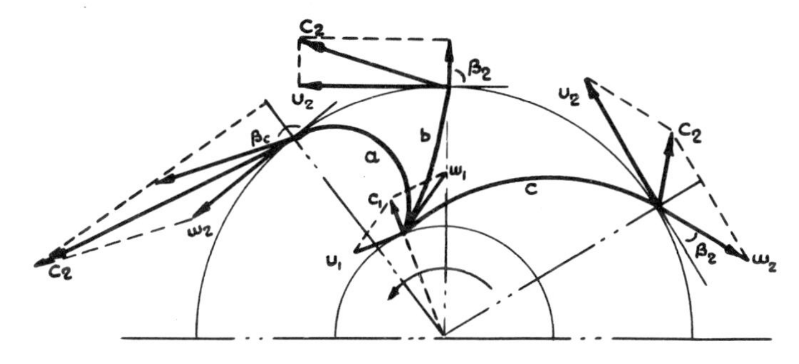

Figure 11.4 Effect of blade form on velocity diagrams (a) Forward curved $\beta_2 > 90°$; (b) Radial $\beta_2 = 90°$; (c) Backward curved $\beta < 90°$

The ratio of the static head developed in the impeller to the total head is defined as the degree of reaction.

An important dimensionless parameter in centrifugal compressor theory is the head coefficient ψ which is defined thus:

$$\psi = 2gH_{ad}/u^2$$

Forward curved vanes ($\beta > 90°$) give higher values of ψ but with increasing values of β_2 the degree of reaction and the static head developed in the impeller decreases. In the limit the static head becomes zero, the whole of the energy put into the impeller being thus converted to kinetic energy. Reconversion to pressure energy in a diffuser is difficult and accompanied by high losses.

The radial vane, corresponding to $\beta_2 = 90°$, gives the maximum static head, which means the greatest reduction in relative velocity of the medium in passing through the impeller. From the strength point of view the radial vane is superior to both the forward and backward curved types owing to the absence of bending stresses – so that extremely high tip speeds, up to about 600 m/s, are possible. Accordingly this form of blade is preferred where minimum dimensions and weight are of importance, as in turbo-superchargers, compressors for jet engines, etc., whilst the forward curved vane is used where large volumes have to be handled at small static pressures.

The backward curved vane ($\beta_2 < 90°$) is, because of its better efficiency and more suitable head/flow characteristic, preferred for compressors for industrial applications. To achieve the optimum efficiency the decrease in relative velocity in the impeller channel should be as small as possible (in the limit $w_1 = w_2$). In practice it is found that a 10% to 20% reduction can be allowed. For an inlet angle of 30° optimum efficiencies are given with an outlet angle between 35° and 50°.

The above is only the briefest sketch of the idealised flow in the impeller of a centrifugal compressor. The actual flow is much more complex and is only amenable to approximate mathematical treatment. The Euler theory assumes uniform flow and therefore uniform pressure distribution across the impeller passage, but in order to transmit power to the fluid the pressure on the leading or front face of the vane must be higher than on the back. There is also an internal circulation or vortex flow in the channel between two vanes. The above theory also assumes incompressible flow, but a critical factor in design is the Mach number, particularly at the impeller inlet where the Mach number is defined as the ratio of the maximum local velocity—usually taken as the relative velocity at entrance—to the velocity of sound. Experience shows that the Mach number at impeller entry should not exceed about 0·75.

The most important design variable is the impeller angle at outlet, and all the design constants etc. depend on its value. Its effect on the impeller characteristics has already been referred to above.

11.3 Reciprocating v. Centrifugal Type for High Pressure

The high pressure centrifugal compressor was specifically developed to meet the demand for a machine capable of handling the large volume of gas involved in the growing size of ammonia plants. In this it entered into direct competition with its rival, the reciprocating compressor, and—broadly speaking—for ammonia outputs between 600 tonnes/day and 1500 tonnes/day the user has a choice between the two. For the latter output two reciprocating compressors of 15,000 hp (11,000 kW) would be required—both working—and this is at present the maximum size available, whereas a single train of centrifugal compressors would present no problems; indeed because of the large volume of gas to be handled the design of a machine of this size would be relatively easier than one for, say, 600 tonnes/day, assuming the same delivery (synthesis) pressure.

Under present-day conditions it is generally considered that the figure of 550 to 600 tonnes/day ammonia output represents the lower limit for a single-stream centrifugal plant. Below this, the reciprocating compressor installation, though slightly more expensive in capital cost, compares favourably in production cost. This is because the efficiency of the high pressure centrifugal compressor falls off sharply when the capacity drops below about 70,000 m^3/h (which corresponds to an ammonia output of about 500 tonnes/day) for a synthesis pressure of about 300 bars.

Control of output to cover operation at part load is much easier, and can provide a much wider range of control, with the reciprocating machine. With a two-compressor plant driven by steam turbines the output of the plant can be varied from 100% down to about 35%, but a single stream centrifugal plant is only capable of control (by speed variation of the turbine) down to about 70% of full output – and this reduction is accompanied by a significant drop in efficiency, which even at maximum output is some 10% or so lower than that of its rival.

A problem which arises with the centrifugal, but not the reciprocating, type is that of surging at reduced capacity. This further reduces the control range and, because of the harmful effect to seals and bearings if surging occurs, requires the provision of anti-surge gear to give automatic by-pass from the delivery to the suction side of the compressor.

Although we have discussed the role of the two types of compressor with reference to ammonia plants the centrifugal compressor is also suitable – provided the output is large enough – for other high pressure processes in the pressure range 200 to 350 bars, e.g. the production of methanol (compression of hydrogen and carbon monoxide) and urea (compression of carbon dioxide). The recent development of a low pressure process for the former considerably increases, of course, the attractiveness of the centrifugal type for this process.

Summarising, we can say the limit for the high pressure reciprocating compressor for pressures up to, say, 450 bars has already been reached in machines of 15,000 hp. Indeed there would seem to be little probable demand for compressors of this type for the traditional high pressure processes for ammonia, methanol and urea manufacture. The minimum size of urea plant suitable for a single-stream centrifugal arrangement is at present above the largest plants in operation, so that the reciprocating compressor is likely for some time to find application either for compressing the CO_2 over the full pressure range or perhaps as a booster compressor following a centrifugal. There are certain other high pressure processes – involving carbonylation reactions – which will continue to use reciprocating machines, because the size of the required compressor is well below the maximum figure quoted above.

The one field in which the reciprocating type will undoubtedly continue unchallenged is that of extremely high pressures – the field of the hypercompressor. Design pressures here have increased rapidly during the past few years and have now reached 3,500 to 4,000 bars, but for industrial machines it is unlikely that there will be much increase, in the near future.

So far as pressure is concerned, a limit would appear to have been reached for centrifugal machines in the designs which have been developed for 400/450 bars, though no plants are yet operating in this range. Development in both reciprocating and centrifugal compressors in the past decade has been so rapid that a period of consolidation is desirable. This is perhaps most necessary in the case of the centrifugal compressor in view of the many problems which have arisen with the first high pressure machines. Certainly their performance so far has in many instances fallen short of expectations.

REFERENCES

11.1 TRAUPEL, W., *Thermische Turbomaschinen,* 2nd ed., Springer Verlag, 1968.
11.2 ECKERT, B. and SCHNELL, E., *Axial-und-Radial Kompressoren,* Springer Verlag, 1961.
11.3. STEPANOFF, A. J., *Turboblowers,* John Wiley, 1955.

12 Liquid Pumps and Intensifiers

12.1

Where liquids at high pressures are involved in the chemical and process industries the volumes to be handled are, in general, comparatively small and are therefore best suited to the reciprocating or plunger pump. Whilst in a few special cases the centrifugal pump has been able to supplant its rival at medium high pressures, its challenge is unlikely to be a serious one – unlike the challenge of the centrifugal machine in the compressor field. At the highest pressures only the reciprocating pump can provide the solution.

In recent years the diaphragm or membrane pump has been increasingly used at pressures up to 1,000 bars though only for very small volumes, for example 1 litre (0·001 m^3) per hour at the above pressure or 0·005 m^3/h at about 500 bars. This type is particularly suitable for handling corrosive liquids and for metering purposes.

12.2 Reciprocating Pumps

The principles discussed in Chapter 9 governing the design of the high pressure cylinders of the compressors apply with equal force to the design of high pressure pump bodies. Modern large pumps of this type for medium high pressures (300–450 bars) are usually horizontal with three throws, though occasionally pumps with six or more throws are to be found where an exceptionally uniform rate of flow in the discharge pipe is required. It is worth noting that, taking the ratio of maximum to mean velocity in this pipe as the criterion, a three-throw pump is better than a four-throw pump, a five-throw better than a six, and so on. There is indeed little advantage in more than five throws even where very steady flow on the discharge side is required.

Vertical designs are also produced by a few manufacturers. In one arrangement the gland is at the lower end of the cylinder with the risk of leakage from the gland passing into the crankcase unless the distance between the two is sufficiently great – that is, longer than the stroke – to allow an "umbrella" to be fitted on the plunger. A better arrangement from this point of view is that in which the gland is at the upper end of the cylinder so that the plunger moves downward in the latter; this however involves a system of tie rods and yokes for each individual plunger.

Because of the need to avoid cavitation, particularly on the suction stroke, mean plunger speeds are very much lower than the mean piston speeds of compressors. The range is so wide – from well below 1 m/s to about 2·0 m/s – that it is not possible to do more than generalise. The nature of the liquid being pumped also has a profound influence on the choice of speed. The expression for deciding whether or not cavitation will occur on the suction side of a pump involves – with other parameters – both the speed and diameter of the plunger, so that both cannot be independently chosen; much more careful consideration must be given to conditions on the suction side of a pump than is required for a compressor.

The valves may be accommodated in the pump body itself or in a separate head attached to the body by elastic studs or bolts. Each arrangement has its advantages and disadvantages, and in each case the valves may be "off-set" – side by side – or the delivery valve may be mounted directly above the suction valve. With the "off-set" arrangement each valve is independently accessible. This is usually not so with the alternative method in which the delivery valve must first be removed to give access to the suction valve, but in some small pumps the suction valve is inserted from the underside. There is no doubt that from the design point of view the separate valve head is better in spite of the need for a joint, subject to fluctuating pressure, between it and the cylinder. The cylinder itself, moreover, becomes a simpler forging especially if, as in some designs, the gland is also housed in a separate forging studded onto the body. In a three-throw pump the valve heads are sometimes combined into a single head which enables the suction and discharge manifolds to be incorporated in the forging itself, leaving only a single connection in each case for the corresponding pipes. Again the advantage of this arrangement must be weighed against the necessity for replacing the whole forging in the event of failure at any point. The most likely point for the initiation of a fatigue failure is the face between the valve seat and body or head where the intensity of pressure is necessarily high.

The type of valve depends on the size (capacity) of the pump, the nature of the liquid and the final pressure. In the very high pressure field, 2,000 to 4,000 bars, where the size of unit is small the choice is virtually limited to ball – sometimes double ball – and thimble type valves, but at medium high pressures disc and mushroom valves are usually found in large pumps. Fig. 12.1 shows a design of valve head with the off-set or side by side arrangement and flat disc valves. The method of holding in the valve seat is similar to that which is commonly adopted for the larger valves of compressors and because the cover studs are subjected to fluctuating load conditions the assembly requires the same careful "elastic" analysis as, say, the "big end" of a connecting rod in a compressor – or pump. In the latter the danger of using shims between the two halves of the rod of a different material and elasticity from the rod itself is well known, but it is not, perhaps, so well appreciated that the choice of material for the joints between the seat and body and between the cover and cage is equally important. Moreover, whilst it is now usual to specify the degree of pre-tightening for the bolts of the motion work of machines this is rarely done, as it should be, for the cover studs of assemblies such as the one in Fig. 12.1.

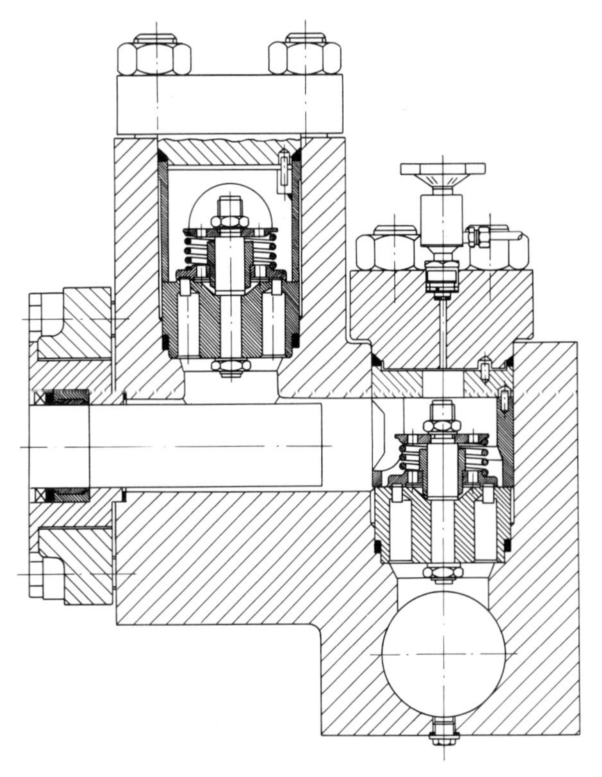

Figure 12.1 Valve head with side by side disc valves *(Pumpenfabrik Urach)*

A design of valve assembly, with a separate head suitable for a pump in the 2,000 bar range, is illustrated in Fig. 12.2. Of particular interest is the symmetry of the head and the breaking down of the assembly into separate components of simple form of which only the head itself is subject to fatigue loading.

At high pressures – and indeed sometimes at medium high pressures – the compressibility of the liquid cannot be ignored and the question of the clearance volume becomes important. If we denote the compressibility by C where

$$C = \frac{1}{V} \frac{\delta V}{\delta p}$$

in which δV is the change in volume of a volume V under a change of pressure δp at constant temperature, it can easily be seen from Fig. 12.3 that the volumetric efficiency η_v of the pump is given by:

$$\eta_v = 1 - \frac{c}{s} \; C \, p \tag{12.1}$$

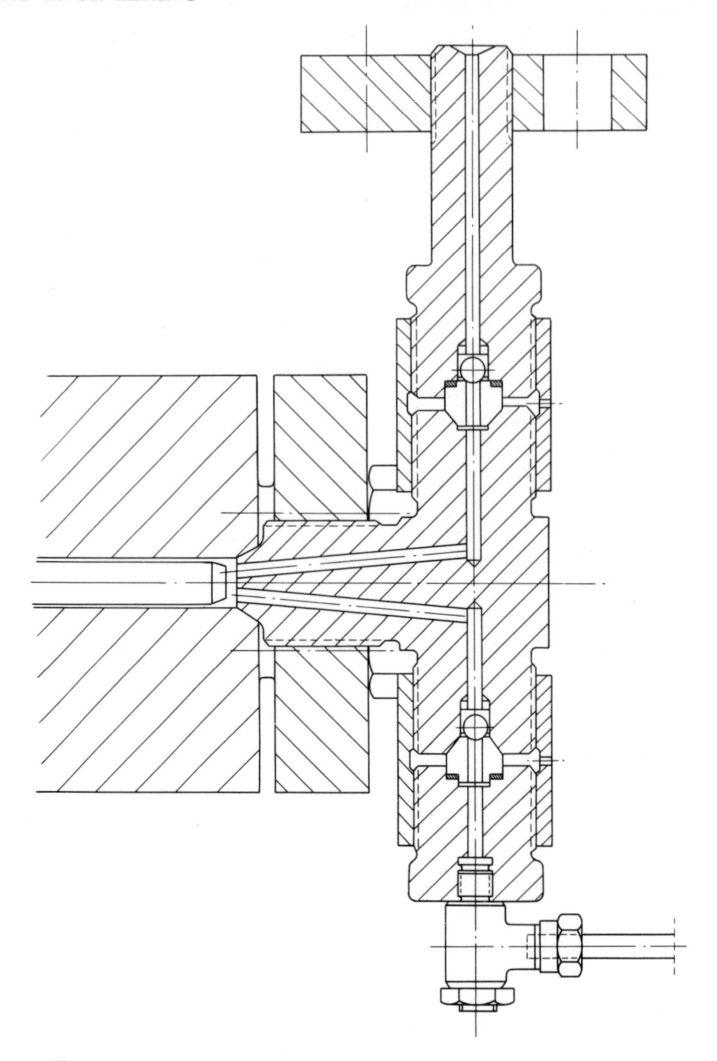

Figure 12.2 Valve head with ball valves: working pressure 2,450 bars *(Friedrich Uhde GmbH)*

where c and s denote respectively the clearance and stroke or displacement volume and p is the delivery pressure of the pump. C usually decreases as the pressure increases and in general a rise in temperature increases the compressibility though, according to Amagat, water has a minimum compressibility at about 50°C.

Eq. (12.1) shows the importance, especially for very high pressure pumps, of keeping the clearance volume to a minimum to avoid too great a reduction in volumetric efficiency. Further, the effect of the compressibility of the fluid can be seen from the fact that for water at 15°C the value of $C \times 10^6$ is 36·3 whilst, at the same temperature, it is approximately 103 for methyl alcohol and about 62 for paraffin oil.

Without question the major problem associated with high pressure reciprocating pumps is that of satisfactorily sealing the plunger. In this connection the

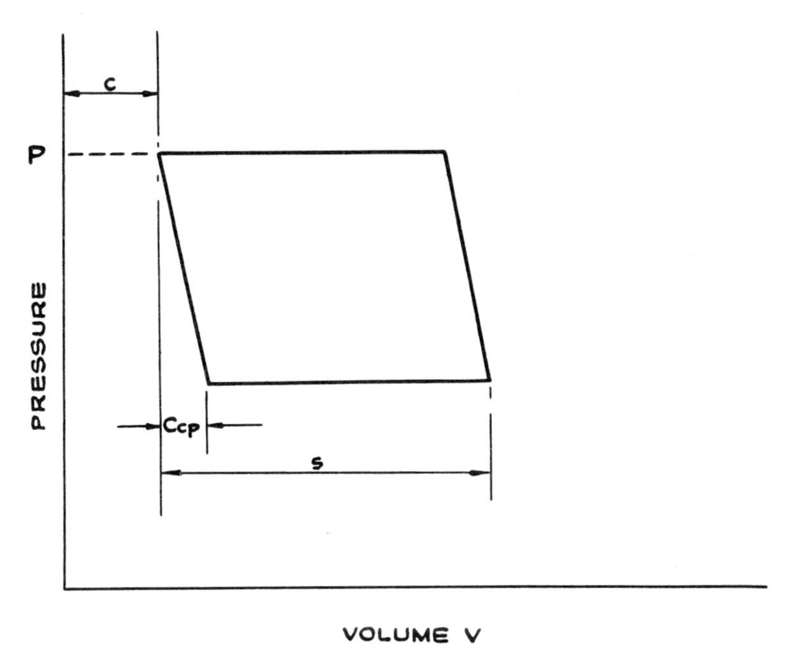

Figure 12.3 Theoretical *P–V* diagram for pump showing effect of compressibility

nature of the fluid – whether or not it is corrosive, has some lubricating property, is clean or dirty – is a vital factor in deciding not only the type of packing but the design of the gland itself. To avoid an excessive rate of wear of the plunger a liquid film must be maintained between it and the packing; if this cannot be provided by the fluid being handled separate lubrication must be arranged. But, no matter how good the design of the gland, its performance depends ultimately on the skill and care with which the individual rings are fitted. The old concept that the higher the pressure the greater the number of rings and the deeper the gland is no longer accepted as a universal criterion. Where, as in the pumping of aggressive or volatile liquids, a deep stuffing box is desirable the design of multiple gland shown in Fig. 12.4 is an excellent solution. The inner set of packing – to which there is good access for assembly – can be tightened up, and adjusted, independently of the outer set. There is also provision for lubrication; in other cases, as for example with the type of liquids mentioned above, a water-sealing circuit may be incorporated.

Although lubricated metallic packing is used to some extent in small high pressure plunger pumps, probably the type of packing most generally favoured in modern pumps is based on PTFE or other similar plastic material with low friction coefficient and excellent resistance to attack from most liquids. There is also a wide variety in the form of ring, from the normal square section to the chevron type and U shape with support elements in which an attempt is made to use the pressure of the liquid to deflect the limbs of the U to provide the seal. But no design of packing provides the universal answer to all problems and in practice a good deal of experimenting is often needed to establish the most suitable

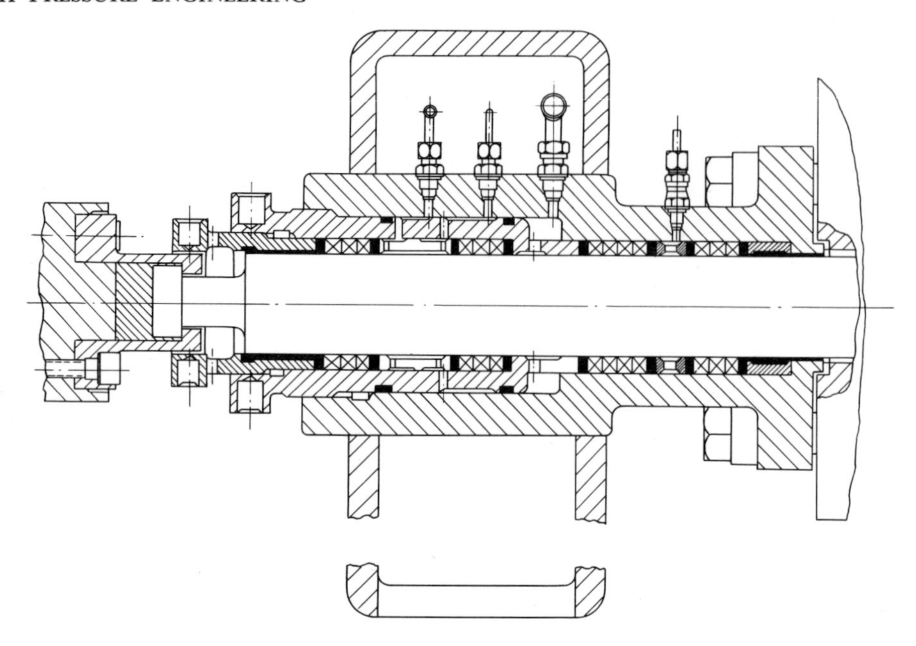

Figure 12.4 Sectional arrangement of 'multiple' gland *(Pumperfabrik Urach)*

form of packing and design of stuffing box. The problem is, of course, most acute in small pumps for very high pressures where the plunger diameter is, perhaps, of the order of 10–20 mm only and even the smallest leakage past the packing is not permissible.

12.3 Centrifugal Pumps for High Pressures

The efficiency of the reciprocating pump is high even at high pressures and significantly above the figure attainable in centrifugal pumps for the large outputs needed (as feed-water pumps) in modern power stations. Until comparatively recently this latter type of pump was considered quite unsuitable for the outputs and even the medium high pressures encountered in most chemical processes. Apart from the efficiency aspect one factor which operated as a deterrent against its use was the high speed at which it would have to run – speeds well outside the experience of chemical concerns. With the tendency to larger units operating without spare machines simplicity and reliability have become more important than efficiency. The situation in the pump field is, indeed, very similar to that which arose within the past decade in the high pressure compressor field and, in the authors' view, the future will see the centrifugal pump taking over many of the duties previously regarded as the exclusive province of the reciprocating pump, especially some of the more difficult pumping problems. Already work is proceeding in this direction but development is only just beginning. However, unlike the centrifugal compressor, in which for high pressures a whole series of new problems had to be solved for pumps there is already a wealth of design data and operating experience at pressures up to about 300 bars and the question is basically one of scaling down.

12.4 Diaphragm or Membrane Pumps

Basically the construction of the membrane pump is similar to that of the compressor of the same type though there are differences in certain details, for example in the valves, the disc type valve of the compressor being replaced by cone valves. The chief advantage of the diaphragm pump is the absence of any gland and accordingly complete freedom from leakage; but it suffers from the same uncertainty regarding life of the membrane as does the compressor. Nevertheless it fulfils a valuable function in handling—to medium high pressures—liquids which must be kept free from contamination of any kind or which are otherwise troublesome. The maximum delivery pressure for pumps—of the order of 350 bars—is much lower than that for compressors.

Hofer of Germany, Corblin of France and Pressure Products in the U.S.A., are three firms who have specialised in the high pressure diaphragm type of pump as well as compressor.

12.5 Small Intensifiers for Ultra-high Pressures

12.5.1 *General*

Pressures of at least 10 kb can now be generated safely and conveniently with relatively simple and inexpensive equipment. Up to about 7 kb (100,000 lbf/in^2) several manufacturers offer units driven by compressed air at normal mains pressure (say 7 bars or 100 lbf/in^2). One of the first of these was Charles Madan & Co. Ltd. of Broadheath, Cheshire, whose version, known as the "Airhydro-pump", is illustrated in Fig. 12.5. This has been fully described by Crossland and Alexander[12.1]. Several of the American companies dealing with high pressure equipment, such as the Pressure Products Industries Corporation, of Hatboro, Pa., Autoclave Engineers Inc. of Erie, Pa. and the American Instrument Co. of Silver Spring, Va., all make fairly similar types of intensifier, while the Harwood Engineering Co. of Walpole, Mass., offers a unit to work up to 200,000 lbf/in^2 (13.8 kb) see §12.5.5 (see below).

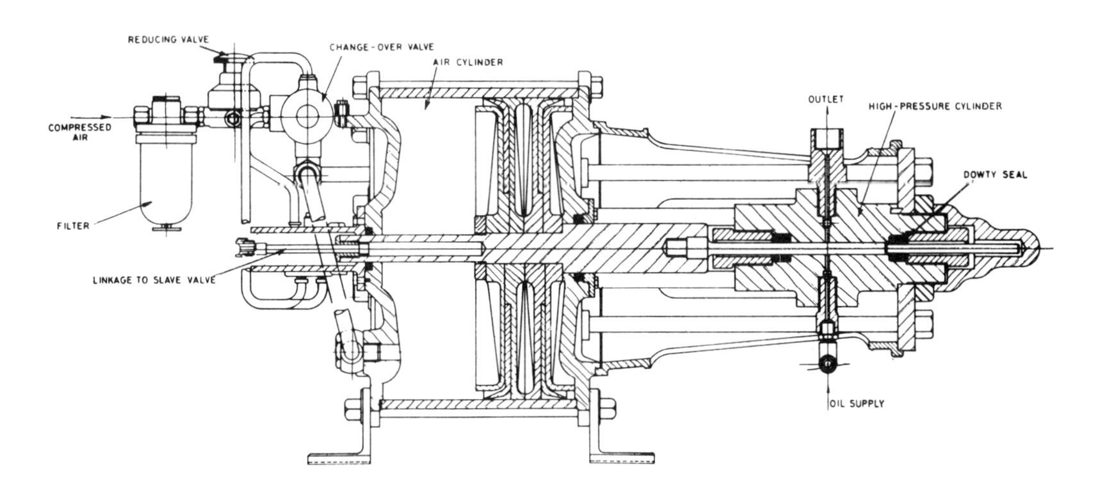

Figure 12.5 'Airhydro' pump *(Charles S. Madan & Co Ltd, Altrincham)*

Professor Crossland and his colleagues at the Queen's University of Belfast have studied the basic problems involved and have published several papers on the subject, see for instance Refs. (12.2) and (12.3). Their work has led to some interesting designs which will be discussed later.

12.5.2 *Basic Considerations*

At the higher end of the range of pressure considered here, i.e. for pressures greater than 15 kb, it becomes difficult to produce tubing strong enough; in fact, this can only be done at all by using material which is too hard to bend when cold. Units for these conditions are therefore usually connected direct to the object which is to contain the pressure, thus eliminating tubing altogether. Ordinary control valves are likewise unnecessary on the high pressure side, because the operation can be wholly controlled from the low pressure supply.

As usual with very high pressure work the principal problems are the sealing of the high pressure plunger and the prevention of fatigue in the cylinders and valves. So far as fatigue is concerned, Crossland[12.3] is doubtful whether indefinite repetitions of pressure in excess of about 4·5 kb (65,000 lbf/in^2) can ever be achieved without failure with existing materials. This is perhaps unduly pessimistic, but there can be little doubt that he is wise to work on the principle that the parts of his equipment most likely to fail in this way should be designed for easy replacement, and so that their failure cannot cause injury to personnel or damage to neighbouring equipment. He also recommends making the pressure-resisting cylinder as geometrically simple as possible, with no counterbores, screw threads, or other changes of shape which inevitably cause stress concentrations. In one design, as we shall see, he uses a simple cylinder with flat flush ends, and this appears to have had a greater fatigue endurance than he expected.

Several types of seal seem to give satisfactory service, although once again it is desirable that they should be easy to replace when working intensively at pressures of this magnitude. Most of the successful designs have stationary seals, although the Harwood 200,000 lbf/in^2 unit is fitted with a Bridgman type seal at the end of the high pressure plunger. Such a seal usually involves making a hole in the end of the plunger, which is thus subjected to extra stresses and becomes liable to crack in consequence. The Harwood Co. appear to have overcome this, however, by suitable choice of materials.

It is worth noting that a unit of this general type, designed and built by I.C.I. in 1933 for chemical research at pressures up to 12 kb, used a simple neoprene bung as the seal, which was thus similar to those used by Poulter[12.4]. This was made slightly oversize and pushed up the bore by the plunger which was ground flat; for a description see Ref. (12.5). The results were very satisfactory and the unit has had a long life, although it was usually necessary to renew the bung after each experiment at the full pressure. It is quite likely, however, that the use of chamfer rings would have avoided this.

Generally speaking, a stationary gland is to be preferred for various reasons; for instance the life is usually greater, the seals are easier to renew, and the stress cycle of the high pressure cylinder is less severe. Thus, if the seal is located at the outer end of the high pressure cylinder, the pressure will be uniform throughout its length, whereas there will be a very big pressure drop across a moving seal and consequently a very complex and mobile stress pattern in the wall around its position.

Crossland has preferred either C.S.N.† rings or the "Hypak G" material made by James Walker & Co. Ltd. of Woking, Surrey. Madan's use the ingenious Dowty seal.[12.6] For the still higher ranges of pressure (up to 25 kb) Thomas, Turner and Wall[12.7] report the successful use of neoprene O-rings with chamfer rings of beryllium copper. It would appear that the success of this seal depends very much on these rings, which prevent the actual sealing material from being extruded into the clearance between the plunger and the cylinder wall (see Fig. 13.7c). These authors have carried out a series of more than 250 runs at over 20 kb with the same seal assembly and final breakdown was due apparently to the breaking up of the beryllium copper.

Care must be taken to see that the high pressure plunger does not fail by instability as a strut. This is more likely with a moving seal since the plunger then carries the maximum load when its whole length is unsupported. Thus if p_{max} be the maximum pressure, d the diameter of the plunger and L its length, the crippling load will evidently be:

$$p_{max} \times \frac{\pi}{4} d^2$$

and, according to the Euler theory, if there were no constraints at the ends this would be given by:

$$p_{max} \times \frac{\pi}{4} d^2 = \frac{\pi^2}{L^2} EI \qquad (12.1)$$

where E is Young's Modulus and I the second moment of area of the plunger. In practice there will be some restraint which can most easily be allowed for by means of a constant factor (let it be μ) on the right-hand side of the equation. Then, on substitution for I in terms of d, we have:

$$L = \frac{\pi d}{4} \sqrt{\left(\frac{\mu E}{p_{max}}\right)} \qquad (12.2)$$

The value of μ is a matter of some conjecture, but if one regards the end behind the moving seal as free (it is usual to reduce the diameter to a moderate clearance except for a short distance immediately behind the seal) and the other end where it is joined to the low pressure piston as more or less fixed, one would expect a value of about 2 for μ. In one case the I.C.I. 12 kb apparatus already mentioned was overloaded owing to faulty operation, and the high pressure plunger did fail in this way. Subsequent investigations indicated that the factor then was probably about 2·6. Thus a value of 2 for μ is likely to be on the safe side.

This can be a serious limitation and, as can be seen from eq. (12.2), no amount of extra strength in the material will help since its modulus will not change so long as we keep to ferrous materials. Fortunately, for very small plungers, tungsten carbide has the valuable property that its modulus, usually about

† Made by Trist Mouldings & Seals Ltd., Bath Road, Slough, Bucks.

6×10^6 bars, is about 3 times that of steel and would therefore allow the safe length to be increased by some 70%. The successful use of tungsten carbide is however limited by the difficulty of getting really sound and uniform material, which in turn is a function of size. However, plungers up to 2 cm in diameter and to, say, 20 cm in length should be obtainable in the necessary quality.

The pressure-transmitting fluid can prove quite a difficult problem at pressures of this order, since many of the liquids normally used solidify and will not flow through the ducts at room temperature. Crossland and Austin[12.2] recommend the use of brake fluids, most of which are mixtures of neutralised castor oil and methyl alcohol. Petroleum ether has also been used and certainly has a high freezing point; it would, however, be very dangerous if ejected as a spray through the failure of any part. Argon gas has also been used, but it has no lubricating action and it too becomes solid under extreme conditions, although there is as yet little published data on this phenomenon.

The friction will be affected by the lubricating properties of the fluid, but this is seldom large and various authors (e.g. Refs. 12.3 and 12.7) report values varying between 3% and 6% of the total force exerted by the low pressure fluid.

For most cases of work at pressures not exceeding 15 kb the brake fluid type of liquid is quite satisfactory. Its viscosity will be increased by the pressure, and time may have to be allowed for the equalisation of pressure along a small bore pipe, but this is not usually a serious matter, and Crossland and his colleagues report successful experiments with the liquid flowing through $\frac{1}{16}$ in bore tubes, even at 10 kb.

12.5.3 *The Airhydropump*

For satisfactory operation from compressed air mains this requires an intensification factor, i.e. ratio of areas of high and low pressure pistons of more than 1 to 1,000. Since, for convenience, the latter is kept down to 14 in, it follows that for a simple solid plunger the diameter could not exceed about 0·4 in, which would be inconveniently flexible. A differential plunger is therefore used (see Fig. 12.5) with diameters of approximately 1 in and $\frac{15}{16}$ in. The seals are fixed and of the Dowty pin type (see Section 13.3), which has proved very satisfactory in maintaining the seal and in its low frictional loss. The need with differential pistons for two seals instead of one is a disadvantage, but in this case it has not proved serious. Another objection is the need for radial ports which introduce severe stress raisers, but again with care taken to avoid sharp corners and undercutting, and with appropriate selection and preparation of the material, long and satisfactory working lives are regularly being obtained.

The operation of the valves to maintain continuity is obtained by means of a rod passing through a gland at the back of the low pressure cylinder. In this way, when the plungers reach the end of their working stroke they are drawn back and the supply of working fluid then refills the high pressure space for the next stroke, which is automatically begun as soon as the withdrawal is complete. For details of the necessary hydraulic circuit see Ref. 12.1.

12.5.4 *The Belfast Intensifiers*

As we have noted, Crossland and his colleagues at the Queen's University of Belfast have developed a simple and effective type of intensifier for operation at pressures up to 16 kb. In this case, as with the Harwood Engineering Co's. design, the intensification factor is much lower, and it is presumed that a low pressure supply of liquid at 500 to 600 bars is available. For experimental work at Belfast they have used a standard "Black Hawk" hand pump, and, for continuous operation, an automatically controlled Towler pump.

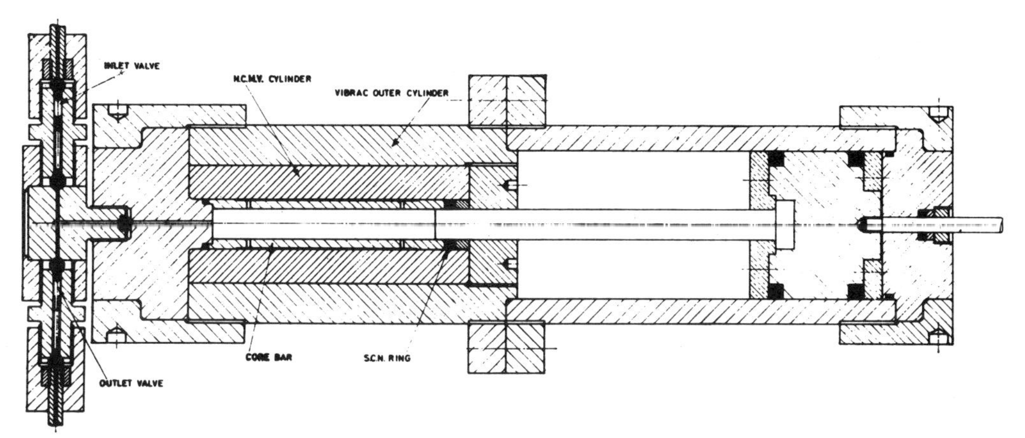

Figure 12.6 14 kb intensifier – Queen's University, Belfast

The first successful design, after minor modifications, is illustrated by a sectional plan in Fig. 12.6. The main cylinder is, as will be seen, completely free from any stress-raising effects, and it is subjected only to the radial action of the pressure, the lengthwise forces being taken by the outer mantle in which it is merely a push fit. The cylinder has a diameter ratio of 2·5 and is made from the English Steel Corporation's NCMV steel, which is capable of appreciable ductility even when hardened to a U.T.S. in excess of 15,000 bars. In Ref. 12.2 it is stated that the cylinders have "a shear yield stress in excess of 50 tons force per sq. inch". Presumably it was in fact considerably more than that figure, i.e. in excess of 7,700 bars, since the pressure to cause the whole wall of a 2·5 to 1 ratio cylinder to overstrain would be about 14·5 kb if its limiting shear stress had that value, and it is reported that one of these units has been working at or near that pressure for a considerable number of strokes†.

The seal, which is of the C.S.N. type with supporting rings of bronze in this case, is as shown in Fig. 12.7. As will be seen from Fig. 12.6, this is easily replaceable, as also is the main cylinder, since by separating the unit at the middle flange, and by removing the high pressure end plug, the whole of the high pressure assembly can be taken apart.

The outer mantle, which is of En 25 steel with a U.T.S. of around 8,000 bars,

† Intensifiers of this type and a considerable range of ancillary equipment are now manufactured and marketed by the Coleraine Instrument Co. Ltd., of 82 Killowen Street, Coleraine, Northern Ireland.

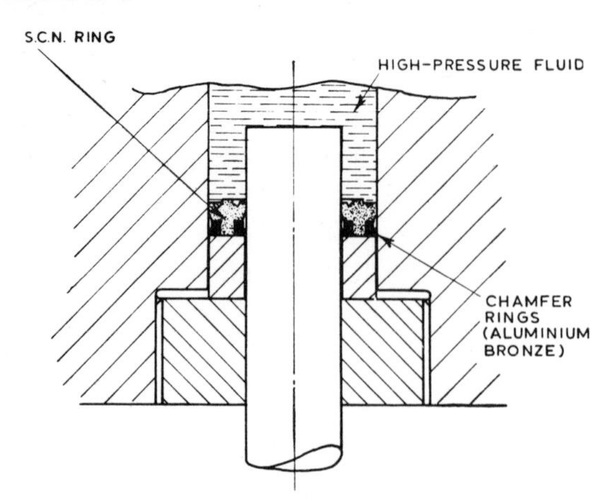

Figure 12.7 Seal used by Crossland & Austin (Ref 12.2)

carries the longitudinal forces and will also give some support to the main cylinder, but this will not be large because the action only begins when the outside of the cylinder swells under the pressure. On the other hand, if the cylinder fails – and material of the hardness used here may fail catastrophically – the mantle prevents any fragments from getting loose.

The T-piece at the high pressure end is also a very highly stressed component which cannot be made without some concern as to its life, especially if the pressure fluctuates. In consequence, it has a sleeve round it to take the load of the plugs holding the connections, but at the highest pressures – and units of this type have been operated at 17 kb – there is some danger of its fragmenting, and, as seen from Fig. 12.6, it cannot be fully protected by the rest of the assembly; for this reason these units are generally placed behind suitable barricades for the highest pressure working.

Experience with intensifiers of the above type shows that the T-piece and valves are the weakest parts and various alternative designs have been studied. For instance, if some way of refilling the high pressure cylinder during the withdrawal stroke could be found, the delivery valve could be very conveniently located in the part of the end plug which projects into the cylinder, thus providing external pressure and keeping the unknown stresses in its neighbourhood mainly compressive. The solution to this is to use a hollow piston and plunger system with a non-return valve, the "suction" valve in fact, at the high pressure end of the plunger. This is perfectly workable provided the high pressure fluid can be used also to provide the low pressure. The arrangement is then as shown diagrammatically in Fig. 12.8. Thus when the piston and plunger assembly is moving to the left the suction valve will prevent the fluid passing backwards through the plunger, and this fluid will be compressed until its pressure reaches that in the delivery system, whereupon the delivery valve will open and the charge will be delivered through it until the plunger reaches its inner dead centre. The delivery valve will then close and, by transferring the low pressure feed to the other side of the low pressure piston, the latter will return and the high pressure space will be refilled with fluid through the duct in the plunger and through the suction valve.

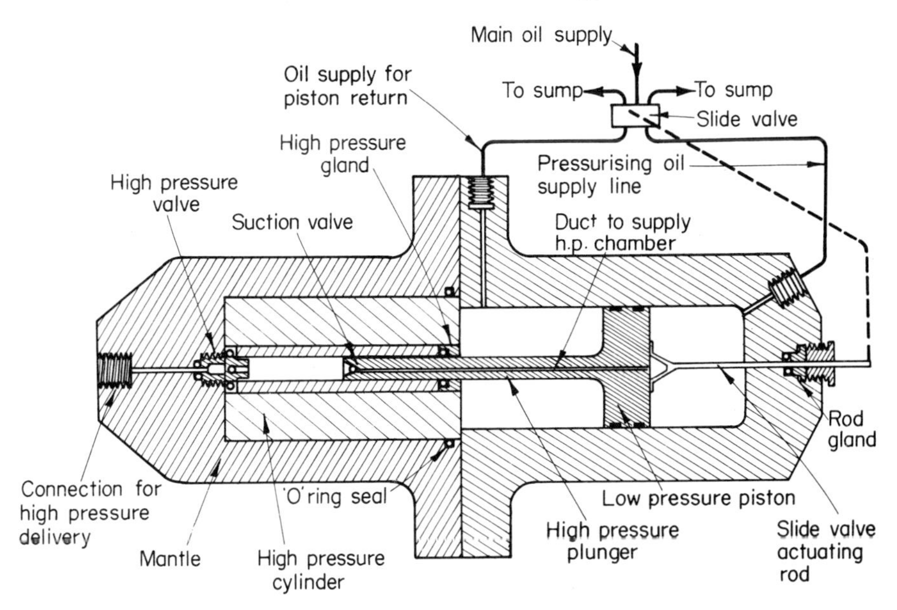

Figure 12.8 Proposed modification to Queen's University, Belfast, intensifier

If, however, it is necessary to use a separate low pressure liquid, the arrangement becomes complicated by the need to have a separate chamber for introducing the high pressure fluid. An ingenious solution of this has been suggested and is on trial at Belfast, as shown in Fig. 12.9, in which a differential plunger is used, but both parts of this are separated and each is held stationary while the high pressure cylinder (which is also the low pressure piston) reciprocates over them.

It would seem that this arrangement could perhaps be improved in detail, for instance by moving the valves to the open ends of the plungers where they would have the supporting effect of the pressure outside them. It would also appear feasible to retain the geometrical advantages of a right circular cylinder by shrinking inside it a component to carry the seals, and assembling the whole in a canister with inserts of lead bronze or some low friction material; and it is clear that there is plenty of room for ingenious designers to improve the embodiment of these ideas still further (see also an I.C.I. design illustrated in Brit. Pat. No. 736,664).

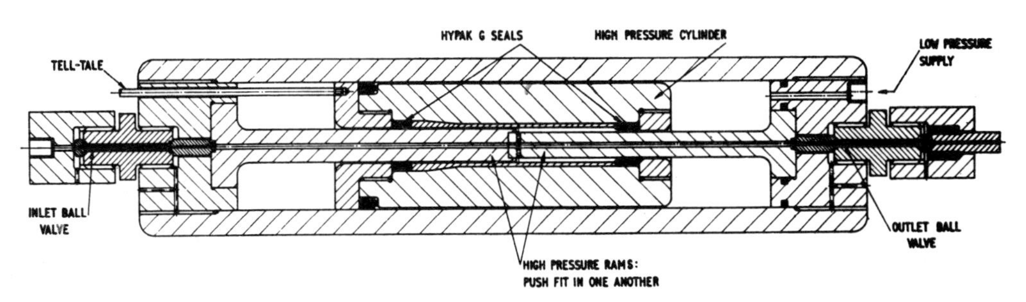

Figure 12.9 Proposed differential piston intensifier, Queen's University, Belfast (Ref 12.3)

12.5.5 *The Harwood Intensifier*

This is probably the only fully tried 15 kb (200,000 lbf/in^2) unit available commercially at present. It is a development of a double-acting injector for introducing catalyst into the reaction space of high pressure polyethylene plants. Fig. 12.10 is a section through this unit and, as can be seen, the low pressure cylinder (which is normally operated by pressurised water in the catalyst injectors) is provided with parts at each end for double acting if a second high pressure cylinder is connected at the left-hand side.

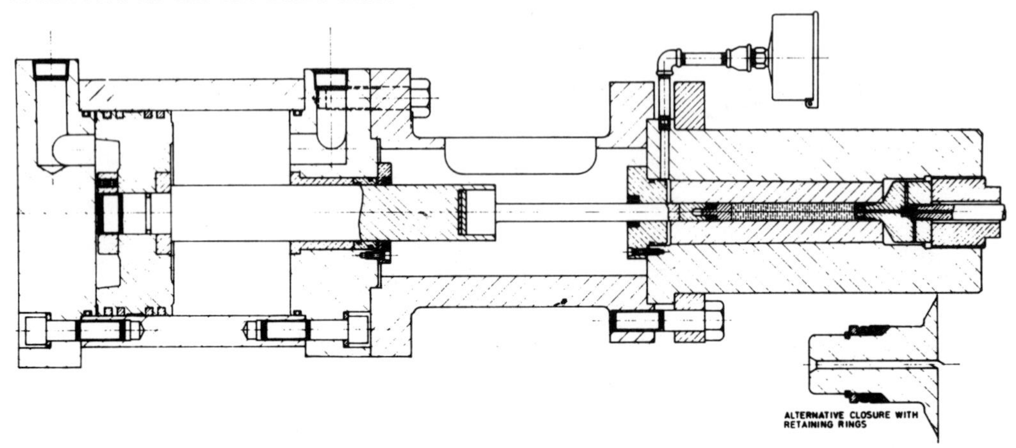

Figure 12.10 Intensifier for 14 kb (200,000 lbf/in^2) by the Harwood Engineering Co. of Walpole, Mass, USA (reproduced from Ref 12.2)

The cylinder is of duplex shrunk construction, but there seems some question as to whether even this will be fully clear of the risk of eventual fatigue failure. On the other hand, we understand that these intensifiers have given a good acount of themselves in service. The rest of the design appears to be along orthodox lines, and—as in the Belfast units—attention has evidently been given to ease of dismantling for inspection, or for replacing worn or damaged parts.

REFERENCES

12.1. CROSSLAND, B. and ALEXANDER, R. L., *Engineering,* **189,** 462, 1960.

12.2. CROSSLAND, B. and AUSTIN, B. A., *Proc. Inst. Mech. Eng.,* (Conference Report). **180,** Pt 3A, 118, 1965–6.

12.3. CROSSLAND, B., SKELTON, W. J. and WILSON, W. R. D., *Proc. Inst. Mech. Eng.,* **182,** Pt 3C, 175, 1967–8.

12.4. POULTER, T. C., *Phys. Rev.,* **40,** 860, 1932.

12.5. *Engineering,* **141,** 192, 1936.

12.6. BINGHAM, A. E., *Proc. Inst. Mech. Eng.,* **169,** 881, 1955.

12.7. THOMAS, S. L. S., TURNER, H. S. and WALL, W. F., *Proc. Inst. Mech. Eng.,* **182,** Pt 3C, 271, 1967–8.

PART III

FITTINGS, MEASUREMENTS, PRESSURE LIMITS AND SAFETY

Notation used in Part III

A	Area of pipe or valve port, etc.
b	Width of contact band in joint
D	Diameter
d	Depth of penetration (of flying fragments)
E	Young's Modulus
F	Force due to friction
g	Acceleration due to gravity
h	Pressure head (of mercury)
K	Diameter (or radius) ratio
k	Constant
L	Length of vessel or pipe
n	Constant (index pressure-volume relation)
P	Reaction in vent pipe during blow-off
p	Pressure
r	Radius
U	Strain energy
u	Radial shift due to strain
V	Volume
v	Velocity
W	Dead weight
α	Constant relating to diameter of dead weight gauge plunger
β	Intensification factor
γ	Ratio of specific heats of gas
ε	Strain
ν	Poisson's Ratio
ρ	Density
σ	Direct stress
τ	Shear stress

Suffixes

1 and 2	Numbers to identify different items
i and o	Inside and outside – applied to pressures, radii, etc. in containers, etc.
o	Also used to denote initial (unstrained) dimension
a and r	Denoting advancing or returning movement in plungers
L and H	Low pressure and high pressure – as applied to the pistons and cylinders of intensifiers, etc.
s	Socket – as applied to the diameter of joint sockets
f	Fragment – applied to safety considerations when containers fail in brittle manner giving rise to fast flying fragments
θ, z and r	Tangential, axial and radial directions (in cylinders, etc.)

13 Joints, Valves and Fittings

13.1 General Considerations

In this chapter we shall endeavour to state the basic principles underlying the design and construction of such parts as joints, valves and fittings. For most of them it has been established that success can be achieved by a number of different approaches, and—particularly in the case of static joints—the choice is now very wide. The size of the part naturally influences the way the principles are applied, but it does not affect the principles themselves.

Moving seals, i.e. at positions where one part moves relative to the other while at the same time the working substance is prevented from leaking between them, are much more troublesome, and problems of this kind cannot always be solved with entire satisfaction. We have considered the sealing of reciprocating parts such as plungers and piston rods in Part II of this book, but the allied problem of preventing leakage past a rotating shaft, e.g. for stirring the contents of a pressure vessel, will be considered here, and also certain methods of dodging the problem in small-scale applications.

Valves for plant control are not usually difficult to design or to operate unless the conditions are particularly severe, but their success in service depends—as do most high pressure components—very much on the way they are looked after. On the whole, stop valves are a good deal easier to deal with than flow control valves where the throttling action is apt to cause erosive wear, even when the working fluid is almost completely free from abrasive particles.

This is perhaps the most convenient chapter in which to mention also a few of the design problems which are common to many applications in high pressure engineering, for instance screw threads and other connecting devices.

13.2 Joints

13.2.1 *General*

It is convenient to divide these into two parts which should be considered separately. First there are the seals, which actually prevent the leakage, and then the couplings, which resist the tendency of the pressure forces to push apart the components to be joined. In most instances too the type of seal will influence the forces to be resisted by the coupling and hence the design of the latter, and we must consider each separately as well as their combined effect.

The number of different joints in successful use today for dealing with high pressures is very large, and here we can only look at a few. For this reason we describe a few typical joints, from which it should be possible to select one that will satisfy the conditions of most jointing problems, up to say 10 kb.

Various methods for designing both joints and seals have been proposed from time to time, but there seems to be, as yet, no universally accepted basis. Experience suggests as a good working rule that the contact pressure between *properly mated* joint faces should never be less than the pressure to be contained. The words italicised here mean that the surfaces must either be flat enough to resist seepage between them, or that they must have been indented – one against the other – so that no micro-channels are left (for instance between neighbouring high spots) through which the leakage can occur. It has been estimated according to Gough[13.1] that the width of such a channel must not exceed about 3×10^{-7} in even to hold water at room temperature and obviously for light gases even closer mating becomes necessary.† Clearly, surfaces of these orders of smoothness cannot be produced by ordinary workshop techniques, and the usual method is either to use a soft non-metallic gasket, or to make the components all of metal – preferably the mating surfaces should be of different hardness – and then to force them together so hard initially that the parts indent one another and thus close up the leakage paths. The success of this must naturally depend on having a properly designed coupling to ensure that this preliminary loading can be reliably carried out.

A further problem that must not be overlooked is the tendency for the internal pressure in the system to cause elastic distortion of the coupling; this unloads the seal, and to overcome the trouble various forms of "self-tightening" joint have been devised. These have the property that the contact pressures on the seal are automatically raised as the internal pressure is raised and therefore continue to prevent leakage until something breaks or becomes grossly distorted.

A few typical high pressure seals will now be considered in more detail.

13.2.2 *Lens Rings and Cone Rings*

The lens ring joint appears to have originated with the Haber process for ammonia synthesis in Germany during the 1914–1918 War. It gets its name from the fact that its flanks have curved surfaces which are usually sections of spheres. The ring can be made either harder or softer than the surfaces it is going to mate with, and the latter can be either flat or recessed with conical hollows. Fig. 13.1 illustrates this joint in its simplest form and shows an example with a flanged and bolted coupling. In this case it is presumed that the ring is hard and the flat seatings relatively soft. The curved surface mating against a flat one must initially have a ring of line contact located almost at the bore, but as the coupling is pulled up this is appreciably widened by the indenting effect of the ring.

The introduction of the working pressure will to some extent reduce the

† This reference also discusses the theoretical aspects of very narrow leakage paths.

preloading due to the initial tightening, the effect depending on the rigidity of the coupling device. Generally speaking this is not a serious matter and a joint which is properly designed in the first place should give very little trouble. If a leak develops after some time in service, it can usually be stopped by letting off the pressure and tightening up the bolts. The practice of slogging up a slightly leaking joint without blowing down the pressure is certainly dangerous and should always be discouraged, even if it has on occasions succeeded in stopping the leak.

The ring must be designed to withstand the radial pressure and, although it receives considerable support from the friction of the abutting surfaces of contact, it is probably safest to make it strong enough to resist the working pressure as a cylinder in its own right. Fig. 13.1 shows (with some exaggeration) an annulus of width *b* over which the indentation has taken place, and thus the force which the coupling has to exert must not be less than:

$$p_i \times \frac{\pi}{4}(D_i + 2b)^2$$

where p_i is the pressure and D_i the diameter of the bore. In practice this contact pressure is usually exceeded by a factor of two or three, but if the joint is working at high temperature, creep in the bolts and flanges may cause a gradual reduction and eventual leakage. It is evident therefore that the coupling must play a considerable part in the successful operation of these joints.

Lens ring joints are cheap to make and require little skill in use in the smaller sizes, e.g. up to about 6 in diameter and to, say, 1,000 bars. Even at lower pressures, however, the couplings have to be heavy, and considerable physical effort is usually required to prevent leakage with the larger sizes.

Cone rings differ from lens rings only in having their flanks straight instead of being slightly curved. They are thus a little easier and cheaper to produce in large quantities and quite as effective as lens rings, except perhaps in places where frequent opening and closing is required, when the curved flanks of the lens ring seem to last rather longer.

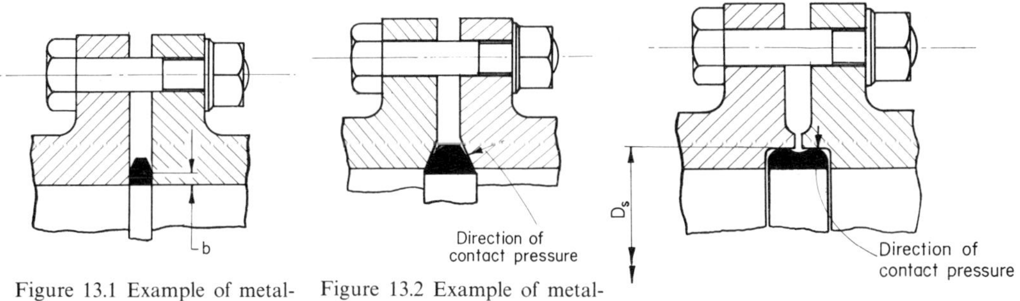

Figure 13.1 Example of metal-to-metal joint

Figure 13.2 Example of metal-to-metal joint

Figure 13.3 Example of metal-to-metal joint

13.2.3 *The Wave Ring Joint*

The I.C.I. engineers at Billingham introduced an important development into the lens ring joint during the inter-war period, when they turned conical recesses in the seatings and splayed out the ring somewhat as is shown diagrammatically in Fig. 13.2. The essence of this is that the pressure tends to expand the ring to a greater extent than the combined effect of the expansion of the seatings and of their being forced apart; and so the ring gets wedged tighter into the recess thus formed for it, and a considerable degree of self-tightening is introduced. This proved very successful at pressures up to 500 bars, but a fair amount of axial load was still necessary, requiring heavy couplings and both strength and skill in assembly for satisfactory use. When I.C.I. turned to the study of chemistry at 2,000 bars and upwards, it was soon found that the available joints for diameters of more than 2 or 3 in were not sufficiently reliable. This led to some further study of the problem, as the result of which the Billingham development of the lens ring described above was extended to its limit, i.e. the conical recesses were made into parallel sockets, and the ring was opened out until it seated by pressing radially outwards against these sockets. Fig. 13.3 shows such an arrangement.

Here we see that the sealing pressure has no component in the axial direction at all and the only force that the coupling has to carry is that due to the working pressure acting on the area of the sockets, i.e.

$$p_i \times \frac{\pi}{4} D_s^2$$

where D_s is the socket diameter as shown in Fig. 13.3. Thus not only is the axial force smaller than in most other types of joint; it is also known exactly. Consequently we can design the coupling without having to make allowances for unknown extra loads. The initial contact pressure has to be supplied by making the ring a slight interference fit in the sockets, but an allowance of 0·002 in/in is usually sufficient for this.

The problem of surface roughness is less serious here because the operation of inserting the ring against its interference usually smoothes out the irregularities in the socket surfaces, which are anyway mostly in the circumferential direction, whereas the most likely leakage path is in the axial direction. Wave ring joints, by reason of this interference, are very useful in instrument work because of the almost perfect alignment which is automatically effected by the interference fit. Another useful quality is that a slight yielding in the coupling—as might be caused by high temperature creep, for instance—does not at once involve leakage. Some tests on this are described in Ref. 13.2.

Against this must be set the extra cost of the fine tolerances on both the ring and the sockets which this joint requires, as well as the risk of damage if any dirt or abrasive matter such as rust gets into the socket as the ring is forced in; this may necessitate boring out the socket, and in large vessels this is an awkward and costly operation. Nevertheless, with experience and with the choice of suitable materials, these joints give excellent service, as well as permitting some convenient features to be introduced into the couplings.

A rather similar joint, commonly used in large vessels, is shown in Fig. 13.4 and, as will be seen, it is a very widely splayed lens or cone ring which thus acquires some of the valuable properties of the wave ring without requiring quite such exact workmanship. The use of shim metal between the mating sur-

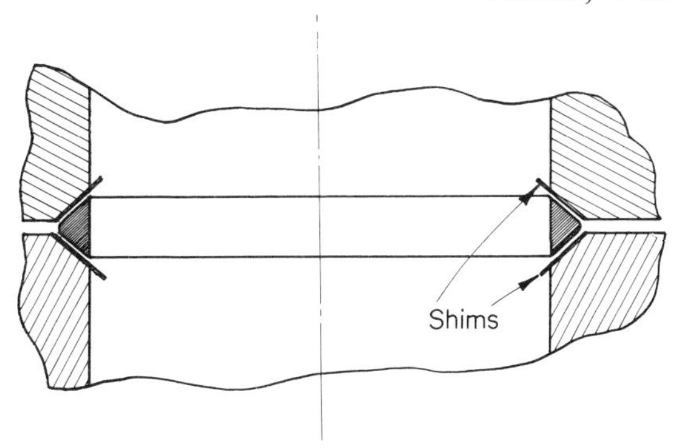

Figure 13.4 Example of metal-to-metal joint

faces also tends to protect the vital areas from damage through scoring or fretting. This is usually known as the delta ring joint.

13.2.4 *The Bridgman Unsupported Area Seal*

This was devised by the late Professor P. W. Bridgman and used extensively in his classical work on the effects of ultra-high pressure on physical phenomena, but it has since been much more widely used, in some instances on a large industrial scale. The essential feature of it is shown in Fig. 13.5, the force on the

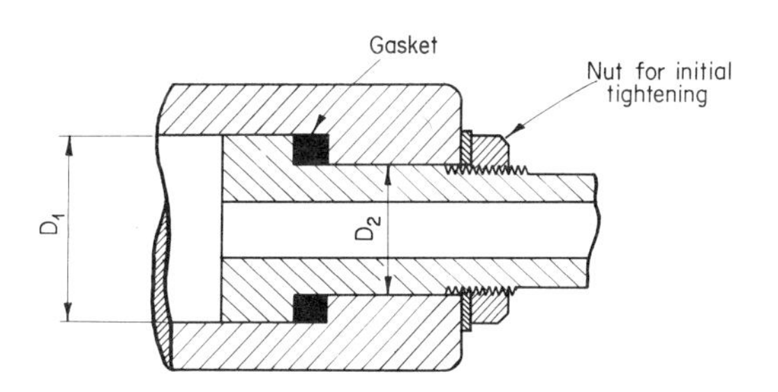

Figure 13.5 Bridgman's 'unsupported area' seals

plug being clearly equal to $\pi D_1^2/4$, and this is supported by the shoulder, the back of which rests on the gasket whose area is $\pi(D_1^2 - D_2^2)/4$. Hence the pressure on the gasket must have an average value of $p_i D_1^2/(D_1^2 - D_2^2)$, but if the material of the gasket is relatively soft it will transmit this pressure laterally as well as in the axial direction, and this will result in all the leakage paths being closed by contact pressures substantially greater than the working pressure they have to contain.

Fig. 13.6 shows how this arrangement can be applied to a moving ram, provided the motion is slow. This can be extremely effective although its life is usually short owing to the wearing away of the plastic packing.

For closing vessels this joint is very convenient, especially as it lends itself to quick-acting couplings involving such devices as the bayonet joint or the breech block mechanism. The most likely cause of trouble, particularly in large diameter applications where the factor $D_1^2/(D_1^2 - D_2^2)$ is likely to be rather high,

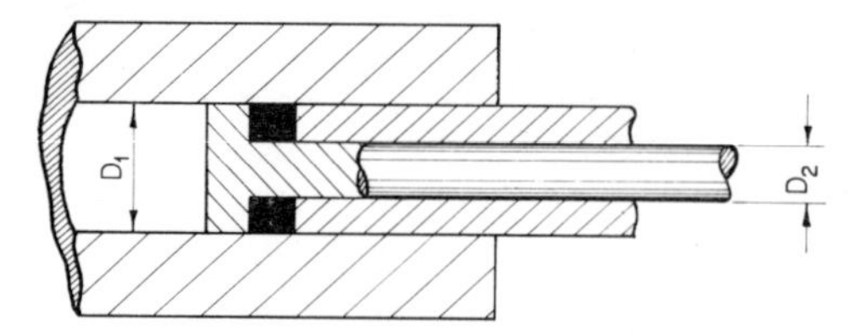

Figure 13.6 Bridgman's 'unsupported area' seal

is that the gasket pressure may be enough to crush the wall against which it is seating. The result is that it indents a band right round the periphery and locks itself in; usually the only remedy then is to machine away the plug.

In spite of occasional difficulties of this sort the Bridgman seal is very widely used, in all sizes from the very small to diameters of several feet, and in pressure ranges up to 25 kb. It can even be used at quite high temperatures (e.g. 350°C) if ductile metals like silver are used as the gasket material.

13.2.5 *O-Ring Seals*

For stationary seals at room temperature—or even up to about 250°C—the O-ring is remarkably handy and effective. Two applications of it are shown in Figs. 13.7a and 13.7b. As will be seen, it requires a ring of elastic material, usually of circular section, to be trapped in an annular hollow in such a way that it presses against the walls of the hollow on all sides except one and in this direction it is exposed to the pressure. It is very easy to make (in its simplest form) and extremely effective up to considerable pressures; the limit depends on a good many factors,

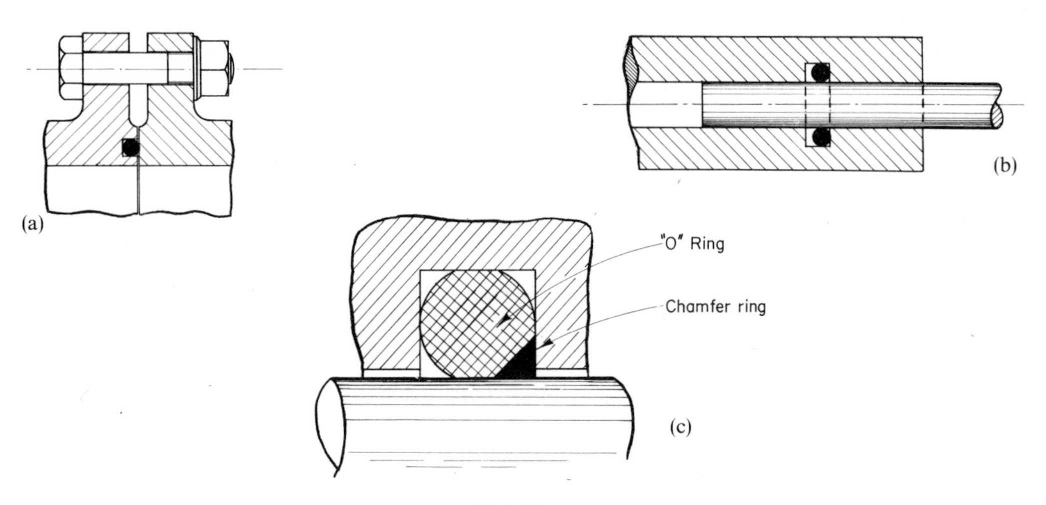

Figure 13.7 'O'-ring seals

including the size, the ring material, and whether the pressure is steady or frequently varying, but at room temperature one can easily seal the end of a 3 in diameter cylinder with an arrangement such as that shown diagrammatically in Fig. 13.7b against 3,000 bars.

For higher temperatures various materials show good prospects, particularly glass-fibre-filled nylon or "Fluon" (PTFE), but the life of the ring can be increased quite markedly by inserting small chamfer rings, such as those shown in the enlarged section, Fig. 13.7c. Beryllium-copper is very effective in this application when bedding against steel surfaces, and in certain applications it is possible to seal for reasonably economic periods of time, even when the internal member is reciprocated slowly. Some interesting experiences with seals of this kind up to 25 kb are described by Thomas et al.[13.3]

The use of O-rings has increased very rapidly in recent years, and analogous developments are now under test, such as the use of gas-filled metal rings, or rings which are open at one side like a capital C, and these may have a profound effect on high pressure technique especially in the range 100 to 1,000 bars.† An interesting combination of the O-ring with the wave ring principle is described by Crossland and Austin[13.4]; in this a pair of O-rings is recessed into the main ring, thus replacing the wave crests.

13.2.6 *Couplings*

These can be divided into two main classes; those which have to be tightened to pre-load the seal, and those for seals like the wave ring which require no initial axial force. There are also certain intermediate types like the Vickers-Anderson coupling which apply a small amount of initial load, but are mainly designed to resist the pressure load.

Of the former type by far the commonest is the bolted flanged coupling, which is a reasonably practical proposition up to 1,000 bars on a 6 in diameter and to about 4 ft at 300 bars. However, the flanges are then very heavy and the numbers and sizes of the bolts are large. Much work has been done on the standardisation of flanges and we shall not therefore discuss this system further here. See consideration by one of the authors (S.L.)[13.16].

A convenient form of connection for smaller sizes of tube or closures for laboratory vessels, etc. is the ordinary union nut, but this becomes awkward to use when the pressure force alone amounts to 5 tons or more, unless one has a wave ring seal or something with similar properties. It can, however, be made quite satisfactory for this by providing a ring of set screws for applying the final nip on the seal. Fig. 13.8 shows such an arrangement.

Fig. 13.9 shows the Vickers-Anderson coupling, from which it will be seen that it consists of a collar split into 3 pieces which are held together by bolts in a tangential direction through holes near the outer surfaces. The components to be joined are provided with shoulders with conical surfaces and the split collar also has hollow cones of the same angle. Thus, as the pieces of the collar are drawn together by the bolts, their conical inside surfaces produce an axial thrust on the shoulders of the main components. In this diagram, incidentally, the coupling is shown in conjunction with a hollow lens ring.

† The Wills Pressure Filled Joint Co. Ltd. of Bridgwater, Somerset, are active in this field.

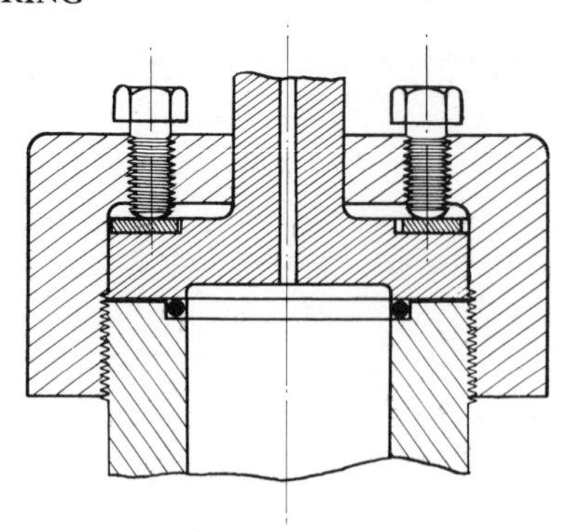

Figure 13.8 Combined union nut and compression screw head

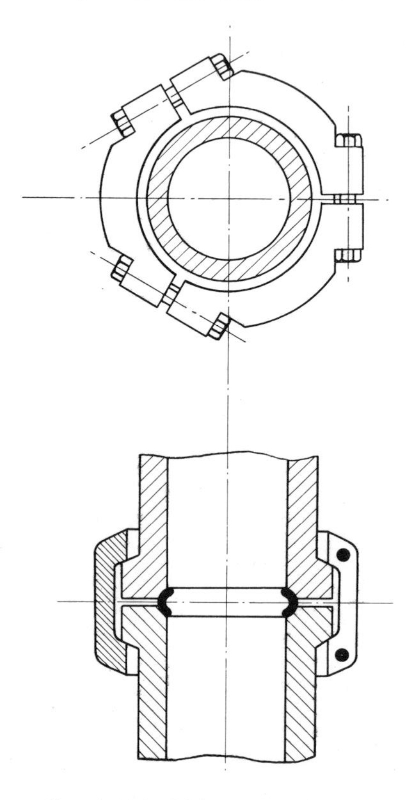

Figure 13.9 Vickers-Anderson joint: closed by pulling split collar over flanges with tapered bearing surfaces

Couplings for wave rings can be of many varieties; the simplest is the ordinary union coupling which well illustrates the great convenience of this joint. The ring is first mated into the socket of one of the components (which in small sizes, up to 3 in say, can be done by simple tapping with a light hammer) and the second component is then drawn over the ring until the end surfaces are abutting. To ensure that no unnecessary load is applied to the union, it is usual to unscrew it a turn and then bring it back by hand. Split couplings of the Vickers-Anderson type can be simplified since there is no longer any need to provide the conical

Figure 13.10 3,000 bar vessel with split coupling engaging buttress grooves†

shoulders for producing axial loading. Instead the engagement can be through a series of parallel grooves of buttress section, as shown in Fig. 13.10, where a man is lifting onto a vessel of 75 mm internal diameter the complete coupling which weighs about 16 kg and resists a pressure load of no less than 200 tonnes. The small bolts shown in the foreground are all that is necessary to keep the collar in place, and its manipulation is further assisted by the fact that the other two joints between the collar pieces are hinges; the hinge pins moreover project upwards and have lugs at their ends, as can be seen in the photograph, and there are small shoulder plates mounted on the top of the cover into which the lugs can be dropped, and which then ensure that the collar is correctly positioned to engage the grooves, both on the vessel and on its cover.

The vessel shown in Fig. 13.10 is provided at the bottom with an ordinary union nut which also forms a suitable base for it to stand on. Since no force is required to make the joint this nut can be rolled on by hand and turned by a bar in the holes provided for the purpose, one of which is clearly seen in the photograph.

For larger vessels the wave ring (and other radial sealing joints) enable various types of split collar coupling to be used. Sometimes it may be advantageous to divide these into a considerable number of separate pieces and hold them in place with rings. In this way the weight of the individual pieces can be reduced more or less indefinitely at the expense of a rather more complicated fitting operation. An interesting example of this has been given by Rolfe.(13.5)

With wave ring joints of large size some difficulty may arise in splitting apart the components unless provision is made for doing this by means of some sort of jacking arrangement. For instance in the head of the vessel shown in Fig. 13.10 the eye bolts serve the double purpose of engaging lifting tackle and also of jacking off the head, since they can be screwed right through in tapped holes until they bear against the top surface of the vessel proper.

† An automatically actuated version of this type of coupling is now marketed by Messrs. Walter Somers Ltd. of Hales Owen, Worcestershire, which enables a large vessel to be fully opened and closed by remote operation.

13.3 Moving Seals

The problem in these is usually to obtain surfaces which are smooth enough to allow the width of the clearance between them to be reduced to a very small value, and to enable this state of affairs to be maintained. A further consideration is to reduce the resulting friction as far as possible, and also to remove the heat that it generates.

In reciprocating seals, e.g. a piston ring in a compressor cylinder, the working surface of the ring is continually rubbing on the wall, but points on the surface of the latter are only intermittently in contact and thus much easier to keep cool. With rotary seals, on the other hand, both parts rub continuously while the unit is in operation.

It is probably impossible to operate a seal or gland without some leak if there is a large pressure difference across it, and the problem is either to provide some other substance, such as a lubricant, to move slowly through the seal and so keep the loss of working substance down to the very small amount which can get dissolved or entrained in the lubricant, or to devise a seal in which some leakage is accepted, because it is small enough. An example of the latter is the arrangement at one time used by the Linde company for compressing oxygen, in which a double-acting piston was so accurately guided in its cylinder that the clearance was small enough for the leak past it to be acceptable, even though the issuing gas was probably moving with the velocity of sound. The possibility of extending this principle to higher pressures is, however, limited because of the sharp rise in the sound velocity in real gases with pressure, and incidentally Linde have now adopted more conventional seals in their oxygen machines. The principle is, however, still of great technical importance in the labyrinth glands of turbines.

Another factor which is apt to be forgotten in designing glands for high pressures is the deformation due to pressure stresses, which may completely alter the small clearances. A good example of this is the simple plunger and cylinder, where an almost perfect fit can be obtained with care and skill, and it has been suggested that the supply of very small amounts of lubricant at the inner end would enable it to pump up to very high pressures with negligible loss of working material. Unfortunately, however, the effect of the pressure is to increase the diameter of the cylinder and to reduce that of the plunger, so that the clearance which was initially so extremely fine usually becomes coarse enough to blow out the lubricant and allow the substance being compressed to follow it.

But first we must briefly consider ways of avoiding the problem which can sometimes be used for stirring chemical autoclaves, etc. on a fairly small scale, i.e. up to 10 litres capacity and to, say, 3 kb.

The methods by which this can be done can be divided into two main groups: those in which the vessel itself is shaken or rocked, and those where some stirring device is installed within it. The latter can be further subdivided into those in which the shaft has to pass through a gland, and those in which the stirring power is actually generated inside the vessel, or by a magnetic flux passing through non-magnetic walls.

The problem of sealing against moving surfaces is, of course, of major importance in reciprocating compressors and pumps, as we have seen in Chapters 9 to 12, but it may be worth adding here that the Morrison type gland (see p. 289) can be very useful in such applications, whether for pumping or agitation. Another gland, which also depends on the principle of the "unsupported area" to maintain the surface pressures, is that made by the Dowty Company of Cheltenham. The chief feature of this gland, which has been described by Bingham[13.6], is the presence of a number of floating pins, usually 4 or 6, spaced equally round the periphery and passing through holes in the packing, see Fig. 13.11. The pressure to be sealed against acts upon the whole annular area of the gland, but the resulting load when transmitted to the packing inevitably raises therein a higher pressure because of the reduction in its area due to the pins. With suitable packing materials, e.g. Neoprene, and with chamfer rings to prevent its extrusion into the clearances, these glands can give good service both with reciprocating or with rotating members, although the rubbing speeds have to be kept fairly low, i.e. to the order of 25 cm/s if prolonged continuous operation is required.

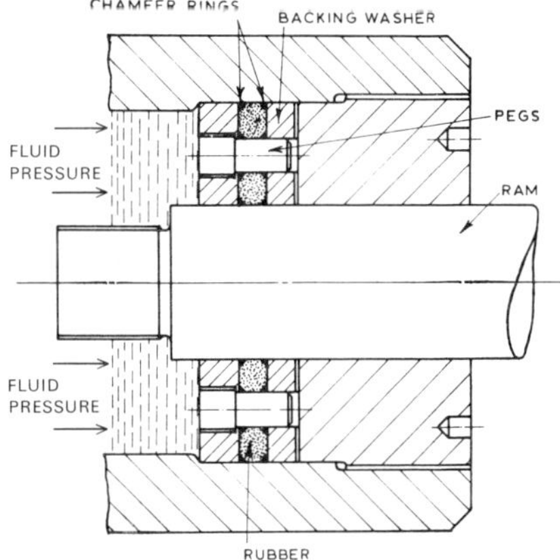

Figure 13.11 Dowty gland *(Dowty Equipment Co. Ltd., Cheltenham)*

13.3.1 *Agitation by Moving the Vessel*

The first category is somewhat naturally restricted to fairly small vessels, but a number of proprietary devices are on sale for laboratory use covering vessels of about 1 litre capacity and operating up to 3 kb. These usually have a long coil of capillary tubing which is able to deflect by the necessary amount without becoming stressed above the fatigue limit, and by this means the contents can be sampled or additions made while the reaction is in progress. The usual layout is to have a cylindrical vessel with its axis horizontal, and to make it reciprocate along its axis, usually with an amplitude of between $\frac{1}{4}$ and $\frac{1}{2}$ of its internal length. By oscillating at fairly high frequencies (200 a minute, for instance) quite good agitation of the contents can be achieved.

For larger vessels, with volumes up to 5 litres and pressures to, say, 2,000 bars, the rocking system becomes easier to operate. Here the cylindrical vessel is mounted in a cradle and rocked through an angle of about 30° on either side of the horizontal about an axis passing diametrically through its mid-point.

The connections then have to be introduced through radial holes on the line of the rocking axis, and the tube leads are usually wound in a flat spiral to take up the angular displacement of the rocking. Again care is required in the design to prevent fatigue failures occurring in the connections. The speed is here necessarily much slower, not usually exceeding 10 a minute, but the extent of the mixing action can be increased by inserting a steel ball with a diameter about $\frac{3}{4}$ that of the bore of the vessel; this then rolls from one end to the other, displacing the contents as it goes.

Generally speaking, both these types are quite effective when there is a liquid phase present in the reacting substances; but, where they are all in the gas phase, it is preferable to use a separate agitating device like an internal rotating paddle.

13.3.2 *Vessels with Internal Stirrers*

The simplest arrangement is to have a stirrer on a shaft, which is introduced through a gland, usually on a vertical centre line, and this seldom gives much trouble with pressures below 100 bars. Difficulties soon increase, however, as we go to higher pressures, but even so there are many successful arrangements in operation.

For pressures of 1,000 bars and over it often pays to use a stirrer shaft which projects through a gland at both ends; the extra complication of a second gland may be more than offset by the fact that we no longer have to deal with the large axial thrust caused by the pressure trying to blow the shaft out of the vessel.

Many types of gland have been tried, but we can only mention a few here. There are, however, certain features which are common to most successful designs and which seem worth briefly noting.

In the first place it is often overlooked that, because in a rotating seal the rubbing of the components is continuous and on the same small areas, the removal of the frictional heat becomes a matter of great importance. Thus, for all but the slowest speeds of rotation, it is advisable to surround the stuffing box with a cooling jacket, and in some cases it may be worth using a hollow shaft so that internal cooling is available on the shaft side of the gland.

Where conditions permit, it is generally found that packing materials with a polytetrafluorethylene base are the most satisfactory. This material is too soft by itself, but when reinforced with glass fibre or with asbestos it becomes firm enough to withstand the force of the gland ring without extruding down the shaft, and its very low frictional properties can be fully utilised. In this connection, however, it must be remembered that it also has a very low thermal conductivity; consequently, although it minimises the amount of frictional heat generated, it also tends to prevent this heat from escaping. Thus overheating can easily occur; one way of preventing this is to introduce thin layers of metal in between thin layers of the reinforced plastic. A possible arrangement is shown in Fig. 13.12.

Another type of gland which seems to offer considerable promise in this application is that developed at Bristol by Professor J. L. M. Morrison.[13.7] This is illustrated in Fig. 13.13, from which it will be seen that it contains a plastic

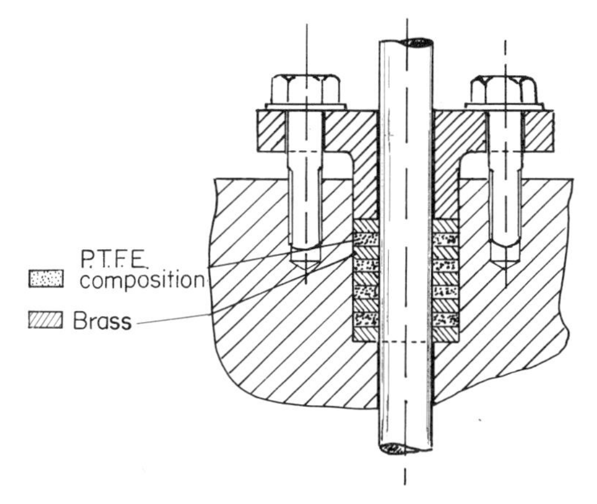

Figure 13.12 Seal for stirred vessel

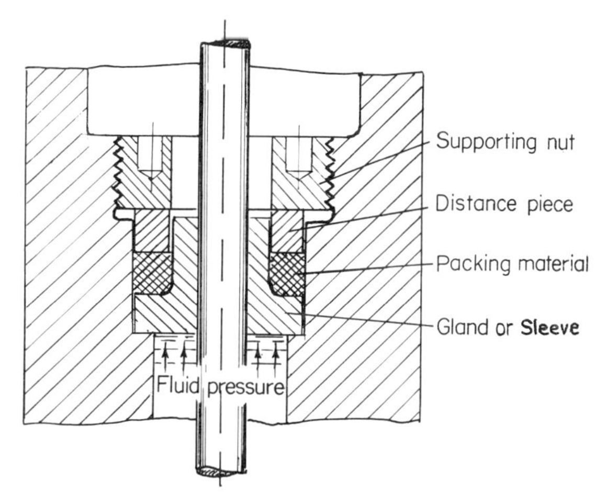

Figure 13.13 Gland developed by Professor J. L. M. Morrison (Ref 13.7)

packing outside the sleeve surrounding the shaft. This packing is loaded as in the Bridgman unsupported area system and thus squeezes the sleeve inwards against the shaft with a pressure that is always proportional to the pressure to be sealed. It is particularly effective if a certain amount of lubricant can be used, and it has the important property of controlling the clearance without causing it to disappear. Thus, if a lubricant be injected near the bottom of the sleeve, it will be driven out very slowly by the pressure; and, because there is no continuous solid contact, the heat generated is considerably less.

The condition of the shaft surface is important in all types of gland, particularly those in which a non-metallic packing bears against it. Hard chromium plate, ground and superfinished, has been found quite effective, and it has the added virtue of being fairly resistant to corrosion. Nitride-hardened surfaces are also good though easily tarnished when highly polished. Various other ultra-hard layers can be deposited, either by welding or flame showering with powdered material – and in some cases by electro-deposition – but any of these needs matching with the packing material into suitable combinations. There are many possibilities, but the arrangement suggested in Fig. 13.12, using reinforced "Fluon" (the I.C.I. brand of polytetrafluorethylene) and a hard chrome-plated or nitrided shaft surface, will usually prove satisfactory for pressures up to 1,000 bars. Above this the thrust problem becomes serious, and the necessity of introducing a ball thrust race complicates the design. However, the Morrison gland is then likely to be the most successful. For pressures much above that level there is usually a case for doing away with the gland and having some sort of internal drive.

13.3.3 *Internal Drives*

These can be of several types, the simplest being a reciprocated paddle operated by an external solenoid round a non-magnetic pressure-resisting extension from the vessel. This is, however, of little use outside the laboratory owing to the very small power that can be transmitted. A rather more elaborate and slightly more powerful effect can be obtained by using a rotary stirrer driven by a rotating magnetic field outside the non-magnetic casing, and this can be produced either with a more or less ordinary a.c. induction motor stator, or by rings of permanent magnets driven round on a cage. By lengthening the casing a considerable increase in power is theoretically possible, but the efficiency is inevitably low and falls rapidly with increase in thickness of the casing, which in turn must be thickened as the pressure is increased.

Various methods of increasing the power of reciprocating stirrers by producing synchronism with return springs have been tried with varying degrees of success, but on an industrial scale it would seem that the only really effective means of setting up good agitation is with a motor actually located within the pressure space.

D.C. machines, or a.c. induction motors can be used, and both are reported to be effective. One small advantage of the former is that it can be operated with an earth return and therefore requires only one insulated pressure electrode, whereas it is advisable to have three such electrodes for the induction motor. Starting torque is often a matter of considerable importance, but effective results can be obtained with either system.

Difficulties could, of course, arise with d.c. motors in conducting media because of their commutators, but apart from this there appear to be no troubles and sparking is not a serious problem since the pressure evidently tends to suppress it. On the other hand, lubricating the bearings of internal stirrers may be difficult, but this depends very much on the reacting substances. With dry gases the amount of lubricant required by good quality ball or roller bearings is so small that an amount sufficient for several days – or even weeks – of continuous running can be supplied when the vessel is assembled. Care may, however, be necessary with hydrocarbon gases, some of which can act as powerful solvents when under pressures of 1,000 bars or more. These problems must of course be considered in the light of all the pertinent factors, which include not only the working pressure and temperature, but also the way these may vary or be controlled for process reasons. Changes of pressure may be particularly detrimental to bearing life because of the dimensional changes in the vessel body which they cause. Other factors to be considered are the length of time between openings up of the vessels, the properties of the working substances (if solids are involved, their abrasive action may be very serious), and any chemical action they may have on the metal of the stirrer and on any lubricant that is used. The problem can in fact be very complex and clearly we cannot take it further here.

13.4 Valves

13.4.1 *General*

The essential problem here is to move a plug into and out of a hole. For a stop valve it must seal up the hole completely, but for control it has only to constrict the hole to a greater or less extent. In order to achieve these requirements it is necessary to have a seal through which the actuating shaft passes and moves, either with a straight line or rotary motion, unless it is possible to have an actuating electric motor immersed in the pressure space, in a similar manner to the stirring gear in some reaction vessels (see Section 13.3.3).

The seal in a stop valve is usually of the metal-to-metal type, with the mating surfaces easily changeable so that they can be replaced if serious wear takes place. The force required to operate a valve may be considerable, especially where the plug of a stop valve is being held shut with the pressure behind it.

The motion is usually transmitted by means of a screw thread, and in small valves this may be cut in the same rod as the moving part of the seal; thus the mating surfaces move against one another with the rotary motion of the screw. These are nearly always direct hand-operated. Fig. 13.14a shows such a valve diagrammatically, and it will be seen that, as the spindle is moved in, work is being done against the pressure, which makes such valves stiff to operate except in the smallest sizes.

A modification which is useful in small hand-operated valves is shown in Fig. 13.14b. Here the spindle has a double cone swelling so that in the full open position there is a positive seal to the valve cavity and leakage through the gland will not cause any loss from the system. There is also the safety consideration that the spindle cannot be screwed out of the thread and blown clear by the pressure behind it. This is an important matter in small-scale research apparatus since an arrangement such as that shown in Fig. 13.14a has nothing to prevent this from taking place. It should be noted in passing that the assembly shown in Fig. 13.14b could only be put together if the spindle were in two parts, but these sketches are schematic only and make no attempt to show constructional details.

Fig. 13.14c shows a fine adjustment valve, the main feature being that the spindle and seating are both tapered in small-angle cones so that the passage through the valve includes a narrow conical annulus which can be adjusted within close limits by comparatively coarse movements of the spindle.

In Fig. 13.14d is shown an arrangement in which opening and closing the valve does not alter the volume of the cavity and therefore involves no work being done on or by the pressure fluid. The gland also acts as a seal on the "unsupported area" principle. This arrangement has the further advantage that the spindle moves in a straight line and does not screw into the seat as it does in the other

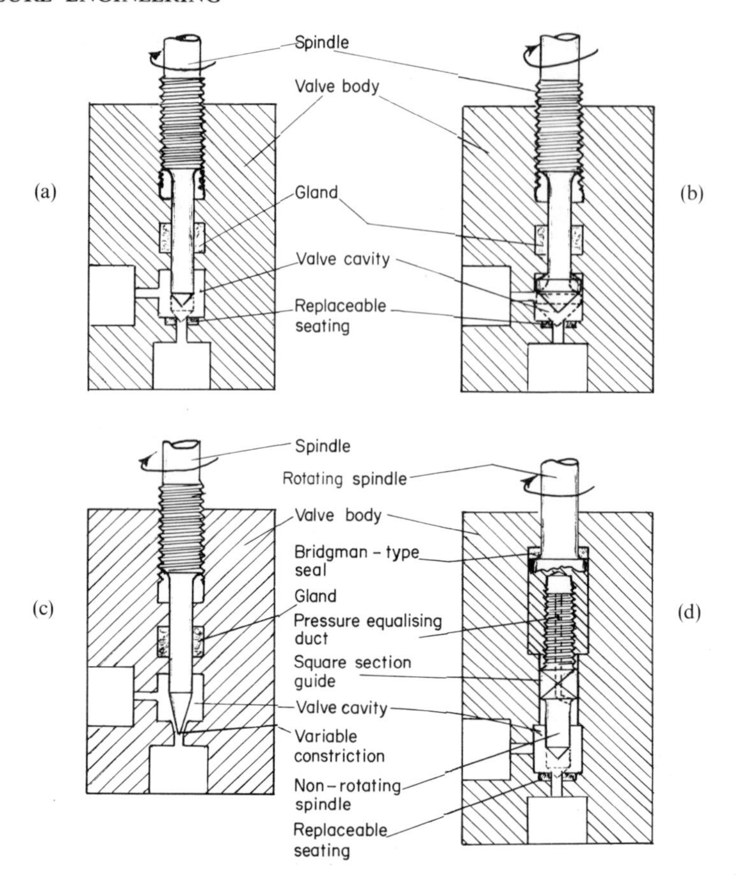

Figure 13.14 Schematic representation of valve types (a) small hand-operated stop valve (b) small hand-operated stop valve with metal-to-metal seal in full open position (c) hand-operated fine adjustment valve (d) valve which does not change its internal volume in opening or closing

types shown in Fig. 13.14. To obtain straight-line motion without rotation in a more conventional valve requires a thrust-race to resist the pressure load and so to transmit the motion from the screw to the non-rotating spindle.

We shall now show in more detail two successful designs of high pressure valve, one small enough for laboratory work, or for use on plant instrument lines, and the other a plant valve of $1\frac{1}{4}$ in bore designed to operate up to 2,500 bars.

13.4.2 *Small Hand-Operated Valve for* 4,000 *Bars*

This example, which is illustrated in Fig. 13.15, is typical of the valves now being turned out in very large numbers, especially in the United States. They are used extensively in small scale research and also for the instrument lines in large industrial plants. This one is made by the Pressure Products Industries Co. of Hatboro, Pa., and also by their British associates, Pressure Products Industries (U.K.) Ltd. of Cheadle Hulme, Cheshire. It sells in England for about £15 for the $\frac{1}{8}$ in bore version.

As will be seen, the spindle is in two parts, referred to as the shank and stem in Fig. 13.15. The upper or driving portion is rotated (by hand) and produces an up and down motion by engaging with the screw thread in the gland plug

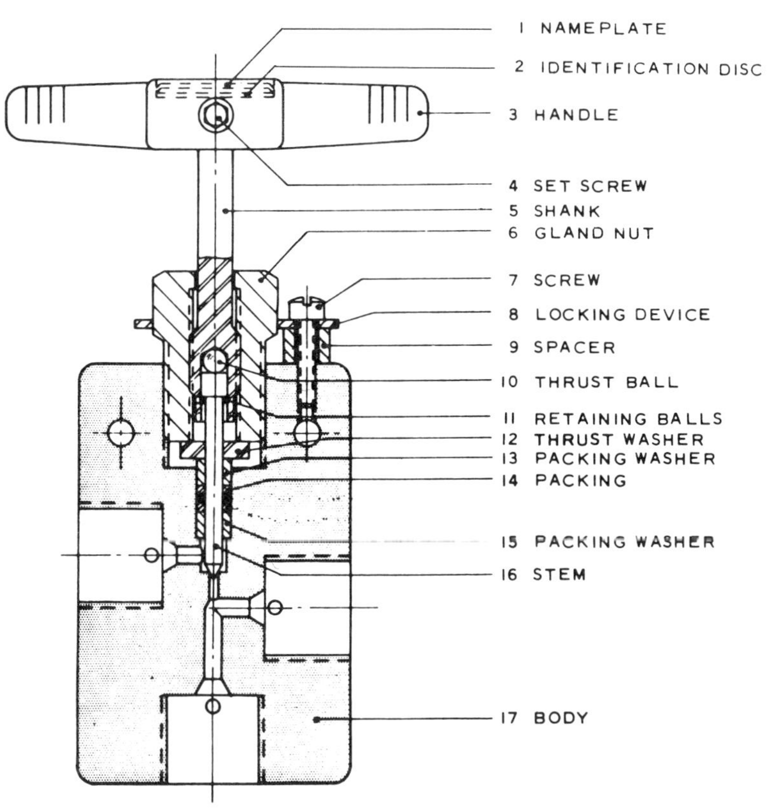

Figure 13.15 Small valve ($\frac{1}{8}$ in bore) for 4,000 bars *(Pressure Products Industries Co. Inc. USA)*

(called "gland nut" in the diagram). This motion is transmitted to the lower portion or stem through a single ball at its upper end, which is enlarged so that a small thrust race can be fitted beneath the enlargement for upward withdrawal. Thus the stem, which is normally made of Stellite, does not rotate, but merely descends onto or rises away from its seat. The latter is integral with the body of the valve, which is made of hard-drawn austenitic stainless steel.

The gland arrangement consists of a close-fitting ring of Teflon (polytetrafluorethylene) reinforced with glass fibre with a metal or reinforced nylon ring above it to resist extrusion. This allows greater working clearances in the hardened steel rings which are fitted above and below the plastic materials. Glands of this type are well proved in service with both liquid and gaseous media at 4,000 bars and can be operated with very low frictional resistance.

The spindle thread is larger in diameter than the hole in the top of the gland plug and cannot therefore be forced out through it; the plug itself is prevented from unscrewing by a locking plate secured by a set screw as shown. The body can be supplied with connections which are either parallel with one another or at right-angles. Thus, referring to Fig. 13.15, connection can be made from the left-hand side either to the right or downwards, but when it is used as a stop valve the pressure side will normally be below the seat, i.e. to the right or towards the bottom, so that the gland is not made to hold against the full pressure longer than necessary. This does not mean that it will not resist it fully, but it increases the life of the packing to avoid this.

The recesses for the connections are normally made to take the standard American joint for small-bore pipes with coned ends to fit into hollow cones in the valve body, as shown in Fig. 13.15. The two holes shown in the top of the body are for fixing purposes.

As will be seen, this type of valve is designed for quantity production, which accounts for its surprisingly low price. In consequence of this, many users are ready to scrap the whole valve and replace it when the seat begins to wear or becomes too deeply recessed. On the other hand, the suppliers market an inexpensive hand tool for trueing up these seats and also the coned recesses in the body connections.

13.4.3 *British Engines Ltd. $1\frac{1}{4}$ in (31·8 mm) Stop Valve for 2,500 Bars*

This is a recent high pressure valve designed and patented by British Engines Ltd. of Newcastle upon Tyne. Its chief feature is the motion, which avoids screw threads altogether and uses instead an eccentric driven through a worm-reduction gear by a hand-wheel. In this way it becomes possible for one man using a hand-wheel of about 50 cm diameter to close the valve against the full pressure, which will be exerting a force of about 28 tonnes.

As will be seen from Fig. 13.16 it is a right-angle flow valve with both the spindle tip and the seating easily replaceable. The packing is of the chevron type, held between suitably shaped steel rings and compressed by a plug screwing into a recess in the main block. This can be pulled up by means of a bar inserted into suitable holes in it. A small hole leading into a hollow above the gland is provided to carry off any leakage and prevent a build-up of pressure at that point should it become serious.

The motion is so arranged that the eccentric is just about on bottom dead centre as the valve shuts; in consequence there is a very large mechanical advantage just when it is most wanted. The eccentric drives a well-guided cross-head to which the thickened upper end of the non-rotating, spindle is flexibly attached. The position of the spindle is indicated by a needle fixed to the eccentric shaft and reading over a dial on the side of the structure which carries the motion, and which is attached to the valve body by 4 massive studs screwed into it at the corners of a square, thus allowing it to be mounted in different positions relative to the connections if so desired.

The main components are made from forgings of high tensile steel and the completed assemblies are all tested to 3,750 bars before delivery.

These valves have been designed primarily for service in high pressure polyethylene plants where many are now in operation in various parts of the world. Their somewhat unorthodox spindle motion is an attempt to overcome the special problem of reducing the very high forces required, so that they can be manually operated without an excessive amount of turning of the hand-wheel. The eccentric arrangement does this very neatly by providing a relatively small mechanical advantage for the movements with the valve well open, and by increasing this as the spindle approaches the seat, when much larger forces are clearly required. From this aspect they are of an unusual type, but the other components, i.e. body, spindle, gland, seat, etc., present an example of good conventional modern

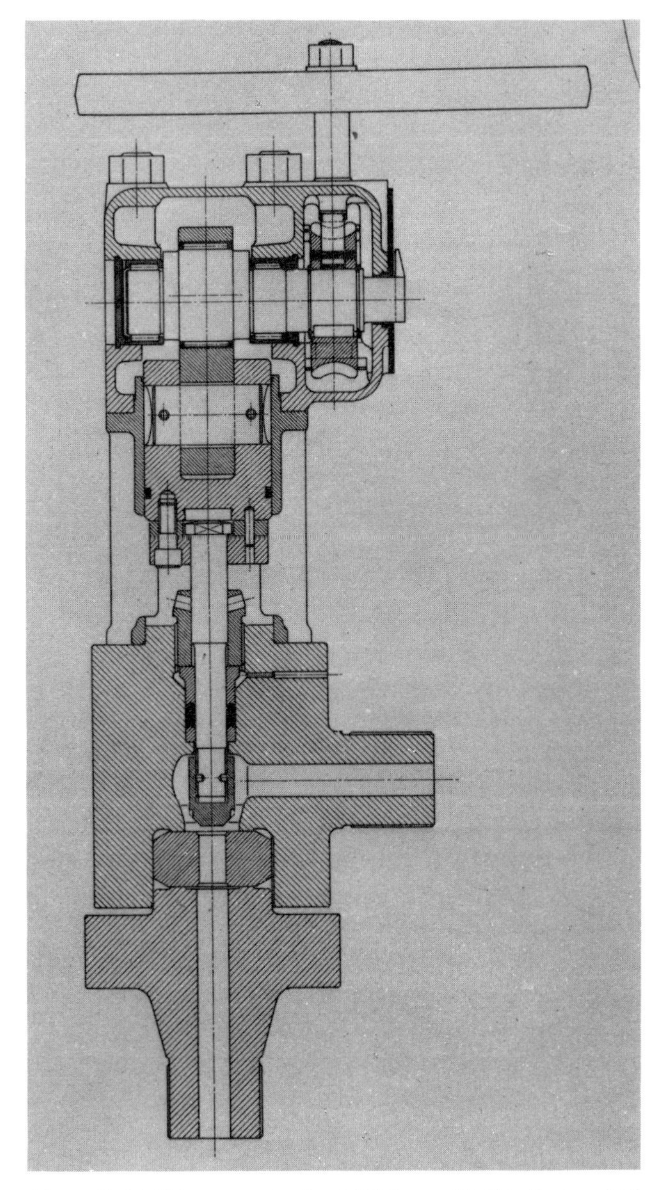

Figure 13.16 $1\frac{1}{4}$ in stop valve for 2,500 bars *(British Engines Ltd)*

practice. Similar valves are now being offered by this company for service at 3,500 bars.

13.5 Pipe Connections

These are increasingly being made by welding, although there are still many places in chemical plants where it is necessary to dismantle pieces of pipe in order to clean them or clear blockages; also, not all pipe materials are really suitable for welding.

With openable connections the main problem is the attachment of the coupling to the pipe. Fortunately, however, in installations where the pressure is above

1,000 bars, the walls will be of considerable thickness and it will generally be quite safe to cut a thread on the outside. Where this will be used, for instance, with a right-hand threaded plug or union assembly, it will be advantageous to use left-hand threads on the pipe so that there is no danger of the partial unscrewing of one thread while another is being pulled up.

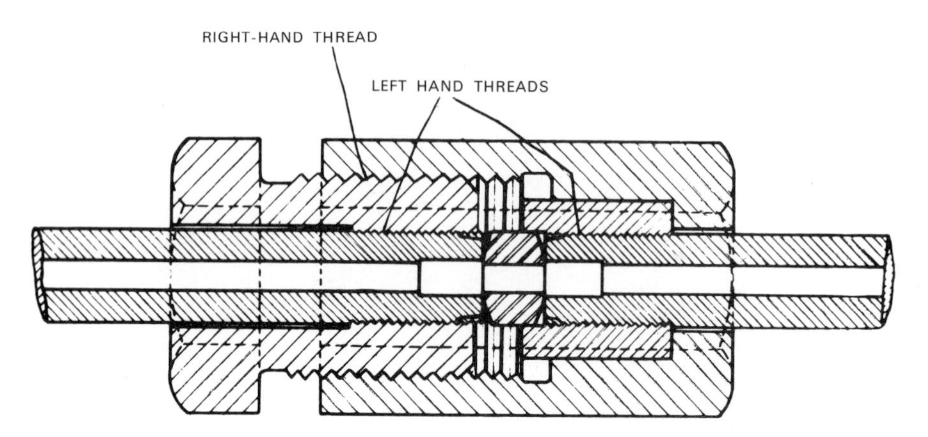

Figure 13.17 Small union connection for 3,000 bar tubing *(as made by Marston Excelsior Ltd, Wolverhampton)*

The actual seal can usually be made on the metal of the pipe, either by shaping it or by using a lens or cone ring. A typical assembly suitable for pressures up to 5 kb in small sizes (5 mm diameter and less) or to 2·5 kb up to 8 mm diameter) is shown in Fig. 13.17.† In this case the tightening can be done by hand tools for pipe bore sizes of 5 mm and less, and the life of such a joint should be at least 100 openings and closings, although it will probably save trouble if the joint ring is renewed after every dozen or so openings. It should also be noted that Fig. 13.17 shows a male and a female type of connection together thus making the pipe joint. Either can be used separately to connect the pipe into a valve or other piece of equipment. In the latter case the more usual practice, especially in the U.S.A., is to dispense with the joint ring and machine a 30° half-angle cone on the end of the pipe to mate with a conical cavity in the other item. Some specifications call for the angle of the latter to be 1° or 2° less so that contact first occurs at the inner end.

For lower pressures – up to about 500 bars – the sleeves shown with left-handed threads in Fig. 13.17 can be merely slipped over the pipe provided there is some arrangement for forcing them into contact hard enough to withstand by friction the force tending to separate the tubes. The Ermeto coupling‡ is an example of this type, although its action is more positive because the sleeve, by reason of its shape, is made to bite into the outer surface of the pipe, as shown in Fig. 13.18. It is obviously important that joints of this type, which dispense with

† A range of fittings of this kind for pressures up to 575 bars is manufactured and marketed by Marston Excelsior Ltd. of Wolverhampton, a subsidiary of Imperial Metal Industries Ltd.

‡ Made by the British Ermeto Co. Ltd. of Beacon Works, Hargrave Road, Maidenhead, Berkshire.

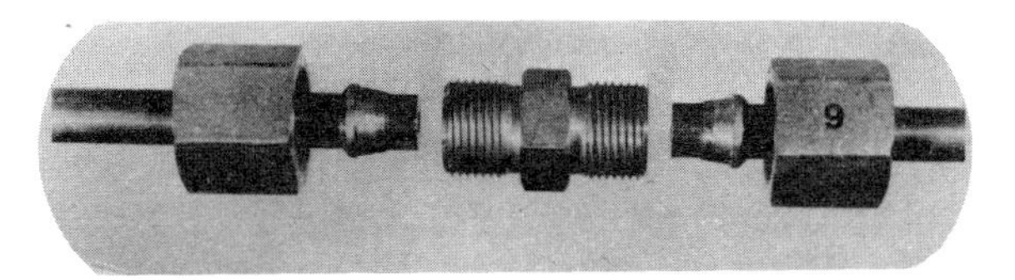

Figure 13.18 'Ermeto' pipe joint *(British Ermeto Co. Ltd, Maidenhead)*

a screw thread to transmit the coupling resistance to the pipe, should have their components made to close tolerances, and this means that the outside diameter of the pipe must be subject to similar dimensional control.

13.6 Screw Threads in High Pressure Equipment

13.6.1 *General*

The sizes and shapes of the various screw threads in use in different parts of the world have mostly been arrived at by methods far removed from any scientific or logical process. For instance, Sir Joseph Whitworth in establishing the standards now known by his name merely averaged the pitches and thread profiles of all the screwed parts he could get hold of. Moreover, he did this round about 1840 when the materials available were limited to cast iron, wrought iron, and a few non-ferrous alloys such as brass and the simple bronzes. Consequently, if his threads were a useful compromise for the relatively low strength materials he was concerned with—and the fact that they are still used so widely shows that they were—this is no reason why they should be right for the vastly different and stronger materials used today in high pressure engineering.

The evolution of the Unified Thread system during the last war appears to have missed a great opportunity to produce a system of standards based on true strength. The main objective of those who compiled it seems to have been to produce compromises wherever possible so as to cause a minimum of short-term dislocation in the industries of Britain, the U.S.A. and Canada, and as most of the work was done while the war was in progress this is not altogether surprising. And it must be accepted that the practical results of using these standards, particularly in the smaller sizes, are good enough for most purposes.

13.6.2 *Strength Considerations*

Symmetrical thread forms with sloping flanks, as in the Whitworth and most other systems, can fail in three ways. Thus, considering a nut and bolt assembly, the bolt may fail in tension across the section reduced in area by the cutting of the thread (what is usually termed failing "at the bottom of the thread"); or it may shear off at the roots of the male thread; or the radially outward component of the flank pressure in the nut may overstrain and expand it sufficiently to let the bolt slip through. In many instances the nut would split before it had expanded sufficiently, but the more usual occurrence is for the nut to expand partially and so reduce the area of metal trying to resist the shearing force until it can no longer do so. These are, of course, considerations of failure under static loading. Where vibrations are involved, with the consequent risk of fatigue, the most likely form of the failure is by the spread of a fatigue crack across the section of the bolt normal to its axis, having been started by the stress-raising effect of sharp edges at the bottom of the thread section. Again, the low strength materials mainly used, and their considerable ductility, have probably been able to relieve the peaks of stress and so to avoid many a failure which would have occurred with the more susceptible materials now coming into use.

Whitworth's dimensions seem to have ensured that failure in tension at the bottom of the thread was unlikely, and his nuts had enough metal outside the thread to offer fairly adequate resistance to the expanding effect of the sloping flanks. On the other hand, his profile with its rounded tips and roots tended to fill the available spaces almost completely. Thus any plastic deformation due, for instance, to localised overstraining, or alternatively the presence of small pieces of dirt, etc., was liable to cause seizing between the surfaces and produce severe damage to both components if attempts were made to force them apart.

13.6.3 *Design Methods for Symmetrical Threads*

Experience has shown that the risk of thread seizure (which from a practical point of view is often as objectionable as actual failure) can be greatly reduced by truncating the tips of both male and female threads, as with the "Unified" profile. It is also desirable to arrange for the flank pressures to be kept under control; in fact they should be appreciably less than the bearing pressures allowed for other contact surfaces. A convenient design basis for threads of this kind, i.e. Whitworth with its 55° included angle or the Sellers and "Unified" with 60°, is to consider the shear in the engaged threads, applying a factor of 60% to allow for various stress-raising effects such as the inevitable imperfections of shape leading to variations in contact pressure from place to place in the threads. This will at the same time control the bearing pressure on the flanks. If then we keep the shear stress down to about $\frac{1}{4}$ of the tensile 0·2% Proof Stress, experience shows that there is little risk of static failure, or of thread seizure, at any rate in materials whose Ultimate Tensile Strength does not exceed 12,500 bars (180,000 lbf/in^2), and where good finishes are insisted upon.

Example. Let us consider a plug of 50 mm diameter, screwed with a 5 mm pitch thread which must resist a total axial force of 80,000 kgf. If the material of the plug and also of the block into which it is screwed has a 0·2% Proof Stress of 6,080 bars (62·0 kgf/mm^2) what should be the engaged length of the thread?

We must limit the shear stress to $\frac{1}{4} \times 62{\cdot}0$ or 15·5 kgf/mm^2.

The effective diameter of the thread will be about 46·75 mm, so if L mm be the engaged length the equivalent area resisting shear (allowing a 60% factor) will be:

$$\pi \times 46{\cdot}75 \times L \times 0{\cdot}6 = 88{\cdot}1 \times L \text{ mm}^2$$

Consequently the force must be 88·1 L × 15·5 and this must be 80,000 kgf, whence $L = 58{\cdot}6$ mm.

It is interesting to see approximately what is the average contact pressure on the projected area normal to the axis of the flanks of the threads. Since the mean diameter is 46·75 mm, the thread depth will be about 3·2 mm, and the projected area will thus be $\pi \times 46{\cdot}75 \times 3{\cdot}2$ or 470 mm^2. Allowing a factor of 75%

for truncating and for the imperfections of the mating, we have a net area for each pitch length of 352 mm^2 and, since the engaged length is 58·6/5 pitches, the total flank area will be 4,130 mm^2, giving a mean bearing pressure of about 19·4 kgf/mm^2 or 1,900 bars.

Experience shows that decently cut threads of this size, loaded in this way, should not give trouble in service, and should be capable of frequent disconnecting without seizure.

13.6.4 *Other Thread Profiles*

In discussing the modes of failure of threads we mentioned the expanding effect of the sloping flanks. This can of course be reduced almost indefinitely by the use of square or knuckle threads, and also by buttress threads, provided that the main load is always in the same direction. In practice it is usual to have a slight inclination on what is generally known as the "flat" side so as to avoid cutter interference when milling the thread. A typical profile is shown in Fig. 13.19. With a thread of this kind the allowance for stress-raising and other effects of the sloping flanks (taken as 60% for the Whitworth or "Unified" profile) can be raised to 80% or 85%. Apart from this the design procedure does not need alteration.

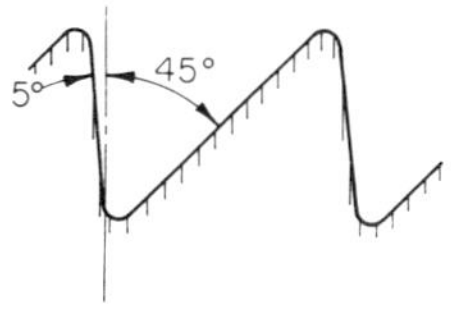

5°
45°

Figure 13.19 Buttress type thread, suitable for high pressure components

Buttress threads certainly have some advantages in heavily loaded connections whether in high pressure work or in other applications. They are stronger and on the whole less liable to seize, but the additional strength often seems to be overestimated; it is doubtful if this ever exceeds a factor of about 1·5 and in most cases it will be safer to consider a factor of 1·33. Again, risk of seizure is not entirely eliminated, especially if a certain amount of the load may be reversed, e.g. by the mere weight of the component. They are certainly more difficult and more expensive to cut and usually need to be coarser than the corresponding Whitworth threads, and this last factor may reduce the strength advantage somewhat. The more awkward machining operation, especially with long female threads, also tends to leave rougher contact surfaces and this in turn can increase the risk of seizing. We can sum this up by saying that buttress threads appear to have been overrated in some quarters; they still possess important advantages, but they should not be specified without a thorough study of the pros and cons in any particular situation.

Symmetrical threads with flanks which are normal to the axis or inclined at a small angle can be treated similarly to buttress threads in regard to design, but they have of course the advantage that they can be loaded in either direction. "Square" threads, i.e. those in which the flanks are truly normal to the axis, are awkward to cut and in practice it is usual to allow a small angle to clear the tool.

The American Acme thread with an included angle of 29°, with flat tops and roots of about $\frac{1}{3}$ the pitch and with a depth of cut about one half the pitch, may prove useful in this class of work, especially where a jacking action, i.e. movement under load, is required.

13.6.5 *Overstraining of Threads*

Even the best modern screw-cutting machinery cannot hope to produce such perfect mating that the flanks of high-duty threads are loaded with true uniformity. In fact with normal practice and with reasonably ductile materials this does not cause any serious trouble because, in the areas of higher than average contact pressure, localised yielding usually tends to distribute the load more evenly, and this process evidently helps to prevent further trouble. However, it is often helpful to lubricate the thread generously before the first pressurising of a new assembly, using for instance a grease containing low friction substances like graphite or molybdenum disulphide.

Sometimes, especially in chemical research, the apparatus has to be constructed from materials which have the required corrosion resistance or absence of effect on the reaction, but which are otherwise unsuitable for such applications. For example, special high chrome or austenitic chrome-nickel stainless steels may be called for, and in such cases it is often worth using rolled threads on the male members. This is usually sufficient to prevent seizure, but in exceptionally severe conditions it may pay to use tap-rolled female threads, although it generally suffices to cut these by ordinary screwing processes. The work-hardening of the flanks and the very good surface finish which rolling gives helps considerably to prevent the risk of seizure and increases the life of the components. The Pressure Products Industries Co. of Hatboro, Pennsylvania, use rolled threads on most of their equipment, as do their British associates, Messrs. P. P. I. (U.K.) Ltd. of Cheadle Hulme, Cheshire.

With materials such as low carbon steel, where the phenomenon of primary and secondary yielding is much in evidence, localised overloading of a thread can spread very rapidly, causing a whole connection to become loose. This may deceive subsequent investigators of some failure in service into thinking that the connection had not been properly tightened, whereas a more probable explanation could be that it was not strong enough in the first place and had become grossly overstrained in consequence.

13.7 Electrical Connections

The methods of insulating a connection through the wall of a high pressure container are fairly well established and present little difficulty up to pressures of 5 or 6 kb. Beyond this there is not much reported experience, but the principles underlying applications to these pressures are probably capable of extension to higher ranges.

The basis of most successful methods is to have a solid conical electrode fitting into a hollow cone of insulating material, which in turn fits into a conical hole in some part of the container wall. The bigger ends of the cones in each case face towards the pressure, which thus tends to force the components more closely

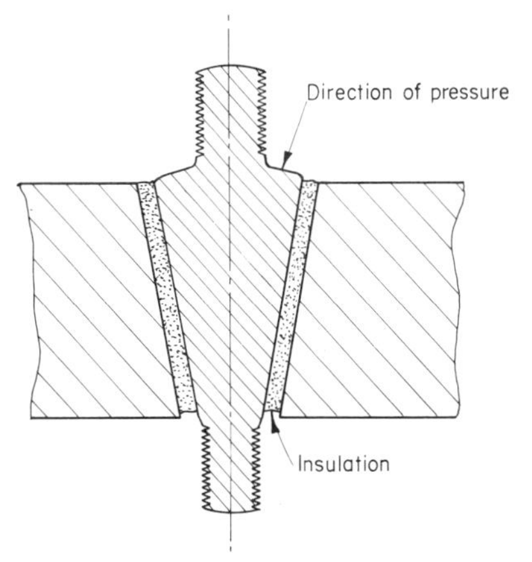

Figure 13.20 High pressure electrode, for high currents

into contact with one another, see Fig. 13.20. It is also advisable to ensure that the electrode will not slip through the hole in the plug if the insulation fails.

We can subdivide further considerations into two categories, namely electrodes where high currents have to be carried – e.g. for supplying internal furnaces or stirrer motors – and those where the currents are small, as in instrumentation, but where the number of circuits may be considerable.

13.7.1 *High Current Electrodes*

The chief problem here is to find a suitable insulating material. It must not only prevent electrical leakage, but must also withstand very high crushing loads at the operating temperature; at the same time it must, of course, prevent the working substances from seeping through or past it. One of the most satisfactory materials for this is glass, since with appropriate metals it can be made to adhere strongly and thus ensure that there is no leakage at the interface. The manufacture of glass-insulated high pressure electrodes is fully described by Welbergen(13.8), who used a steel containing 25% of chromium and a hard glass of the pyrex type. The steel cone was lapped into the glass, which in turn was lapped into a conical hollow in a component of the same steel. The parts were then assembled and placed in a furnace with a weight so arranged as to load the electrode in the same direction as the pressure under operating conditions. The hole was then heated to the softening point of the glass at which it was held for a few minutes before being slowly cooled.

According to Comings(13.9), the General Electric Company of America achieve much the same result by using an electrode which is initially parallel and which is pushed through a parallel glass tube into a parallel hole in a suitable plug; this latter is turned to a conical shape on the outside, and the whole is then swaged at a temperature at which the glass is plastic until the outside becomes cylindrical. The result is to transfer the cones to the electrode and its glass insulating sheath.

Other materials which have been used for insulation in assemblies of this kind are quartz, soapstone, pyrophyllite, and several synthetic substances such as formaldehyde resins reinforced with glass or nylon. An important advantage with glass, however, is its resistance to temperature, but it must be remembered that most of the suitable electrode materials have a relatively low conductivity and—when carrying high currents—they tend to heat up considerably. With industrial units this may require water-cooling and in such cases the electrode should be made long enough to project well on the low pressure side, so that a jacket can be fitted round it. In small scale applications in the laboratory the necessary cooling can usually be supplied by playing a small air jet onto the electrode.

13.7.2 *Multi-Conductor Connections for Low Currents*

Small glass-insulated electrodes of the type described in the previous Section can be grouped into a plug carrying as many as 5 or 6, but these are expensive to make and take up more space than may be available. An alternative procedure for instrument connections, at least up to 3,000 bars, is to pass bare wires through a conical annulus filled with epoxy resin, somewhat as shown in Fig. 13.21. Care must, of course, be taken to see that the wires do not get shorted while they are being assembled and while the resin is still plastic.

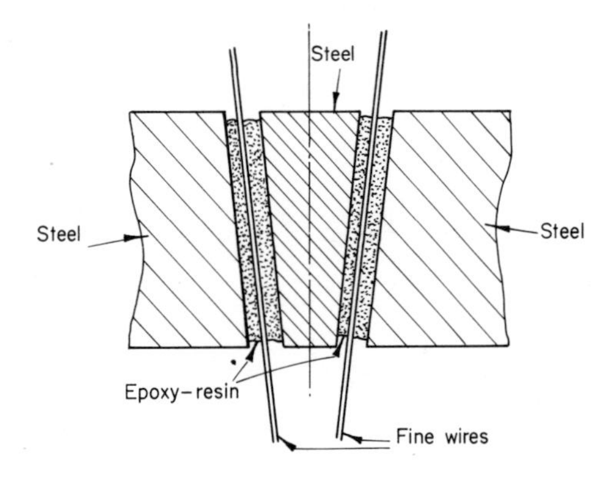

Figure 13.21 High pressure electrode, for multiple instrument leads

13.8 High Pressure Windows

Thick-walled glass tubes for observing liquid levels in boilers and autoclaves have been used for many years, but only for relatively small pressures, e.g. about 50 bars maximum. With suitable selection of the material and with careful annealing, however, hard glass tubes can be safely used for considerably higher pressures. But for the ranges we are mainly concerned with in this book, it is necessary to use other forms of window.

It should be appreciated that glass can be very unpredictable in its behaviour under stress. For instance, Bridgman[13.10] states that tubes may withstand a considerable pressure several times without any sign of damage, and yet fail subsequently at a much lower pressure. It is therefore clear that safety demands that direct observations should never be made unless special shields of, for instance, thick "Perspex" are used to protect the personnel concerned.

One method of achieving greater pressure resistance with glass is to arrange for the pressure to act on the outside of a tube. This naturally involves somewhat complicated optical systems to transmit an image of what needs to be observed to some position where it can be seen. Stryland and May[13.11], and Gilchrist[13.12] have described ingenious methods of dealing with this side of the problem. The last-named has also described some tests to destruction of tubes under external pressure.

For tubes of this kind the only practicable materials are glass and quartz, and Gilchrist's tests show that, with a quartz tube of 10 mm external diameter and 3·3 mm wall thickness (i.e. a diameter ratio of 1·5), a pressure of more than 3,000 bars could be safely carried on the outside surface. Failure would then almost certainly be by instability, as discussed in Chapter 3, Section 3.4, and Gilchrist estimates that this would occur in such a tube at about 5,000 bars.

The use of a tubular sighting device, with the pressure either on the outside or on the inside, involves making a satisfactory seal with the metal walls of the vessel or pipe system of which it is a part. A common way of doing this is by means of a simple packing gland, but great care must be taken if this is tightened by a screwed plug owing to the danger of causing torsional stresses in the glass or quartz. This will greatly reduce its capacity to resist the pressure stresses, and it is therefore preferable to pull up such a gland with a flanged plug secured by bolts, as shown in the upper part of Fig. 13.22. In some cases, however, a screwed

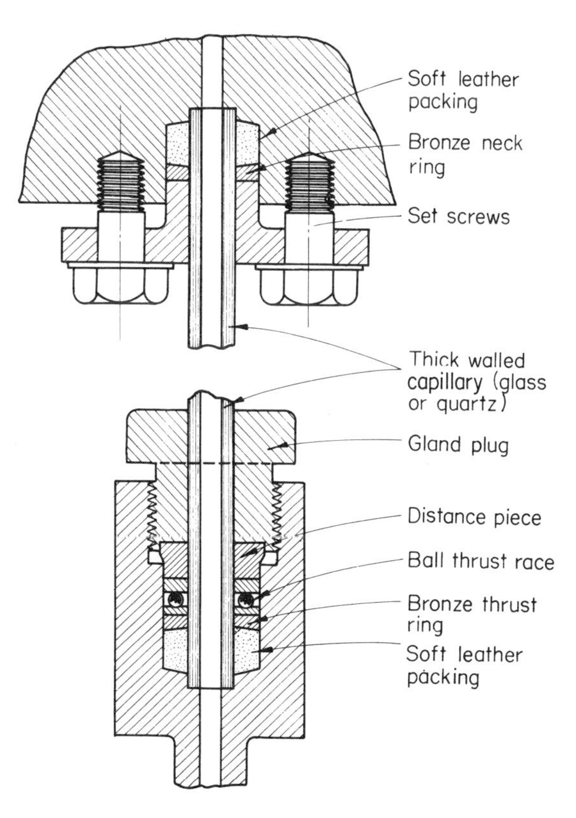

Figure 13.22 High pressure gauge glass; upper gland shown with set screws, lower gland with gland plug and thrust race; methods of preventing the twisting of the glass tube

plug is particularly desirable for other reasons, and Michels[13.13] has described a successful gland in which a small thrust ball race is introduced between the plug and a neck ring, so that virtually no twist can be transmitted to the glass. This arrangement is shown in the lower part of Fig. 13.22.

When the pressure involved reaches several kilobars, visual observation can only be through small ports. A very satisfactory design for doing this was described by Poulter[13.14], in which the actual window is a circular disc of glass with plane faces overlapping a circular hole in a convenient part of the wall. Fig. 13.23 shows such an arrangement; as will be seen, there is contact between the glass and metal around the annular overlap, in which the pressure will always be greater than the pressure to be sealed. Both of the surfaces in contact should be lapped flat to no more than two or three optical fringes, and initial sealing may be assisted by smearing a little Canada Balsam on them.

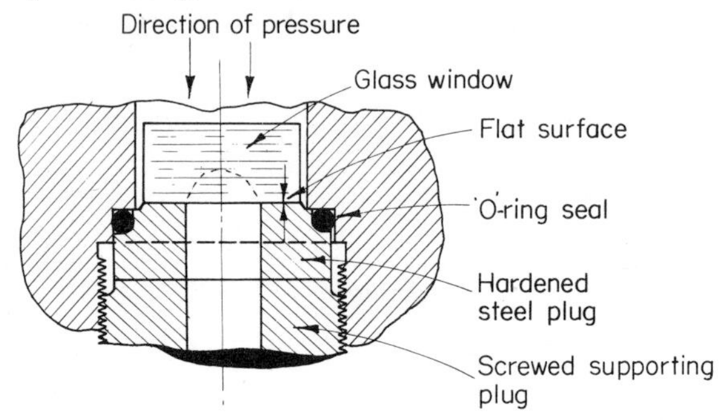

Figure 13.23 Poulter type high pressure window (Ref 13.14)

The actual window may be made of glass, quartz, or synthetic sapphire. Gilchrist[13.12] has modified Poulter's design by adding O-rings both round the window and round the plug on which it is supported; these are shown in Fig. 13.24.

Windows of this kind, in which the thickness of the transparent material is about the same as the diameter of the hole in the plug, have been used up to 30 kb by Poulter. Their lives are somewhat unpredictable under such extreme conditions – as might be expected – and it is worth noting that glass windows, when they fail, usually do so along a surface of maximum shear into two parts,

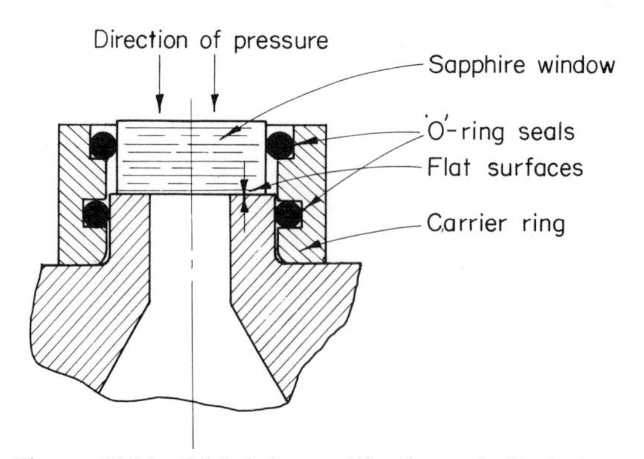

Figure 13.24 Gilchrist's modification of Poulter's window (Ref 13.15)

one being domed in shape and the other mating with it as shown by broken lines in Fig. 13.23. Moreover the surfaces thus exposed are smooth and look as if they had been carefully polished like a lens. Poulter[13.14] illustrated this very effectively by photographing small print through it to show this lens effect. Some years ago a failure of this kind occurred in the Northwich laboratories of I.C.I., though at a pressure of not more than 4 kb. The resulting "lens" was kept in a show case and its magnifying qualities demonstrated to occasional visitors, but after several years – and apparently without any outside interference – it was found to have disintegrated into a little heap of fine powder.

For extreme pressures spectroscopic investigations have been made on very small samples using diamond anvils which serve also as windows. Much ingenuity has been devoted to the development of apparatus of this kind, but it hardly falls within the scope of this book; readers who may be further interested in such work will find an excellent description, by Jamieson and Lawson, in Wentorf's symposium.[13.15]

REFERENCES

13.1. GOUGH, H. J., *Proc. Inst. Mech. Eng.*, **132,** 209, 1936.

13.2. MANNING, W. R. D., *Proc. Chem. Eng. Group, Soc. Chem. Ind.*, **19,** 110, 1938.

13.3. THOMAS, S. L. S., TURNER, H. S. and WALL, W. F., Conference on High Pressure Engineering, London 1967, Paper No. 11, *Proc. Inst. Mech. Eng.*, **182,** Pt 3C, 271.

13.4. CROSSLAND, B. and AUSTIN, B. A., Joint Conference on Machines for Material and Environmental Testing, *Proc. Inst. Mech. Eng.*, **180,** Pt 3A, 118, 1965–66.

13.5. ROLFE, B. W., Conference on High Pressure Engineering, London 1967, Paper No. 18, *Proc. Inst. Mech. Eng.*, **182,** Pt 3C, 239.

13.6. BINGHAM, A. E., *Proc. Inst. Mech. Eng.*, **169,** 881, 1955.

13.7. MORRISON, J. L. M., CROSSLAND, B. and PARRY, J. S. C., *Proc. Inst. Mech. Eng.*, **170,** 697, 1956.

13.8. WELBERGEN, H. J., *J. Sci. Instrum.*, **10,** 247, 1933.

13.9. COMINGS, E. W., *High Pressure Technology*, McGraw-Hill 1956, 114.

13.10. BRIDGMAN, P. W., *The Physics of High Pressure*, 1st Edition, Bell, London 1931, 57.

13.11. STRYLAND, J. C. and MAY, A. D., *Rev. Sci. Instrum.*, **31,** 414, 1960.

13.12. GILCHRIST, A., *Soc. Chem. Ind.* Symposium on The Physics and Chemistry of High Pressure, London 1962, published by the Society in 1963, 219.

13.13. MICHELS, A. M. J. F., *Ann. Phys.*, **72,** 285, 1923.

13.14. POULTER, T. C., *Phys. Rev.*, **35,** 297, 1936.

13.15. JAMIESON, J. C. and LAWSON, A. W., Paper No. 4 in *Modern Very High Pressure Techniques*, edited by R. H. Wentorf, jnr, Butterworth, London 1962, 70.

13.16. LABROW, S., *Proc. Inst. Mech. Eng.*, **156,** 66, 1947.

14 The Measurement of Pressure

14.1 General

This problem presents very little difficulty until we get well into the kilobar range. Ordinary commercial gauges such as the Bourdon tube type are remarkably reliable and accurate over long periods of operation if they are given reasonable maintenance, and if their calibration is checked from time to time. Even when regularly subjected to pressures as high as 3 kb, Bourdon tubes can be highly satisfactory,† but at 5 kb – even under special conditions of manufacture and service – they are approaching their useful limit.

Bourdon gauges are one of a number of types of what are usually termed "secondary" gauges, since they depend on elastic deformation caused by the pressure and need to be calibrated from time to time against a "primary" gauge, which is essentially one in which the pressure supports either a column of mercury or a dead weight acting over a known area. The former are obviously limited; a mercury column to the top of the Eiffel Tower or the Empire State Building would only give a pressure head of about 400 bars, and – although Amagat actually did have a column on the Eiffel Tower and even tried to get a still bigger head in the shaft of a coal mine – it was clear that some other method was called for.

The closed mercury manometer was an attempt to get over this difficulty and still retain the convenience of a mercury column. It consists essentially of a number of mercury columns side by side, with air used to transmit the pressure from the bottom of one to the top of the next. This has some advantages in that the temperature control is much simpler, but for any considerable pressure the number of mercury levels that have to be measured for each and every reading becomes tiresomely large, and this sort of apparatus is now seldom used. The supporting of dead weights by the pressure on a plunger, now generally referred to as a "dead-weight" gauge, or pressure balance, is much more convenient, and for pressures below about 1 kb it is very simple to operate and gives very accurate

† We have had experience of gauges, made by the Budenberg Gauge Co. Ltd. of Broadheath, Cheshire, which stood up to regular usage of this kind for several years without trouble.

results. At high pressures however the deformation of plunger and cylinder becomes sufficient to introduce the need for considerable corrections which cannot, as a rule, be calculated with sufficient accuracy and thus involve calibration checks and other complications, and it ceases therefore to be a true primary (or absolute) gauge.

For calibration at very high pressures, i.e. over 10 kb, we must have recourse to certain fixed points, mostly phase changes or allotropic discontinuities, several of which are now known with considerable accuracy. As a result of these it is possible to measure 25 kb with an accuracy better than $\pm 1\%$. Beyond that the uncertainty is a good deal greater, but work on other fixed points is gradually reducing this.

In practice, both industrially and in small-scale research, secondary gauges are almost always used, preferably with a good primary system available for calibration. For steady pressures two basic types of secondary gauge are available, depending either on elastic deformation (the Bourdon tube is, of course, a gauge of this sort), or on the effect of pressure on electrical resistance.

14.2 Primary Gauges

From a practical point of view the use of a mercury column is hardly of interest; although this appears at first sight to be almost ideal, it suffers from a considerable number of objections if really accurate results are to be obtained, even in the 100 bar range, and we shall therefore concern ourselves in this section only with dead-weight gauges of various types.

Essentially these are all vertical plungers in close-fitting cylinders, loaded with suitable weights which they support by the pressure to be measured acting underneath, as in Fig. 14.1. Thus, if we neglect friction between the plungers and cylinders, and also the effects of the elastic deformation, it is evident that:

$$p = W/\frac{\pi}{4}D^2 \tag{14.1}$$

where p is the pressure, W the total applied weight (including that of the plunger itself) and D its diameter.

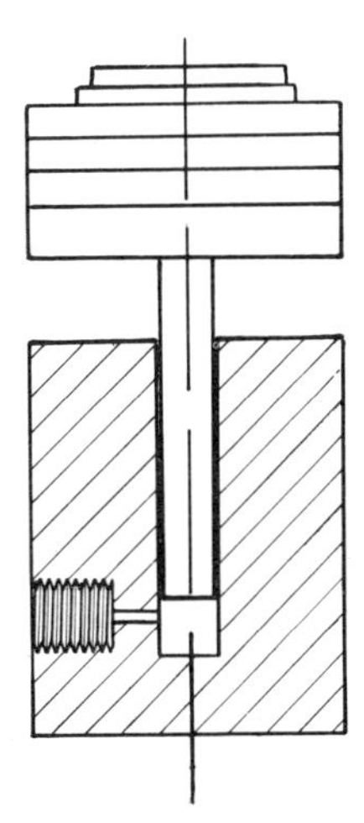

Figure 14.1 Simple dead weight pressure balance, schematic diagram

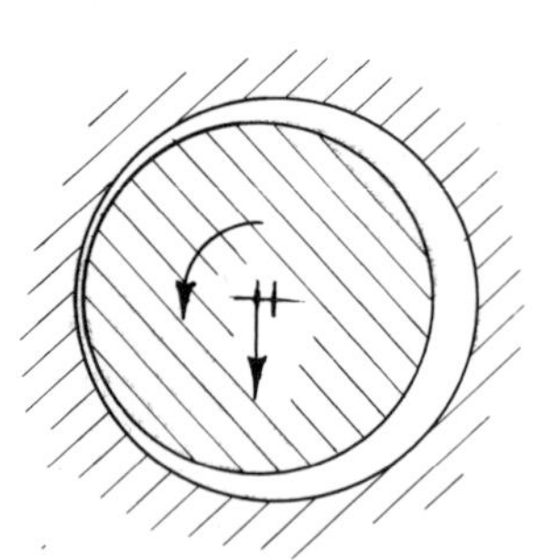

Figure 14.2 Diagram to illustrate precessing motion of piston in balance

The effect of friction cannot be totally eliminated, but it can be greatly reduced by steadily rotating the plunger, especially if the fluid in the system has reasonable lubricating properties. Under these circumstances it has been shown by Michels[14.1] that metallic contact between the plunger and the walls of the cylinder is intermittent. If the system was geometrically perfect and the plunger was perfectly central, i.e. the clearance the same all round the circumference and throughout the immersed length, the situation would be metastable. In practice geometrical imperfections obviously preclude this and the axis of the plunger will normally be eccentric relative to that of the cylinder. The effect of this in a viscous liquid is to produce a force on the plunger at right-angles to the eccentricity, see Fig. 14.2, which in turn causes the axis of the plunger to describe a tight spiral about the axis of the cylinder, until the two surfaces touch. This at once upsets the distribution of viscous forces and introduces a radially inward push on the plunger which once again floats freely until this action is repeated. Michels (loc. cit.) demonstrated this very neatly by insulating the system and connecting the cylinder and a slip ring on the plunger into an electric circuit with means for recording the resistance at any moment. By this means he showed that, so long as the speed of rotation was above a minimum (which was usually no more than 15 or 20 rev/min), there was only intermittent metal-to-metal contact every three or four revolutions or so and the duration of the contact was never more than about 1/20 of the duration of a complete revolution.

A further difficulty is encountered if the pressure-transmitting fluid is very thin, i.e. of low viscosity, since there will then be considerable leakage through the clearance round the plunger; but if it is thick, it will introduce a considerable error through the resistance to the vertical movement of the plunger due to its viscosity. Also viscous liquids such as thick oils tend to "freeze" under pressure at relatively low intensities.

For very high pressures one comes up against a problem of design; if the plunger is to be of such a diameter that the manipulative problems of lapping it accurately into the cylinder are reasonable, the dead weight will be so large and clumsy that the instrument will be difficult to operate: if, on the other hand, the weight is moderate and easy to handle, then the plunger becomes inconveniently thin, like a knitting needle. Michels produced a design which, to some extent, avoided this difficulty by using a differential plunger and cylinder. In this the plunger had a small shoulder on it and there was a step in the bore of the cylinder.

The plunger thus protruded right through the cylinder at top and bottom and was loaded by weights hung beneath it, while the upper end was used for applying the rotary motion, Fig. 14.3. This was a much more convenient arrangement, although it involved a very difficult boring and lapping operation to get a good and truly coaxial fit between two cylindrical surfaces of slightly different diameter. It also resulted inevitably in a considerable increase in the area of the leakage channels. However, it has proved possible to make the shoulder about $\frac{1}{4}$ mm in radial width with the plunger of about 1 cm in diameter. The area of the shoulder is then no more than 0·080 cm^2 so that a weight of 200 kg (which is reasonably manageable with this type of instrument) will balance about 2·5 kb.

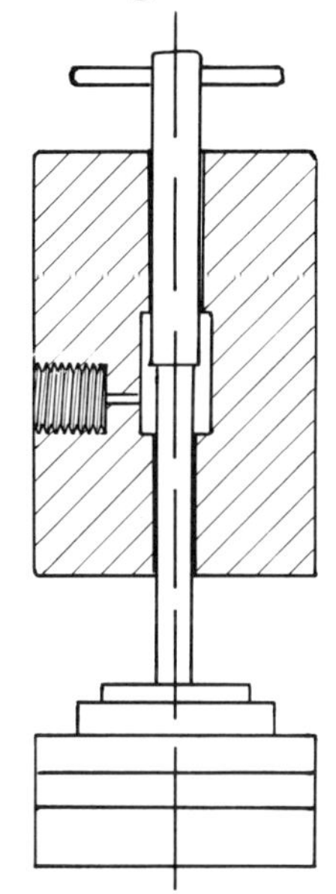

Figure 14.3 Michels's differential piston dead weight balance

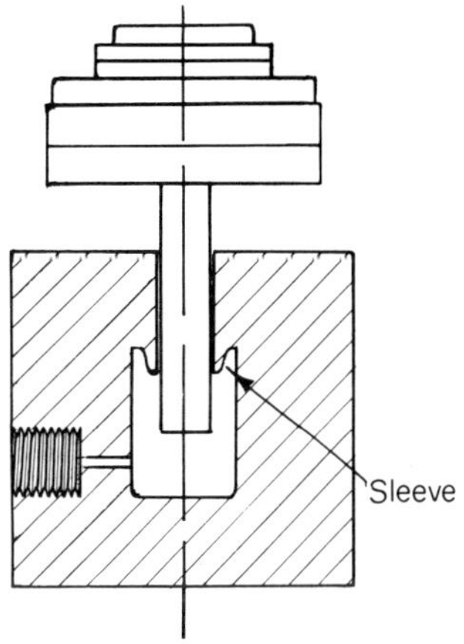

Sleeve

Figure 14.4 Bridgman's lip seal modification to simple dead weight balance

Examples have been made to work up to well over 3 kb, but the rate of leakage then becomes so fast as to make the manipulation required in taking a reading decidedly troublesome. Various devices have been tried with the object of controlling this leakage; Bridgman[14.2], for instance, put a sleeve round his plunger projecting into the pressure space, see Fig. 14.4, so that this sleeve would be under hydrostatic pressure and thus not expanded by the pressure. In the experience of the authors, however, this is not very satisfactory because it tends to close up and grip the plunger, making it too tight to rotate. Morrison's gland, see Fig. 13.13, would seem to be worth trying in this connection, but probably the most successful device is that developed by Johnson and Newhall[14.3] and incorporated in the units manufactured by the Harwood Engineering Co. of Walpole, Mass., U.S.A. This comprises an annular jacket surrounding the actual cylinder bore and connected to a separate pressure supply so that the clearance can be controlled by adjusting this jacket pressure, see Fig. 14.5. Another ingenious arrangement described by these authors enables the weight to be supported on a thrust

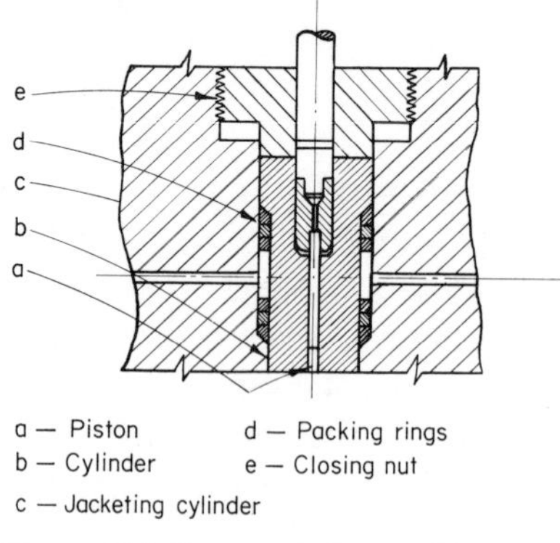

Figure 14.5 Pressure balance with pressurised jacket *(Harwood Engineering Co. (Ref 14.3))*

race except during the actual pressure measurements when it is lifted off and supported momentarily by the very small plunger. Standard units of this kind are available for 100,000 and 200,000 lbf/in^2 (approximately 7 and 14 kb respectively).

14.2.1 *Calibration of Dead-Weight Balances*

Michels[14.1] showed how the effective diameter of his differential plungers could be calculated, taking into account the changes in dimensions due to the elastic distortion of the components. He also discussed the effects of drag in the transmitting liquids and other possible sources of error, and concluded that there would still remain some uncertainties. He therefore devised a means of calibration, using a mercury column in the tower of a church in Amsterdam, and a piezometer which he had perfected for other projects. Essentially the latter is a pressure vessel having within it a glass tube closed at its upper end and with several platinum contacts fused into it. This tube is filled with gas over mercury and the space outside the glass and above the mercury is filled with the pressure-transmitting oil.

Now the application of pressure forces the mercury up inside the glass, compressing the gas as it goes, and—provided the temperature is kept constant—the pressure at which the mercury shorts any particular one of these contacts will be the same to a high standard of accuracy. For calibrating a pressure balance Michels used a system whereby he could switch the mercury column in and out of his system, so that for one reading the head of mercury was in series with the dead-weight balance and for the other the balance was direct onto the piezometer. Thus, if h be the head of mercury and ρ its density, D the effective diameter of the plunger, W_1 and W_2 the weights to balance the pressure p in the piezometer with and without the mercury column in the system respectively, then:

$$p = \frac{W_1}{\pi D^2/4} + h\rho = \frac{W_2}{\pi D^2/4} \tag{14.2}$$

This assumes that the effective diameter is not substantially altered by the difference in pressure at the balance cylinder due to the mercury head. A closer approximation can be obtained by assuming that the effective diameter varies linearly with the applied weights, i.e. $D = D_o(1+\alpha W)$ where D_o is the initial diameter and α a constant; then:

$$p = \frac{4W_1}{\pi D_o^2(1+\alpha W_1)^2} + h\rho = \frac{4W_2}{\pi D_o^2(1+\alpha W_2)^2} \tag{14.3}$$

Since W_1, W_2 and h are measured, D_o can be obtained from one experiment using eq. (14.2), but for the more refined method of eq. (14.3) two experiments at different values of p, i.e. at different contacts in the piezometer, are necessary.

More recently Dadson[14.4] has described an ingenious method of calibration using plungers of different materials so that the elastic constants are varied. According to his work the correction varies linearly with pressure.

For a comprehensive discussion of the errors in gauges of this kind, readers are referred to the work of Bett, Hayes, and Newitt[14.5] in which an ingenious method is described for using a single mercury column up to much higher pressures than its own head by an additive system which, though laborious, was shown to be capable of great accuracy and sensitivity. It was shown that a pressure of one kb could be measured by a dead-weight gauge to an accuracy of 1 part in 1,265. This paper also contains a comprehensive and critical review of the literature on the subject.

14.3 Secondary Gauges

14.3.1 *Bourdon Tube Gauges*

These are so well known and universally used that we need only give a brief note on them here. The essential feature is a tube, closed at one end, flattened to a more or less elliptical section (in some cases with straight parallel sides along the longer axis of the cross-section) and then bent into a curve in the plane of the shorter axis. When pressure is applied to the inside of a tube of this form it tends to unbend, so if the open end is fixed and the closed end connected to some simple rack and pinion mechanism it is easy to transform this movement into the rotation of a needle in front of a graduated dial.

For really high pressure work the tubes are generally made by boring out a piece of solid rod of carefully selected and heat-treated alloy steel. The middle part of this is then flattened and bent into the required curve while the still circular ends are screwed so that one can be closed by a cap and the other connected to a base plug which supports the whole assembly. A useful safety precaution is to mount the dial on a disc of fairly hard steel and put the tube behind it. The needle is then mounted on a spindle passing through a hole at its centre so that it moves over the scale with the steel disc between the tube and the observer. Safety glass or "Perspex" should be used if a dust cover is needed on the dial, but at the back the tube should be uncovered with, for preference, a vent system in case it fails.

For very high pressures the Bourdon tube can be heat-treated after flattening and bending, and some authorities advise a low temperature heat treatment, usually at about 300°C, after subjecting to an initial pressure test. This is said to reduce the hysteresis in readings between rising and falling pressures and is probably worth while if the gauge is to be operated for long periods near the maximum of its scale.

Most manufacturers recommend using gauges which are designed for twice the working pressure which will be required in normal running, but in the authors' experience these are very robust instruments and will usually stand long operation at or near the maximum figure on the dial without serious loss of accuracy. Even then the commonest trouble is the development of a zero error, but if the needle is moved on the spindle so that it again reads zero when there is no applied pressure re-calibration will usually show negligible differences from the original.

The most severe service for Bourdon tube gauges is with rapidly fluctuating pressures, as on high speed compressors, and in such circumstances it is advisable to fit a check valve or some sort of constriction in the connecting pipe to prevent this pulsation effect from reaching the tube, where it could easily do serious damage by fatigue or – if it picked up a synchronous frequency in the gauge – by simple overstraining.

It may be worth adding one small point. Many manufacturers still fit a stop at the zero point, and it is usually desirable to remove this; otherwise, any sudden release of pressure will cause the needle to swing back and hit the stop, often breaking itself in the process, and in small-scale research units it is very difficult to bring the pressure down slowly enough to avoid this.

14.3.2 *Eccentric Tube Gauge*

If a straight circular bar is drilled longitudinally with a hole that is off centre, and if this is then subjected to internal pressure, it will bend. The stress distribution under these conditions is complicated, but it is fairly obvious that – in the axial direction – it will be higher where the wall is thinner, and consequently the strain will be greater there. Thus, if one end is fixed, the other will move; and, if the tube is not overstrained in the process, this movement will be proportional to the pressure.

The earliest applications of this effect seem to have used an optical system of magnification, but the Budenberg Gauge Co. of Broadheath, near Manchester, are now marketing a well-tried gauge with mechanical amplification to work up to 10 kb, see Fig. 14.6. The possibility of extending this to even higher pressures by some sort of reinforcement that gives support to the tube in the plane normal to the axis without stiffening it in the other directions would seem worth studying, perhaps using shrunk-on discs as in the "laminar" method described by Birchall and Lake.[14.6]

14.3.3 *Strain Gauge Transducers*

A number of other elastic deformation pressure gauges have been tried from time to time, but space only permits us to describe one general type. In this the expansion of a tube is measured by strain gauges attached to its surface. Using electric resistance gauges enables the signal to be carried well away from any danger area to a suitably placed instrument panel or control room, and the device can therefore have a number of practical advantages on an industrial plant.

One of the basic difficulties, however, is to eliminate the effects of temperature

Figure 14.6 Eccentric tube gauge *(Budenberg Gauge Co. Ltd, Altrincham)*

change and most modern applications of this principle use two resistance strain gauges, one subjected to strain and the other not, but connected differentially so that any change in temperature is likely to affect each equally and consequently to have no influence on the strain measurement. A system which was first developed by the Baldwin–Southwark Corporation during the Second World War and originally placed on the market in 1945 is shown schematically in Fig. 14.7a. It is made from a rod of high tensile steel, bored out for about half its length. Two similar strain gauges are then bonded to the outside surface, one over the hollow portion and the other over the unbored solid. By connecting these in opposition the only resistance change will be due to the tangential strain on the outer surface of the hollow portion, which – neglecting the supporting effect of the solid portion – will be given by:

$$\varepsilon_\theta = \frac{p_i}{E(K^2-1)}(2-\nu) \qquad (14.4)$$

where p_i is the internal pressure, K the diameter ratio, and E and ν the values of Young's Modulus and Poisson's Ratio respectively.

Now, by reference to Section 2.3, it is easy to see that if τ_{al} is the limiting shear stress that can be allowed:

$$(\varepsilon_\theta)_{max} = \frac{1}{E}(\tau_{al}-p_i)(2-\nu) \qquad (14.5)$$

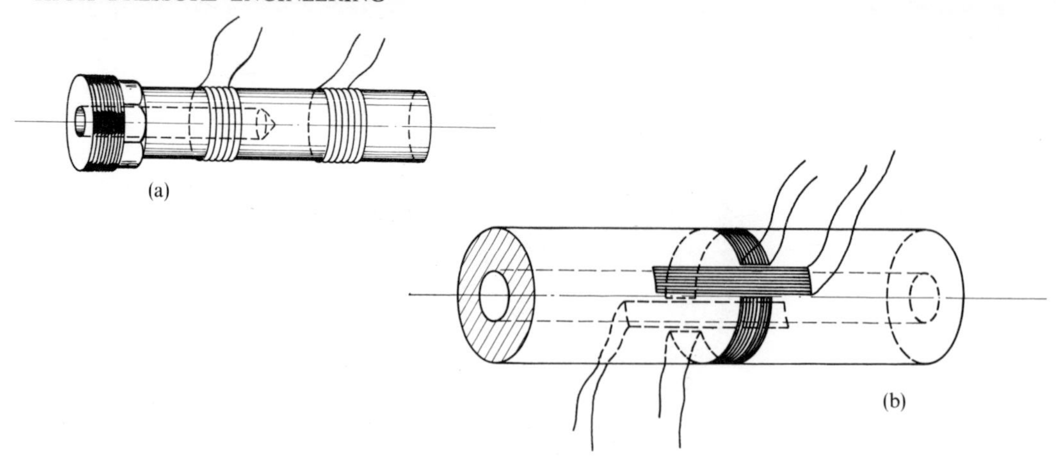

Figure 14.7 Strain gauge transducer arrangements: a) usual dead end type; b) proposed through pipe arrangement

Consequently we want the lowest values of E and ν and the highest of τ_{al} to give the greatest sensitivity. Table 14.1 shows the strains which would be given by transducer tubes of various materials when used to measure a pressure of 3,450 bars (50,000 lbf/in²). In this case we have taken the limiting shear stress as $1/\sqrt{3}$ times the tensile elastic limit.

Table 14.1

Transducers for Maximum Pressure of 3,450 bars (50,000 lbf/in²)

Material	*Ultimate strength*		*Elastic limit in shear bars*	*Young's Modulus bars* $\times 10^{-6}$	*Poisson's Ratio*	$\varepsilon_\theta \times 10^3$	*K*
	lbf/in²	*bars*					
0·15% C steel	61,000	4,200	1,450	2·07	0·30	—	—
En 25 steel	126,000	8,680	4,150	2·055	0·28	0·58	2·43
Cr, Mo, Co high strength stainless	215,000	14,850	7,140	2·09	0·32	2·97	1·39
18% Ni, 9% Co, 5% Mo, maraging	292,000	20,200	10,500	1·86	0·30	6·44	1·22
Hard-drawn phosphor bronze	105,000	6,350	3,670	1·033	0·37	0·35	4·04
1·8% beryllium copper	170,000	11,750	5,770	1·29	0·35	2·98	1·58

The first entry in the Table, a low carbon steel, is seen to be useless for this service because it would be overstrained by this pressure however thick it were made. All the others would contain it, but—as will be seen—the weaker ones require very thick walls, i.e. a high diameter ratio (K), and in consequence the strain at their outside surfaces would be small. In fact the values of ε_θ in the 7th column are really the figures of merit from this point of view.

The maraging steel is particularly good because in addition to its enormous strength it has a relatively low value for Young's Modulus, although these calculations give a very thin wall, and in practice it would be advisable either to increase the thickness somewhat, or else to have the actual transducer tube surrounded by a bullet-proof protecting box.

The sensitivity of these instruments would be very much increased if the strain gauges could be stuck on the inside surface, but it would be difficult to do this without making the tube much bigger, and there remains some doubt about the satisfactory life of a strain gauge in this service, surrounded by a liquid at high pressure. Probably the best way to obtain an enhanced sensitivity would be to have a cylindrical transducer with the pressure on the outside and the gauge on the dry bore. This also means increasing the size and probably the external type is good enough for most users.

Beryllium copper is much used for this purpose by the leading manufacturers of these instruments and, as the Table shows, its properties are very favourable for such service. It is, however, a difficult material to obtain at these high strengths without some local variations in hardness, and in view of this uncertainty, a protecting box is desirable here also.

Clearly many changes can be rung on the use of electric resistance strain gauges for the measurement of pressure, but space does not allow us to discuss them in these pages. One possibility does, however, seem worth mentioning. Nearly all the pressure gauges in normal use involve a dead end, and in chemical processes where a solid product is formed there is always the chance that the line to the gauge will get blocked. With this device, however, there is no reason why an actual part of the pipe system should not be made to measure the pressure, as shown in Fig. 14.7b. If similar strain gauges are applied, one in the circumferential and one in the axial direction, then – since the strain in the former will be considerably greater than that in the latter – we can again eliminate temperature effects by connecting the gauges in opposition. The strain measured would then be:

$$\varepsilon_\theta - \varepsilon_z$$

which is related thus:

$$E(\varepsilon_\theta - \varepsilon_z) = \frac{p_i}{K^2 - 1}(1 + \nu) \tag{14.6}$$

Comparing this with the relation of eq. (14.4), we see that it is smaller in the ratio $(1+\nu)$ to $(2-\nu)$, or – since ν has values varying between about 0·30 and 0·35 in the materials most suitable for these elements – between 22 % and 31 %.

14.3.4 *Pressure Measurement by Change of Electrical Resistance*

The effect of hydrostatic pressure on a wire carrying an electric current is to increase its resistance. This varies considerably from metal to metal, but it is never very large and is always apt to be swamped by the temperature effects. The suggestion that it should be used for pressure measurement seems to have originated with La Fay, who published in 1909[14.7] the results of a series of measurements with wire of Manganin, an alloy with a very small temperature coefficient of resistance. Bridgman[14.8] investigated this and initially had some trouble because of the variation of the pressure effect from batch to batch of Manganin wire. This seems to have been due partly to small variations in composition (Manganin normally contains roughly 10% nickel, 5% each of iron and manganese, and the rest copper) and also to cold work, and he got over these difficulties by obtaining a considerable quantity of the wire from the same drawing, and then annealing each measuring coil before he used it. Unfortunately, however, his remark that "its temperature coefficient of resistance is so low that no special precautions are necessary in the way of keeping temperature constant" has been found to be far too optimistic for accurate work.

Michels and Lenssen[14.9] made a further study of the problem and devised an instrument in which two similar coils of Manganin wire were inserted in two identical cavities in a large block of steel; both of the latter were filled with oil,

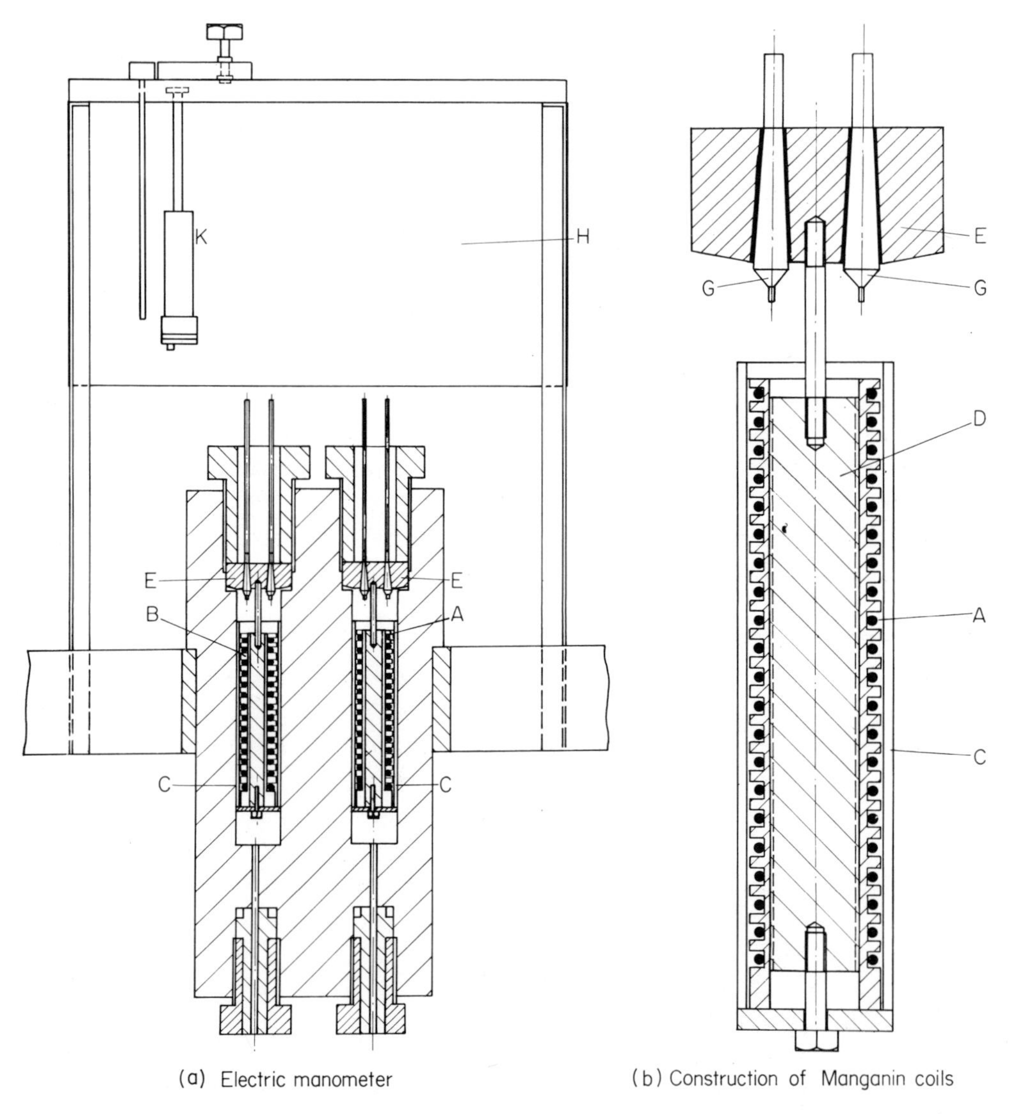

Figure 14.8 Manganin wire electric resistance pressure gauge: arrangement by Michels and Lenssen (Ref 14.9)

but only one was subjected to pressure. The coils were then connected into opposite arms of a Wheatstone bridge circuit, and the temperature effect was thus virtually eliminated. As the change in resistance with pressure is virtually a linear effect without hysteresis, calibration only requires one or two fixed points. It may be desirable to repeat this occasionally if the instrument is extensively used owing to the tendency for the zero to drift. In such cases it is also a good thing to re-anneal the wire, which can be done very easily by merely raising its temperature to 140°C. Fig. 14.8 is reproduced from Ref. 14.9 and shows a section through the instrument at (*a*) and details of the coils at (*b*).

Michels and Lenssen also call attention to the need to allow sufficient time after a sudden change in pressure for the change in temperature of the oil due to adiabatic compression or expansion to be dissipated. This effect is much larger than might be expected, although – with large blocks of metal to absorb the temperature changes – it is not serious unless the variations are frequent or continuous; it does, however, mean that any gauge working on this effect will not be satisfactory for measuring or controlling rapid fluctuations of pressure.

Various other alloys have been studied, mainly in the hope of finding one with a still smaller temperature coefficient. Gold with about 2% of chromium is a possibility here although its pressure effect is also much smaller. Darling and Newhall[14.10] have looked into this with encouraging results, but so far it does not seem to have displaced Manganin.

This type of pressure-measuring device also has the merit that it can be used up to very high pressures with considerable accuracy, provided the means are available for its calibration. The biggest problem then is likely to be the electrodes, although the recent developments using epoxy resins in a conical annulus (see Section 13.7.2) have gone a long way towards solving this. We know little about the effect of pressure on the resistance of such pressure electrodes, but their compactness and simplicity make it comparatively easy to add an extra pair so that the resistance of the actual Manganin coil can be measured by a potentiometer circuit in which no current is flowing.

14.4 Pressure Measurement in Intensifiers

In equipment for the small-scale generation of very high pressures the intensifier principle is almost invariably used, and whilst the high pressure side may go up to intensities which are difficult to measure this does not apply in the low pressure system. Unfortunately we do not, as a rule, know how much is lost in friction and thus an estimate of the high pressure generated merely by measuring the low pressure and multiplying it by the intensification factor (i.e. the ratio of the areas of the low and high pressure pistons) will be in error by this unknown factor.

With plunger packings of the type used by Crossland, Skelton and Wilson[14.11] or by Thomas, Turner and Wall[14.12] the friction is a relatively small percentage of the total forces and it is therefore possible to get a closer approximation if we have the means to detect movement of the plunger system. Thus if after applying a low pressure p_L to the low pressure plunger whose diameter is D_L to give a pressure p_H in front of the high pressure plunger whose diameter is D_H, we decrease the low pressure until at a value p_L^1 the plunger system starts to move back, it is evident that if F_a is the total frictional force while the system is raising the pressure and F_r that when it is moving in the opposite direction, the equilibrium demands that:

$$p_L \frac{\pi}{4} D_L^2 = p_H \frac{\pi}{4} D_H^2 + F_a \qquad (14.7)$$

and

$$p_L^1 \frac{\pi}{4} D_L^2 = p_H \frac{\pi}{4} D_H^2 - F_r \qquad (14.8)$$

Now if the friction is fairly small (say not more than 5 % of the total force acting), it will not introduce any great error if we assume that:

$$F_a = F_r \qquad (14.9)$$

Then, by adding equations (14.7) and (14.8), we get:

$$(p_L + p_L^1) \frac{\pi}{4} D_L^2 = 2 p_H \frac{\pi}{4} D_H^2$$

which simplifies to:

$$p_H = \left(\frac{p_L + p_L^1}{2}\right) \beta \qquad (14.10)$$

where β is the intensification factor.

Using this procedure it is a simple matter to obtain a rough calibration relating the high and low pressures.

14.5 Fixed Points

The calibration of secondary gauges in the pressure range beyond about 10 kb depends on certain known changes of phase. The first is the pressure freezing of mercury at 0°C. This was suggested by Bridgman (loc. cit., p. 73), but later work has shown that the value he gave was considerably in error. Various experimenters have since worked on this using different methods, and it appears that that of Dadson and Greig(14.13) is the most reliable. The value they give is:

7569·2 kilobars

Their paper also reviews previous work and discusses the probable error in their own experiments.

For higher pressures, transitions in solids from one allotropic form to another have proved the most useful, although their determination usually depends on extrapolating such effects as the pressure change in resistance of Manganin or some similar wire. However, Kennedy and La Mori(14.14) have shown that there is a change in pure bismuth at 25·410 kb (±0·095) and in pure thallium at 36·690 kb (±0·110). These workers used a dead-weight piston gauge acting directly onto the solid and detected the change by reason of the sharp change it brought about in the electrical resistance of the specimen.

Further points have been determined, but with less accuracy. For instance Balchan and Drickamer[14.15] suggest a change point in barium at 59 kb and a second change in bismuth at 90 kb, their accuracy being reckoned at about $\pm 3\%$.

REFERENCES

14.1. MICHELS, A. M. J. F., *Ann. Phys.,* **72,** 285, 1923.
14.2. BRIDGMAN, P. W., *The Physics of High Pressure,* 1st Edition, Bell, London 1931, 65.
14.3. JOHNSON, D. P. and NEWHALL, D. H., *Trans. Am. Soc. Mech. Eng.,* **75,** 301, 1953.
14.4. DADSON, R. S., *Nature,* **176,** 188, 1955.
14.5. BETT, K. E., HAYES, P. F. and NEWITT, D. M., *Phil. Trans. Roy. Soc.,* **247,** 59, 1954.
14.6. BIRCHALL, H. and LAKE, G. F., *Proc. Inst. Mech. Eng.,* **156,** 349, 1947.
14.7. LA FAY, A. *C. R. Acad. Sci (Paris),* **149,** 566, 1909.
14.8. BRIDGMAN, P. W., *op. cit.,* 8.71 et seq.
14.9. MICHELS, A. M. J. F. and LENSSEN, M., *J. Sci. Instrum.,* **11,** 345, 1934.
14.10. DARLING, H. E. and NEWHALL, D. H., *Trans. Am. Soc. Mech. Eng.,* **75,** 311, 1953.
14.11. CROSSLAND, B., SKELTON, W. J. and WILSON, W. R. D., Conference on High Pressure Engineering, London, Sept. 1967, Paper No. 8, *Proc. Inst. Mech. Eng.,* **182,** Pt 3C, 175.
14.12. THOMAS, S. L. S., TURNER, H. S. and WALL, H. F., *ibid.,* Paper No. 11, 271.
14.13. DADSON, R. S. and GREIG, R. G. P., *Brit. J. Appl. Phys.,* **16,** 1714, 1965.
14.14. KENNEDY, G. C. and LA MORI, P. N., Contribution to *Progress in Very High Pressure Research,* Wiley, New York 1961, 304.
14.15. BALCHAN, A. S. and DRICKAMER, H. G., *Rev. Sci. Instrum.,* **32,** 308, 1961.

15 Other Measurements—Temperature, Flow and Sampling

15.1 Measurement of Temperature

15.1.1 *General Considerations*

In most industrial applications the general practice is to use thermocouples for the measurement of temperature in high pressure plants, since they are easy to fit and service. Whatever sensitive element is used, however, will be subjected to almost perfect black body radiation conditions from the inner surfaces of the pressure container; and, if the pressurised substance is at a different temperature from that of the walls, the element may well be more affected by this radiation than by the convection from the substance. This is not, of course, peculiar to high pressure systems, and the point of mentioning it here is merely to remind the reader that it needs more consideration than it all too often receives.

A quick response to temperature changes is obviously desirable for control, especially where these may be rapid and unpredictable. This clearly demands a measuring element with the smallest possible heat capacity, and wherever possible bare wires should be used. If the maximum temperature will permit—about 200°C with a suitable design—the bare wires can be carried through the container wall in a conical annulus of epoxy resin (see Section 13.7.2), but this limit can be considerably exceeded if the neighbourhood of the resin is cooled, for instance by an air jet, or by a small water jacket. For higher temperatures, glass or quartz insulation becomes necessary, and the design problems become more involved, though not intolerably so.

We may however be confronted by a situation in which the pressurised substance would affect the bare wires of the element by corrosion, or their readings by electrical conduction. We are then forced to use some kind of sheath, which inevitably introduces a time lag into the response of the measuring instrument being used, although this can usually be kept within acceptable limits.

The sheath is unlikely to present any very formidable design problem, since it is a short blank-ended tube with pressure on the outside. Usually a considerably lower diameter ratio can be used in this than in the pressure-containing cylinders and pipes, the chief consideration being its resistance to inward collapse and, as eq. (3.33) shows, when the length is comparable with the diameter the stability is markedly increased.

15.1.2 *Thermocouples*

Most of the commoner pairs of thermo-element materials, such as iron with constantan, chromel with alumel, platinum with platinum plus 10% rhodium alloy, can be used for bare wire junctions within the pressure space. At this point it may be worth giving the rather obvious, though often overlooked, warning that much care is needed to ensure that these wires are not fouled by any of the internal fittings of the container in a way which might cause an electrical fault, or in extreme cases—as with a moving stirrer—torn off. But the great advantage of using bare wires is their very rapid response to temperature changes in the pressurised substance. Mention has already been made of the liability for the readings to be influenced by the radiation from the container walls; the most effective way of dealing with this is probably to surround the junction with a thin shield which will reflect some of the radiation and be light enough to respond to changes in the temperature of the substance as it flows past.

Even where corrosive conditions are encountered, it may be possible to use a tubular type of thermocouple, in which the junction is made between a tube of one metal and an inner wire of the other. This will have a somewhat greater heat capacity than the bare wires, and therefore a less rapid response, but it will still be a good deal better in this respect than the use of a separate sheath. Also, by making the outer component from a metal resistant to the working substance while the other remains unexposed to it, the couple can be safely used in the pressurised space. In constructing such a thermocouple it is advantageous for the annular space between the tubular component and the central wire to be filled with a relatively soft insulating material, such as Pyrophyllite, which will become compacted and transmit the pressure so that both components are subjected to substantially the same pressure.

The influence of pressure on the thermo-electric effect has been investigated by a number of workers, including Bridgman[15.1], who showed that this varies greatly from material to material. Generally, however, with solid metals it is small, i.e. of the order of 5×10^{-12} volts per deg C per bar. The sign of the effect also varies with different combinations. More recent studies of the problem, notably that by Boyd and England[15.2], largely confirm Bridgman's results, as well as extending them to much higher temperatures and pressures (over 1,000°C and over 40 kb).

It is worth a passing notice that mercury—and probably most metals when in the liquid phase—exhibits a much greater pressure effect when forming a component of a thermocouple than do the solids; it is of the order of 100 times greater. Thus the suggestion, which has been reported from time to time, that where one has, for instance, a mercury seal inside an experimental reaction vessel it should be used as part of the couple for measuring the temperature is not likely to be very reliable if readings are required at a number of temperatures over a considerable range.

For most industrial purposes, the pressure effect can be judged with sufficient accuracy by inserting a couple in a small vessel which can be brought up to the required pressure with the appropriate working substance and held steady for some time, while a similar couple in a similar vessel is at atmospheric pressure, but at the same temperature. By connecting the couples differentially, the magnitude of the effect can be gauged. It would, however, be of great value if an absolute measurement of temperature at high pressure could be devised, and Garrison and Lawson[15.3] have suggested that this might be achieved by means of the thermal "noise" phenomenon in resistors, since this appears to be independent of absolute pressure.

Where it is necessary to have a sheath inserted into the pressure container, the response will naturally be considerably slower. On the other hand, since the thermo-element will not be under pressure at all, there can be no question of the pressure affecting the temperature reading; and with such a system any method of temperature measurement can be used, and often ordinary mercury-in-glass thermometers can prove very handy and efficient.

15.1.3 *Resistance Thermometers*

Platinum resistance thermometers can be used in high pressure service, although the extra complications involved are seldom justified in industrial applications, except in unusual circumstances. For research studies on a small scale they can be very valuable, especially when the coil can serve both as a heater and a temperature measuring instrument. It must be remembered, however, that in any system of this kind we are faced with the problem of separating the pressure effects from those due to temperature, and such corrections could become a source of uncertainty with temperatures in excess of about 300°C. Fortunately, in the case of platinum in the pressure range up to 12 kb and between 0 and 100°C, the effect of the pressure is to decrease the resistance by about 2·2%, and the effect appears to be linear with pressure; these figures are taken from a table given by Strong.[15.4]

One such application for the small-scale study of gas reactions at constant pressure comprised a thin-walled platinum capillary fitted within a steel high pressure tube and insulated from the rest of the system. The platinum tube was coned out at each end and pressed over quartz cones which were in turn pressed over steel cones integral with the end plugs. The platinum was heated by direct current and the experimental gas was forced along the annulus between the platinum and the steel tubes before returning inside the former, and then leaving the system through a let-down valve, from which it could be sampled and analysed. The temperature of the platinum was measured by means of a potentiometer circuit, using separate electrodes for its connections. Fig. 15.1 indicates diagrammatically the arrangement, and it should be noted that a comparatively large electrode is needed to carry the heating current.

This example is introduced to illustrate the sort of temperature control that can be used with the aid of platinum resistance thermometers.

15.2 Measurement of Flow

15.2.1 *General Considerations*

This is a problem that can present considerable difficulties in large-scale plant,

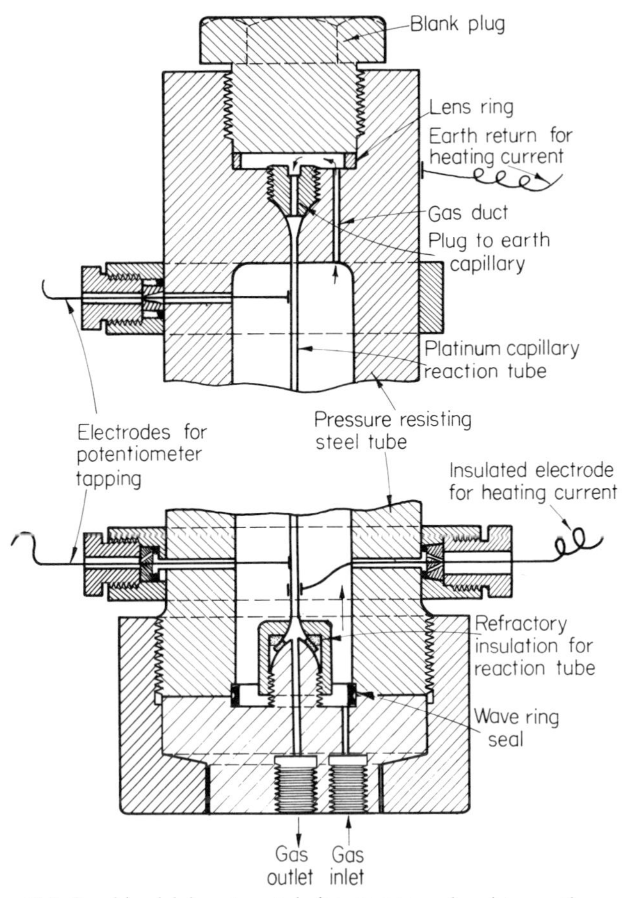

Figure 15.1 Combined laboratory tubular reactor and resistance thermometer

although in recent years several reasonably adequate methods of dealing with it have been developed. Essentially these fall into three main categories, namely mechanical, hydrodynamic and thermal.

The first includes determining the total delivery of a compressor from its suction and delivery pressures and temperatures, its speed, and its volumetric efficiency. Under steady running conditions this suffices for many of the plant control requirements, although it is obviously no help when the flow of pressurised fluid has to be distributed between a number of consuming units such as reactors; moreover, it is in situations of this kind that the problem is most important.

The other mechanical solution takes the form of small axial flow turbines fitted into non-magnetic pipes and sending out signals through the walls by means of permanent magnets in their rotors.

Hydrodynamic applications are based on measuring the pressure drop at a constriction, a problem much complicated in high pressure plants by the fact that it is then a matter of measuring the small difference between two large quantities. Methods are, however, available for measuring the difference direct. Another hydrodynamic method is with the well-known rotameter principle, in which the fluid is made to flow up a vertical pipe which is slightly coned so that its area gradually increases; in it is a small spinner which is carried up on the stream until it floats more or less stationary, the position depending on the dimensions and the rate of flow. Normally, at low pressures this is observed visually through glass, but at high pressures it is necessary to make the rotameter tube of a non-magnetic metal and to sense the position of the spinner by electromagnetic means.

The thermal method depends on putting heat into the fluid at a steady rate and measuring the temperature further downstream. This method is particularly suitable for judging the flow variations in branch pipes. Another thermal possibility depends on the rate of loss of heat from a platinum wire carrying a constant current while exposed to the stream of fluid. This carries away more or less of the heat thus generated according to the rate of flow, and thereby affects the temperature of the wire, which in turn affects its resistance.

15.2.2 *Mechanical Methods*

In many chemical plants it is common practice for high pressure reactors each to be fed by its own compression unit, and the feed rate can then usually be controlled with sufficient accuracy from the speed of the machine. As we have seen in Chapter 9, the delivery rate of a reciprocating compressor is a function of the speed the suction and delivery pressures and temperatures, the swept volume and the volumetric efficiency. With well maintained machines the last item will vary very little and, under normal running conditions, the delivery rate will be almost directly proportional to the speed of the machine. In fact, speed control of this kind may well prove to be the best means of controlling the whole process, but to correlate this quantitatively will generally require careful preliminary calibration.

The problem is more complicated when the compression is carried out by rotary machines, and, owing to their very large capacity, such installations are usually made to feed more than one reactor unit, thus calling for flow measurement in each feed pipe. The output of a rotary machine is also a function of speed and pressures, etc., but—as explained in Chapter 11—it is a good deal more involved and depends on many other factors than arise in the case of reciprocating compressors. On the other hand, the mechanical meters which can be installed in the actual pipes bringing the fluid feed to reactors are small and simple, being in fact little axial flow turbines which are rotated by the flow, and whose speed is directly related to its magnitude. The rotors are mounted in lengths of austenitic steel (non-magnetic) pipe, and their rotation can be picked up by suitable electromagnetic sensing arrangements.

A device of this kind was developed some years ago by the Potter Aeronautical Company, of Union, New Jersey, U.S.A. and is now made and marketed in the U.K. by Kent Instruments (Stroud) Ltd. of Stroud, Glos, see Fig. 15.2. The casing forms part of the pipe system and can be easily inserted in it. The bushings which bear the rotor are held in place within the casing by means of thin cylindrical rings in sets of three, which spring into place and thus ensure correct alignment. The rotor is so designed aerodynamically that the venturi effect tends to push it downstream and so to balance the downstream thrust on the blades; it is

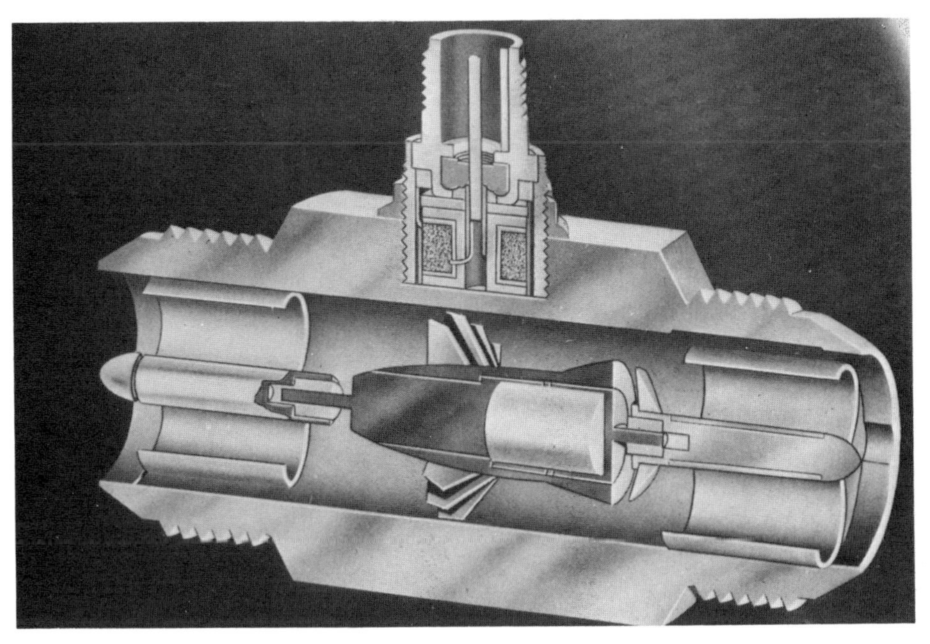

Figure 15.2 Potter flowmeter. *(Pottermeter Division, Kent Instruments (Stroud) Ltd, Gloucestershire)*

thus free to wander in an axial direction until these forces balance one another. The sensing element generates an alternating current whose frequency is proportional to the rotational speed, which in turn is proportional to the flow of the fluid. The current is then rectified and the resulting direct current voltage measured, either with a potentiometer system, or directly with a voltmeter. The standard instrument is suitable for pressures up to about 2,400 bars (35,000 lbf/in^2) and for fluid densities up to 1·6. It is also designed for high temperature work and units for 450°C are available.

15.2.3 *Hydrodynamic Methods*

A stream of fluid moving along a parallel pipe with a constriction in it will produce a pressure drop between the up and down stream sides, just as it does at the usual industrial pressures; the problem is to measure this drop when it is the relatively small difference between two large pressures. One method of doing this is by means of a mercury lute contained in a steel pipe bent into the arc of a circle. The two ends of this are closed and connected to points above and below the constriction in the pipe by coils of fine-bore capillary tubing which can be flexed considerably by very small forces. If then the lute is pivoted about the centre of the arc so that it is free to rotate in a vertical plane, it is clear that a difference of pressure applied to the two mercury surfaces will cause the mercury itself to move round the arc till it is in equilibrium with the pressures, and this in turn will cause the arc of the tubing to rotate to balance the mercury. Fig. 15.3 shows diagrammatically how such a system can be arranged, in this case with a pointer to indicate the movement. According to Perry[15.5] this was known as the Bosch flowmeter, having been first used apparently in the original Haber

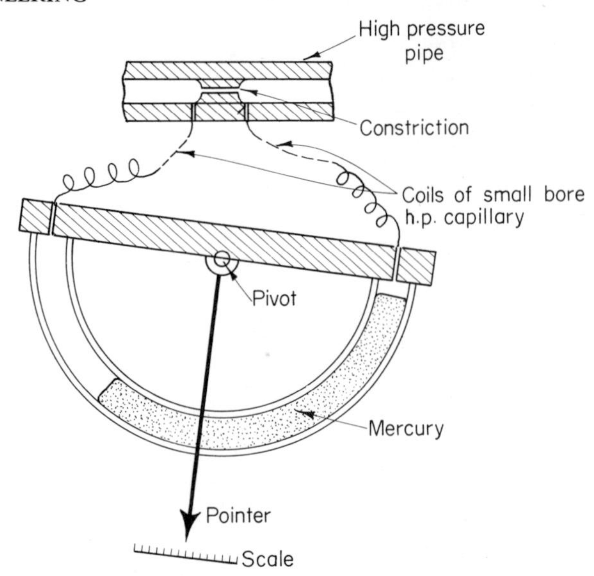

Figure 15.3 Bosch constriction flowmeter

Bosch synthetic ammonia plants. The correlation between the actual flow and the deflection of the lute tube can probably be calculated with fair reliability if the properties of the fluid are accurately known, and if the characteristics of the lute system have been determined. On the other hand, it will always be safer to calibrate such instruments, since there may be some doubt as to the discharge coefficient under varying pressure conditions.

An alternative method is to have a stationary U-tube filled with mercury and to measure the displacement of its surfaces by means of platinum wires stretched along the centre of the straight vertical arms and passing out through pressure-tight electrodes at the top of each. As the mercury moves the wires are shorted by it to a greater or less extent, and the variation in the resistance of a circuit passing through either of these wires to earth indicates the level of the mercury surface. A variation quoted by Perry (loc. cit.) makes one arm of the U-tube of much greater diameter than the other, so that the movement of the mercury is virtually all confined to that in the thinner arm, and only one platinum wire is required.

It is desirable in any system of mercury lutes, especially if the tubes are of metal so that one cannot see through them, to put in a stop valve at or near the lowest point; otherwise, it is all too easy when opening up to the pressure to start oscillations which may blow the mercury into the pipe system.

The extension of the rotameter principle to pressures as high as 700 bars (10,000 lbf/in^2) is said by Perry (loc. cit.) to have been used in experimental work and Tongue[15.6] mentions it as a potential basis for plant instruments, but gives no further details. The essential feature here is a vertical tube of slightly tapering bore so that its cross-section increases linearly as one rises through it. A small spinner is inserted in the tube, and, when fluid flows up it, this lifts and with steady flow takes up a more or less stationary position which depends, among other things, on the vertical velocity of the fluid at that point. In low pressure applications the spinner's position is noted visually, but for high pressure work it becomes necessary to do this by some indirect means, such as having a permanent magnet in the spinner and a non-magnetic material such as an austenitic steel for the tapered tube.

Flowmeters based on any of these effects would need calibration if they were to be relied upon for proper plant control, and this operation is likely to be by no means easy or cheap to carry out. Probably the simplest arrangement would be to put a by-pass in the delivery system of a compressor of known performance, and—when required—to insert the instrument to be calibrated into this by-pass and push all the output of this compressor through it. The feasibility of arranging such a method of calibration would of course depend very much on local conditions, but the need for something of this kind ought to be borne in mind during the design stages of any high pressure industrial project.

15.2.4 *Thermal Methods*

We are concerned here with two main methods. The first, and best known, consists in measuring the increase in temperature brought about in the fluid substance as the result of putting in heat further up stream. It has the virtue of being simple to install, as well as being able, to some extent, to produce an average figure in the case of a pulsating flow. For absolute measurement calibration is essential, since much of the heat put in will be lost by conduction in the metal of the pipes, etc.

The method is, however, particularly useful when it is required to divide the flow evenly between several branches. For instance, if these are all in pipe of the same size, they can be passed through a common jacket in which steam—or some higher boiling liquid—is condensing, and by which a short length of the outside of each branch pipe would be heated to almost exactly the same temperature, see Fig. 15.4. The sensing temperature elements can then be located in each line at the same distance from the jacket, and their readings will indicate the proportion of the total flow which is going down each branch. Evidently, if the temperatures are all the same, it is a reasonable assumption that the flow in each is the same; and this is usually the desired condition in systems of this kind.

Another method, which has been used successfully in small-scale research equipment and was described by Reynolds[15.7], is a development of the hot wire anemometer. It consists of a platinum wire, carried between insulated electrodes, in the pipe through which the flow to be measured is taking place. A constant current is passed continuously through this wire, and its resistance is measured. The wire will thus be receiving energy at a constant rate, but its temperature—and consequently its resistance—will vary with the cooling effect of the fluid moving past it. In this way the resistance of the wire will be a measure of the rate of flow, but this will need calibration if the results are to be obtained in, for instance, the mass flow per second of any particular fluid in a particular pipe system.

15.3 Sampling

This is an important consideration in the control of many processes, but with the

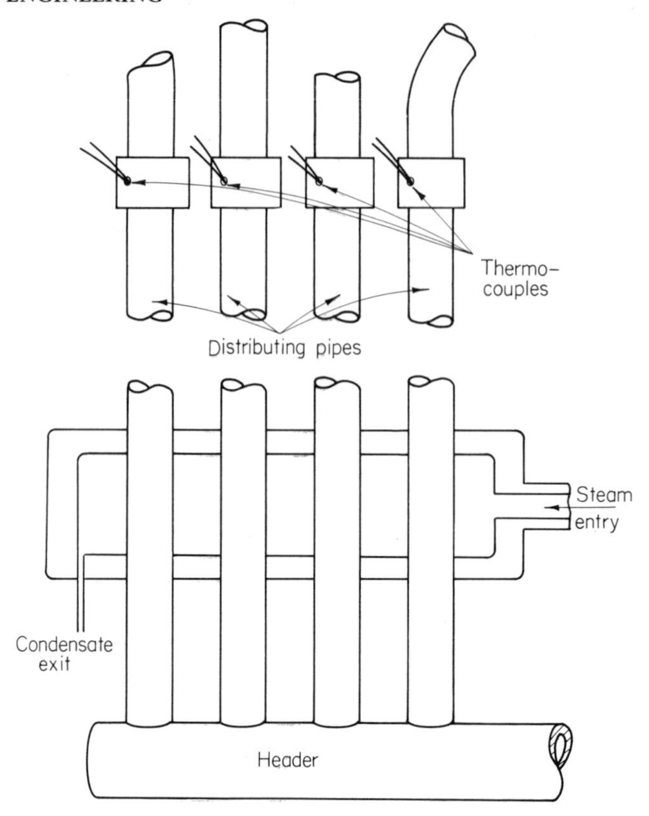

Figure 15.4 Schematic arrangement of thermal flowmeter

large volumes and flow rates normally handled in modern industrial plants the main problem is to locate the points from which the samples are drawn so that these are representative. To a large extent this is a matter of plant design and lay-out, but it is usually possible to duplicate the sampling points in such a way that by comparing the composition of the samples drawn from each the most appropriate locations can be found and the results interpreted with confidence.

The sampling problems in small-scale research studies may, however, be considerably more difficult. Often the trouble is that the sample is an appreciable proportion of the total amount of the reacting substances, and this has to be removed without seriously affecting the conditions in which the reaction may be continuing. A vessel, which was designed to enable samples to be extracted from a constant volume liquid reaction, was used by Gibson, Fawcett and Perrin[15.8] and is illustrated in Fig. 15.5. As will be seen, the reaction space A is contained in a bell over mercury, and the only connection with this is through the valve V. The space into which the valve opens has two connections, B and E, the former leading to the sample receiver C and the latter to an evacuating system (not shown). To extract a sample, C and D are first washed out and evacuated, and the valve is then slightly opened so that fluid can escape slowly into the tube B, while all the time the pressure is being maintained by injecting more of the pressure transmitting oil through the connection P. The illustration shows a sample receiver suitable for containing material which may flash off some of its constituents when the pressure is reduced.

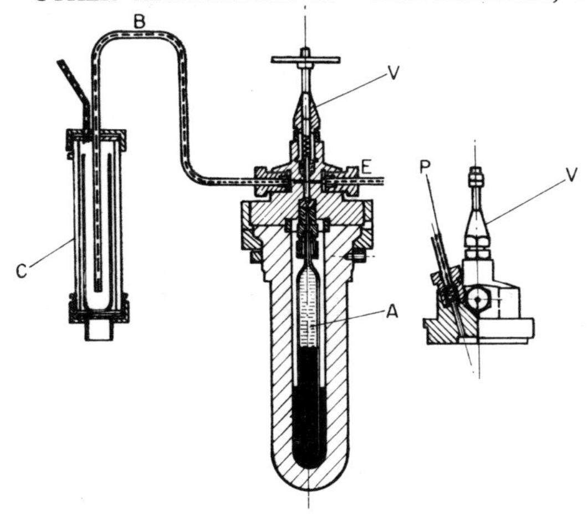

Figure 15.5 Laboratory sampling system for 3,000 bar reaction vessel (Ref 15.8)

REFERENCES

15.1. BRIDGMAN, P. W., *The Physics of High Pressure,* 1st Edition, Bell, London 1931, 295.
15.2. BOYD, F. R. and ENGLAND, J. L., *J. Geophys. Res.,* **65,** 741, 1960.
15.3. GARRISON, J. B. and LAWSON, A. W., *Rev. Sci. Instrum.,* **20,** 785, 1949.
15.4. STRONG, H. M., *High Temperature Methods at High Pressure* Chap. 5 in Symposium on *Modern Very High Pressure Techniques,* edited by R. H. Wentorf, jnr, Butterworths, London 1962, 111.
15.5. PERRY, J. H., *Chemical Engineers' Handbook,* 3rd. edition, McGraw-Hill 1950, 1254.
15.6. TONGUE, H., *The Design and Construction of High Pressure Chemical Plant,* 2nd edition, Chapman & Hall, London 1959, 238.
15.7. REYNOLDS, R. G., *Ph. D. Thesis,* Imperial College of Science and Technology, London (Dept. of Chemical Engineering), 1964.
15.8. GIBSON, R. O., FAWCETT, E. W. and PERRIN, M. W., *Proc. Roy. Soc.,* **A150**, 223 1935.

16 Ultra-high Pressures and Present-day Limits

16.1 General

We have seen in the earlier chapters that, at room temperature and for small numbers of applications, the simple elastic cylinder may be able to contain perhaps 8,000 bars, and that this can be about doubled by compounding procedures such as shrinking or wire-winding. Also autofrettage may give a pressure limit of the same order.

However, a major step forward appears to be obtainable by using a ring of sector-shaped components in the wall, especially if the sectors are made of a very hard material, like tungsten carbide, and accurately figured so that the stress distribution is symmetrical within them. The feasibility of attaining a cylinder with a pressure resistance comparable with the crushing strength of tungsten carbide (say 55 to 60 kilobars) therefore seems reasonable. Moreover, since this inevitably means a high diameter ratio in order to accommodate the sectors, with an internal sleeve and the externally supporting mantle, there is even the possibility that simple cylinders of very special materials, e.g. vacuum re-melted steels of very high tensile strength, could have an ultimate bursting strength of this order. In practice, a sector-reinforced cylinder would require an outside diameter at least 10 times its bore, and if the material of a simple cylinder would overstrain at constant shear stress until the whole wall was plastic the limiting condition of stability would require that:

$$p_i = \tau^* \ln K^2 \tag{16.1}$$

where τ^* is that constant shear stress – see footnote on p. 40.

In actual fact this assumes not only that there is no strain-hardening, but also that the strains themselves are small enough to be neglected; that is to say that the diameter ratio K at the instant immediately preceding failure is the same as it was initially. In practice it is seldom that either can be neglected, but their effects are opposite and such data as we have – and they are very scanty for high values of K – suggest that the actual bursting pressures may be higher than would appear from eq. (16.1). In other words, the strain-hardening effect is greater than that due to the decrease in K.

On this basis and with $K = 10$ and τ^* up to perhaps 9,000 bars, we have p_i in the region of 40 kb. But there are factors to be mentioned later which may tend to strengthen these cylinders, so that the possibility is worth considering that the bursting resistance of very thick monobloc cylinders may approach that of those built up with tungsten carbide sectors to the same overall diameter ratio. Clearly, if that were so, they would be very much cheaper.

We already have some data for small thick cylinders up to 25 kb as reported by Thomas, Turner and Wall[16.1], and this suggests that they may have quite an appreciable endurance of this pressure. To what extent this can be further extended remains to be seen however, see a recent paper by one of the authors (WRDM)[16.17].

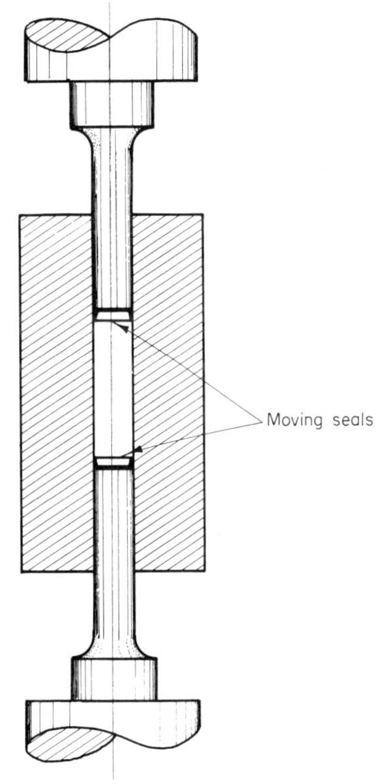

Figure 16.1 Schematic arrangement of opposed plunger apparatus for ultra-high pressure generation

For this type of pressure unit the simplest arrangement is somewhat as shown diagrammatically in Fig. 16.1, where two plungers are driven vertically opposite one another into a simple cylinder open at both ends. With moving seals as shown the pressure is confined within the central portion and the relatively lightly stressed parts of the cylinder outside the seals appear to provide some reinforcement to the part which is actually holding the pressure.

In this type of apparatus, however, the plungers are subjected to almost purely uni-directional compression, and this is inevitably accompanied by large shear stresses. Thus we have the most severe type of stressing from the point of view of tungsten carbide plungers, and the pressure limit is unlikely to exceed 55 kb. On the other hand, this appears to be attainable fairly easily with sector-reinforced cylinders.

It would appear therefore that the first limit we are up against is the crushing strength of tungsten carbide in this type of apparatus, but an arrangement of this sort is needed if an appreciable capacity for movement to change the internal volume is to be provided.

16.2 Use of Tapered Components

Bridgman[16.2] found that the tungsten carbide plungers could be made to withstand much greater compressive stresses if they were tapered; moreover, the wider the angle of taper, the greater was the compression that could be obtained. Thus, incidentally, the sector is likely to be a very favourable shape for increasing the resistance to crushing.

Unfortunately, however, with tapered plungers it is no longer possible to provide ordinary seals to enclose the space to be compressed. Bridgman got over this difficulty in a small relatively flat volume by surrounding his specimen with a ring of the mineral Pyrophyllite† (an aluminium silicate of the talc type with chemical formula ($Al_2Si_4O_{10}(OH_2)_n$), see Fig. 16.2. This had the remarkable property of compressing with the specimen, but at the same time having sufficient coherence to prevent the latter from bursting radially outwards. Even so, it is probable that the actual compression system in the specimen was by no means uniformly hydrostatic. Nevertheless the idea of a material acting as Pyrophyllite was apparently doing, i.e. being a compressible gasket which had sufficient internal coherence to contain very high pressure, and sufficient external adherence to remain attached to the metal of the apparatus, was evidently an important

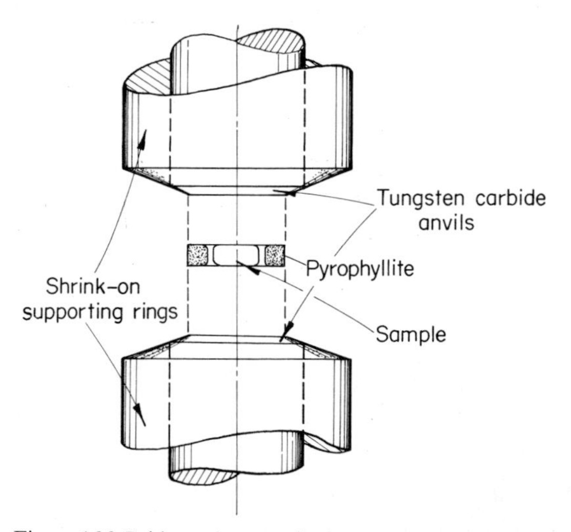

Figure 16.2 Bridgman's tapered plunger arrangement

† Bridgman originally used a material called Pipestone which is similar though less satisfactory than Pyrophyllite, and the latter is now used by most experimenters in this field.

advance, and Hall's "Tetrahedral Anvil" apparatus[16.3] was successful largely because of this.

In essence this device consisted of four separate hydraulic presses set with their axes each making an equal angle with each of the others. Each of these drove plungers towards one another and each plunger was tipped with a tungsten carbide "anvil" in the shape of an equilateral triangle. Thus, as they moved together, they would – if properly adjusted – make contact along their edges to enclose a space with the shape of a regular tetrahedron. The anvils themselves were tapered so that the strengthening effect was obtained, and sufficient power was provided to work up to at least 100 kb.

The ability to reduce the volume of the internal tetrahedron was provided by filling it with a carefully machined Pyrophyllite tetrahedron which was initially a little larger than that cut off by the anvils when their edges touched; thus they would first squeeze into the Pyrophyllite, extruding some of it outwards as a "rag", but imprisoning the rest and generating therein a very high pressure. Having established this principle, Hall put his specimens into holes machined in the Pyrophyllite tetrahedron, and also added means of resistance heating by providing his specimens with metallic conductors in contact with the anvils, and he was also able to measure temperatures by means of thermocouple wires which were led out through the edges and insulated within the "rags", another valuable property of the Pyrophyllite being its excellence for both electrical and thermal insulation.

Using materials such as bismuth and barium which had known transition points, Hall was able to calibrate the apparatus and show that the relation between the load on the rams and the pressure on the specimen was linear at least up to 100 kb.

Tetrahedral anvil units are now being manufactured by several firms, and they evidently provide a reliable method of reaching the 100 kb pressure range. They are however expensive and require considerable skill in setting up; and more recent developments include a means of obtaining a comparable effect with an ordinary hydraulic press. In the arrangement described by Lloyd, Hutton and Johnson[16.4], three rams, each tipped with triangular anvils, are contained in a hollow cone with their outer ends seating against the inside of an outer cone, which is placed on the up-stroking ram of the press, while the fourth anvil is held in the press frame pointing vertically downwards along the axis. Fig. 16.3 shows this diagrammatically.

A further application of the use of Pyrophyllite both as a gasket and as a pressure-transmitting "fluid" is the so-called "Belt" apparatus used by the General Electric Co. of Schenectady, U.S.A., for their synthetic diamond work. This has been described in some detail by Hall[16.5], and essentially it consists of a pair of opposed plungers with anvil extensions shaped as shown schematically in Fig. 16.4. These are made from tungsten carbide and surrounded by a number of hard steel supporting rings forced on against a slight taper. The working space is then held laterally by a ring, the "belt", also of tungsten carbide at the inside, supported by shrunk-on outer steel rings. Again the pressure space is filled with Pyrophyllite into which the specimens to be pressurised are inserted.

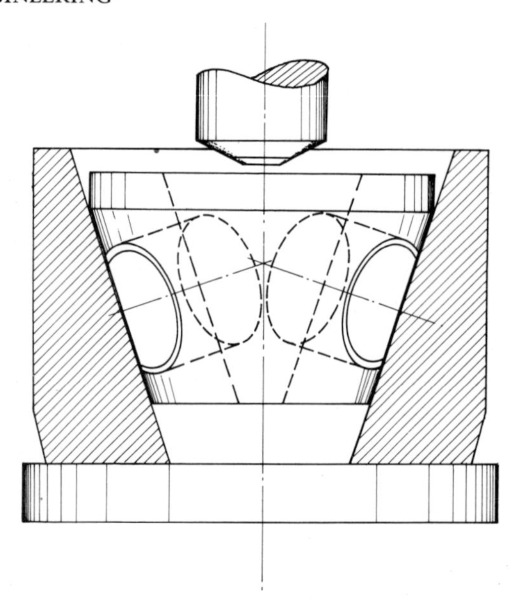

Figure 16.3 Tetrahedral anvil apparatus; adaptation for vertical press, as described by Lloyd, Hutter, and Johnson (Ref 16.4)

The chief advantages of this type of apparatus are the simplicity and the greater compression, i.e. available change of volume in the trapped space.

16.3 The Synthesis of Diamonds

No consideration of ultra-high pressure techniques would be complete without some mention of this development which has stimulated so much ingenuity during the last 15 years in a field which had previously been very little studied. Many attempts have been made to achieve this synthesis, from the time when Lavoisier showed, in 1792, that diamond was an allotropic form of carbon. The first of these to achieve apparent success was that of J. B. Hannay in 1880[16.6] by heating "carbonaceous mixtures" in iron tubes. There is a good deal of mystery surrounding his work and the few small specimens deposited in the British Museum have since been examined by X-ray spectroscopy and found to be more like natural diamonds than any of the synthetic products since examined in this way. In any case our present knowledge of the phase diagram of carbon makes it seem unlikely that diamond could be produced in the sort of apparatus then available, and this fact inevitably casts doubt also upon the claims of Moissan[16.7] who in 1894 reported achieving the synthesis by dissolving carbon in molten iron and shock cooling and solidifying the solution by pouring into large quantities of water. The lumps of iron were then dissolved in acid leaving small hard crystals which were claimed as diamonds. Subsequent study showed that these crystals contained silicon and may well have been silicon carbide.

The main sources of natural diamonds were in partially decomposed igneous rock, which had evidently been subjected to very high pressure and temperature, and this led to the idea of pressure being an important factor. The question was how high must this pressure be. This particularly engaged the attention of the

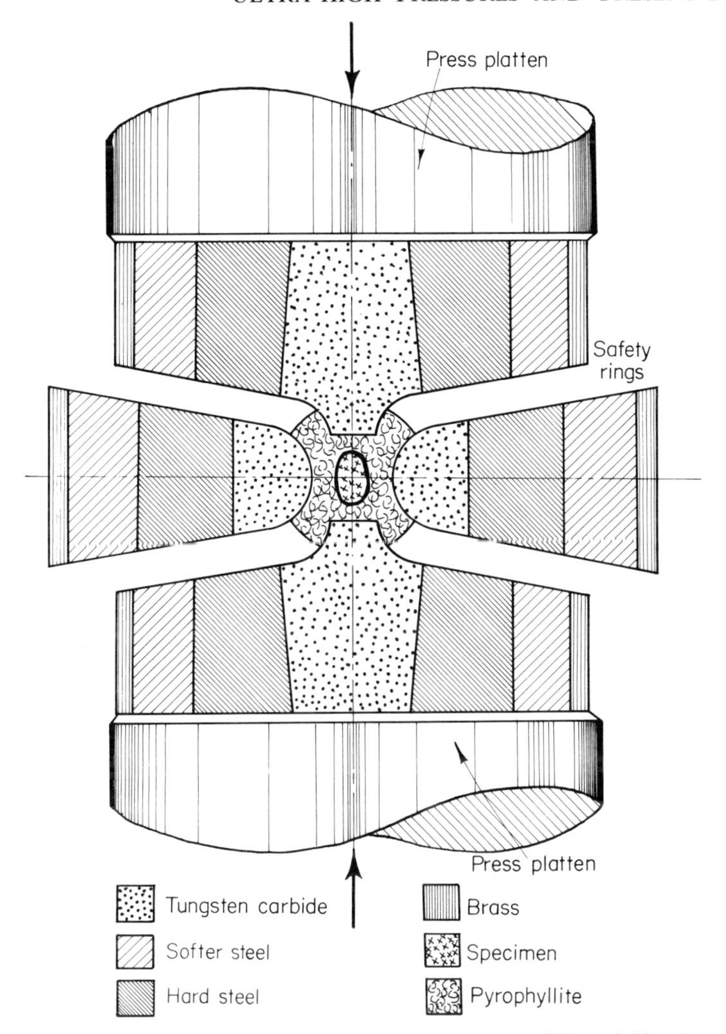

Figure 16.4 Schematic section through 'Belt' apparatus, *(General Electric Co. Schenectady, USA*

late Sir Francis Simon, who had been studying the problem from the thermodynamic point of view. From earlier work of his own and various others he concluded that pressures of at least 10 kb would be required, but he also reckoned that, even if that would take the material into the region where diamond was the stable phase, it might require some other effect to initiate the transformation. In 1950 therefore a small apparatus was developed on his behalf (by the Plastics Division of I.C.I.) which was small enough to insert into what was at the time the only atomic pile available in England (at Harwell). Then, with the co-operation of the United Kingdom Atomic Energy Authority, the high pressure space of the apparatus was filled with graphite and subjected to about 17 kb and held in the pile at about that pressure for 9 days, but without result. The available neutron flux was not very high, but the real explanation was found later when Simon and his colleagues at Oxford redetermined some of the thermodynamic data experimentally, which led to a revised idea of the phase diagram, see Berman

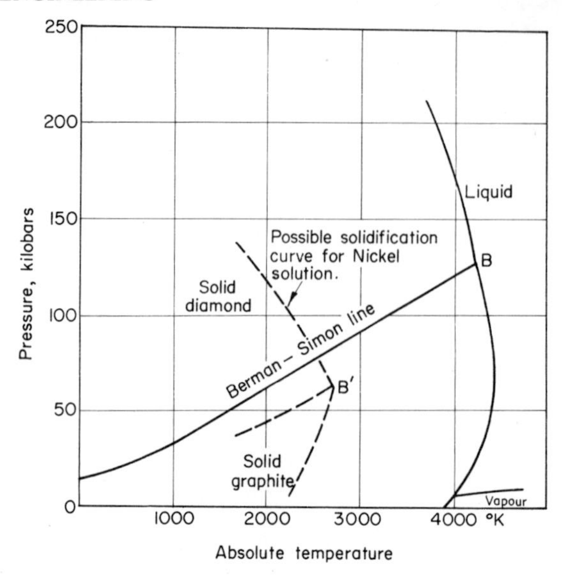

Figure 16.5 Phase diagram for carbon

and Simon.(16.8) Their curve is reproduced in Fig. 16.5 and is reasonably reliable up to about 50 kb and 1,500°K. The rest of this diagram involves a fair amount of conjecture, but it would seem to be enough to explain the subsequent commercial processes now in successful use.

The first of these to be announced was that by the American General Electric Company. Their first successful experiment was actually carried out in their laboratories at Schenectady, N.Y., on December 16th, 1954. The process was patented, but the details were kept secret on grounds of national security although an announcement was made in a short article in *Nature*(16.9) by Bundy, Hall, Strong and Wentorf in July 1955. This announcement had two immediate effects: it stimulated the De Beers organisation in South Africa to initiate an intensive study aimed at providing them with a commercial process, and it led to the Swedish A.S.E.A. company revealing that they had actually achieved a similar success a year earlier, although for some unspecified reason they had hitherto kept quiet about it. Their first public announcement appears to have been in the A.S.E.A. house journal, but in 1960 Liander and Lundblad published an excellent account(16.10) of the work which first achieved success in 1953. The date of the first proved diamond synthesis is not given, but it must have been at least a year before the General Electric whose success did not come till right at the end of 1954.

The A.S.E.A. used equipment similar in some respects to the Tetrahedral Anvil apparatus except that it had six plungers instead of four and therefore enclosed a cube instead of a tetrahedron. The reacting charge was a mixture of Fe_3C and graphite and the temperature was produced by surrounding the charge with a thermit mixture, which was electrically ignited after putting up to pressure. Later electrical heating was employed with a heating element running through the middle of a spherical charge, the whole being packed round with powdered talc.

This latter probably played the same part as did the Pyrophyllite in other varieties of apparatus, but where thermit heating was used it may only have been needed as a gasket, the resulting expansion of the charge producing the extra pressure when the expansion was resisted by the plungers.

A successful diamond preparation was also carried out by Pugh and others at the National Engineering Laboratory at East Kilbride, Scotland, in 1961 and reported in detail.[16.11, 16.12] In this, a modified tetrahedron apparatus was used, having the adaptation to a single ram press described in Ref. 16.4. The conditions in their successful preparations were about 75 kb and 2,000°K. The graphite was in the form of thin sheets and the charge was made up by sandwiching alternate sheets of graphite and pure nickel; these were fitted with tabs of tantalum metal to make electrical contact with the anvils through which the electric current for heating the sample was transmitted. Approximately 10% was converted in a reaction time of about 20 minutes, but it is evident that the actual conversion takes much less time than that, since De Carli and Jamieson[16.13] have reported achieving the direct conversion of graphite to diamond with pressures generated by explosives and evidently thus of very short duration.

Reference to Fig. 16.5 shows that the most likely way of producing diamond is to cool liquid carbon at a pressure above the point B, which is the graphite/diamond/liquid triple point. The lowest pressure that has been suggested for this point is well over 100 kb, although it is quite clear that diamonds have been produced at much lower pressures. The explanation must therefore be either catalysis or some solvent action by the various added substances (usually transition metals of the Periodic Table). The latter would be explained if we assumed that the corresponding lines for a system containing say 10% of nickel in the liquid phase would be somewhat as shown in broken lines in Fig. 16.5, with the triple point now at B′, the pressure then being lowered to perhaps 70 kb.

Synthetic diamond manufacture is going on in many parts of the world in addition to Sweden, the U.S.A. and South Africa where the earlier plants were established. France, Russia, Japan and the Irish Republic are amongst those now manufacturing, and recently claims have been announced for processes operating at much less severe conditions. It seems unlikely, however, that these will be fully substantiated unless the phase diagram shown in Fig. 16.5 is seriously in error, and so far there is no indication of this. The only other possibility is of some form of growth as epitaxial thin films on diamond substrates; this is the subject of a patent[16.14] by Eversole of the Union Carbide Corporation.

We can therefore conclude that the available techniques of pressure generation are amply sufficient for diamond manufacture although this is facilitated by the fact that the raw materials are relatively incompressible and their electrical conductivity makes internal heating comparatively easy.

16.4 Possible Limits

From what has been discussed in the foregoing sections it seems evident that the pressure limit for conditions for which appreciable change of volume is needed is around 55 kb. Where, however, the pressure can be generated by very small changes of volume – or by internal heating – equipment based on tapered anvils can be employed and the pressure limit then may be as high as 200 kb. Explosively generated pressures may go even higher, perhaps to 400 or 500 kb, but their duration is only of the order of a micro-second.

In considering how these limits might be extended, the really major advance would require a material able to withstand a higher uniaxial compression, or – alternatively – some means of laterally supporting parallel cylindrical plungers and at the same time leaving them free to move. Lateral pressure would reduce the intensity of shear stress for a given axial load, and this has been achieved to a limited extent with a stepped plunger device described by Giardini and Tydings.[16.15] This, however, has the limitation that it again requires Pyrophyllite to transmit the pressure and to prevent the escape of the material being pressurised.

Another possible line of development could perhaps come from the fact that the tungsten carbide anvils operating in, for instance, the tetrahedral type of apparatus undergo a progressive deformation similar to the strain deformation of metals, and apparently with considerable strengthening when re-loaded in the same direction. The possibility of thus preparing plungers by progressively

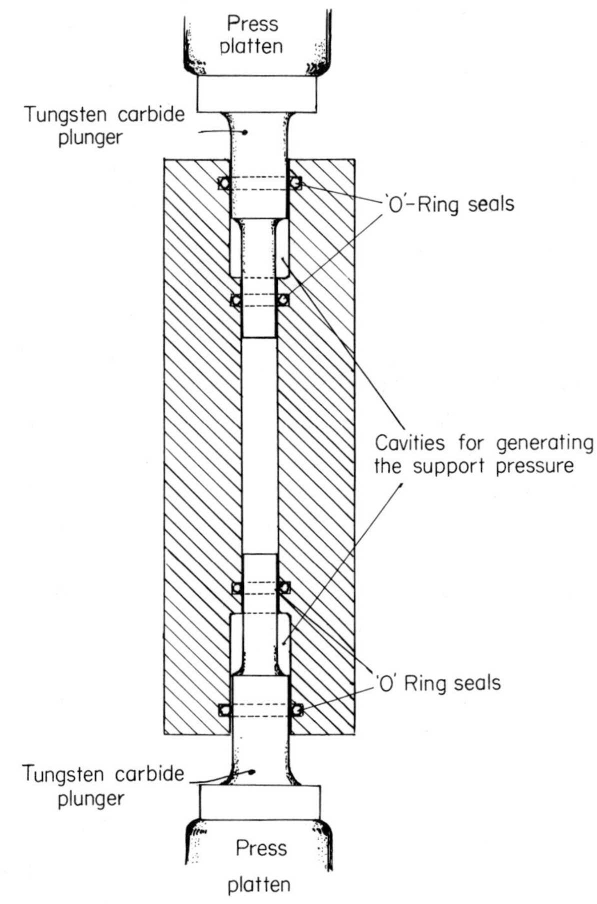

Figure 16.6 Schematic arrangement of double seal plunger support

increasing the compressive stress applied to them may be worth considering. Fig. 16.6 shows schematically another possible means of giving lateral support. It depends however on being able to make seals of the O-ring type to hold against these very high pressures. A single-ended version of this has been successfully used by Boyd and England.[16.16]

With a device of this kind it would seem reasonable to expect the maximum pressure to be substantially additive. Thus if the pressure in the annular cavity could be worked up to 20 kb, one might hope to take the inner pressure to 70 kb. The correct proportions would obviously require a fair amount of trial and error before they were established, but the system would have the advantage that the supporting pressure would rise and fall more or less in proportion as the working pressure rose and fell.

The particular scheme outlined in Fig. 16.6 has plungers entering at both ends since an ordinary blank plug would need supporting laterally at these pressures. Alternatively it is possible that a hemispherical end could be used since these are much stronger than the comparable cylinder, as we have seen in Chapter 4.

REFERENCES

16.1. THOMAS, S. L. S., TURNER, H. S. and WALL, W. F., Conference on High Pressure Engineering, London, Sept. 1967, Paper No. 11, *Proc. Inst. Mech. Eng.*, **182,** Pt 3C, 271.

16.2. BRIDGMAN, P. W., *Proc. Roy. Soc.*, **A203,** 1, 1950.

16.3. HALL, H. T., *Rev. Sci. Instrum.*, **29,** 267, 1958.

16.4. LLOYD, E. C., HUTTON, U. O. and JOHNSON, D. P., *J. Res. Nat. Bur. Standards,* **63C,** 59, 1959.

16.5. HALL, H. T., *J. Chem. Educ.*, **38,** 484, 1961.

16.6. HANNAY, J. B., *Proc. Roy. Soc.*, **30,** 188, 1880.

16.7. MOISSAN, H., *C. R. Acad. Sci. (Paris),* **118,** 320, 1894 and **123,** 206, 1896.

16.8. BERMAN, R. and SIMON, F., *Z. Elektrochem.*, **59,** 333, 1955.

16.9. BUNDY, F. P., HALL, H. T., STRONG, H. M. and WENTORF, R. H., jnr., *Nature,* **176,** 51, 1955.

16.10. LIANDER, H. and LUNDBLAD, E., *Ark. Kemi,* **16,** 139, 1960.

16.11. PUGH, H. LL. D., LEES, J. and BLAND, J. A., *Nature,* **191,** 865, 1961.

16.12. PUGH, H. LL. D., LEES, J., ASHCROFT K. and GUNN, D. A., *Engineer*, **212,** 258, 1961.

16.13. DE CARLI, P. S. and JAMIESON, J. C., *Science,* **133,** 1821, 1960.

16.14. EVERSOLE, W. G., U.S. Patents Nos. 3,030,187/188, 1962.

16.15. GIARDINI, A. A. and TYDINGS, J. E., U.S. Patent No. 2,995,776, 1961.

16.16. BOYD, F. R. and ENGLAND, J. L., *Yearbook Carnegie Inst.*, **57,** 170, 1958.

16.17 MANNING, W. R. D., *High Temp. High Pres.*, **1,** 123, 1969.

17 Safety

17.1 General

The importance of this consideration needs no emphasising, although the dangers of high pressure operations are often exaggerated. A high pressure unit will certainly contain more potential energy than one of comparable size which works at low pressure, but this only becomes a hazard if the energy releases itself; and in most cases these risks can be kept to an acceptable level as the result of modern knowledge and experience.

In this final chapter the authors attempt briefly to indicate the main sources of danger and how they can be guarded against, so far as this is a matter of engineering technique. In one most important aspect, however, safety is a matter of the proper training and discipline of all personnel, including the Management. We can only indicate here the imperative need for this; how it is to be achieved clearly lies outside the scope of this book.

High pressure engineering as we know it today has been developed very largely by the chemical industry, which has in consequence acquired a large store of knowledge and experience of these problems and dangers. More recently, however, the metallurgical and ceramic industries have become interested, in particular with such techniques as hydrostatic extrusion and the pressure compacting of powders (isostatic compacting), and it may well be that some of these lessons will have to be studied by the newcomers in more detail.

Small-scale research at very high pressures is also on the increase and, whereas until recently this was mostly confined to chemical investigations, it is now being used by geologists, biochemists, nuclear physicists, and many others. Fortunately several firms are now offering small-scale equipment of proved reliability at very reasonable prices, and this is certainly a welcome development, although it is likely to result in some of the purchasers not being adequately trained in its use.

The problems of safe operation are primarily to keep the impounded energy under control, and these can be subdivided very generally into matters of design and construction on the one hand and into those of operation and maintenance on the other. In most instances enough information is available to achieve these requirements without any major departures from ordinary practice, but there remain some projects where the background knowledge of the reaction may not be sufficient to guarantee the behaviour of the process, and the pressures may then build up to unknown levels and at unknown rates. Such cases require either the provision of pressure-relieving devices for getting rid of the contents of the plant in a very short interval of time and without danger or damage to the neighbourhood; or, in the worst cases, of providing some kind of safety barricade so that even the bursting of the containers does no harm.

The use of explosives in peace and war, and especially the problems of their manufacture and handling, have provided much valuable information which can be adapted to the rather similar problems posed by the energy in high pressure systems. Several instances have been reported of collaboration between explosives experts and those concerned with high pressure engineering; see for instance, W. G. High.[17.1] This has been of particular importance where protection against bursting of pressure containers has to be provided, since it appears that the resulting energy release can be matched quite closely by firing the appropriate quantity of a given explosive.

17.2 The Principal Causes of Accidents with High Pressure Equipment

These can first be subdivided into two main categories, namely failure of equipment or failure of operation. In the latter must be included, not only mistakes by personnel, but also the incorrect functioning of control systems. The results may be the building up of excessive pressure resulting in relief valves opening and blowing off the contents or – in extreme cases – bursting some part of the system. There is also the possibility of the wrong valve being opened, and such mistakes can sometimes admit the high pressure to parts of the system only intended for low pressure working. This can also liberate quantities of the working fluid (which may be combustible or toxic) into spaces where men are working.

Failures of the human element are usually due to inadequate training, or perhaps to the selection of the wrong people in the first place. On the other hand, bad design may make such things more liable; for instance, a badly arranged control panel can be at least partly to blame if the wrong valve is opened in an emergency.

Failures of equipment in the early life of a plant are generally due to bad design, or to faulty materials or workmanship. Similar failures after some time in service are almost invariably due to faulty maintenance, and it is important to realise that this must include the detection and correcting of any deterioration in the equipment, due for instance to corrosion, creep or fatigue. The importance of proper maintenance in high pressure equipment is not always appreciated, especially where practical experience is limited to low pressure systems.

The results may be anything from the utterly catastrophic failure of a large vessel to the cracking of a small joint ring, allowing the slow (and perhaps undetected) escape of the working substances. The former is obviously disastrous, but the consequences of the latter may be almost as severe if, for instance, it allows quantities of combustible gas to accumulate in a confined space and form a large pocket of explosive mixture which may be subsequently ignited.

Probably the worst condition would arise if a vessel, made of potentially brittle material, developed a fatigue crack (accelerated perhaps by corrosion) to such a size that it spread through the walls at high velocity. In such cases fatigue might well have started a number of similar cracks, and these might spread simultaneously, thus dividing up the shell and separating fragments which would be propelled at very high velocity by the liberated high pressure fluid. Normally, however, vessels for really high pressure service should be made from material of the highest quality, treated throughout to ensure that it is truly ductile. Then we could reasonably expect that the spread of a crack would be slow enough to be detected, given the proper conditions of maintenance (see Section 7.3). Such vessels, if subjected to serious over-pressures, would show gross plastic deformation before they failed, and the actual rupture would take the form of a few fissures (perhaps only one) through which the fluid would escape.

The only other condition which would be likely to give rise to brittle fracture in a vessel of this quality would be a detonation in its contents. This means an explosion which travels at very great speed (of the order of 5,000 m/s), so that the pressure load travels faster than the resisting stess wave in the steel, i.e. faster than the speed of sound in it. Such an occurrence would be very unlikely, but with untried chemical reactions one cannot be sure, and precautions against it may be necessary.

17.3 Design Considerations

In Part I of this book the chief problems of designing high pressure equipment have been considered, and a few suitable materials of construction and their properties have been discussed. The use of these methods for design presumes that the worst conditions within the process are known, but this is not always so. The pilot-plant trials of a new chemical process may reveal unexpected side reactions, and those concerned must always be on the look-out for such developments and be ready to deal with them. An example of this occurred with the high pressure polymerisation of ethylene when it was discovered that, under certain conditions, the gas could undergo violent decomposition with rapid rises of temperature and pressure. Fortunately, in that particular instance the pressure rise was comparatively slow, and the relieving devices fitted to the first pilot plant (that of I.C.I. at Northwich in Cheshire where the material was originally discovered) were able to blow off the excess pressure without damage to the plant or to its surroundings, though not without some alarm in the neighbourhood.

The lay-out of high pressure plant can have a considerable influence on its safe operation. As far as possible the controls should be concentrated at one point, so that those in charge can see at a glance how it is running and make any necessary adjustments from where they stand. Pipes and valves carrying pressures up to about 500 bars hardly need to be behind barricades unless the substance they are conveying is very unstable, but for higher pressures, especially above 1,000 bars, it is desirable to shield the pipes so that the failure of a joint cannot cause any damage from the reaction of escaping material. This can conveniently be provided by a rolled-steel channel to which the pipe is easily secured, while the web of the section protects those behind it. This can also be helpful in preventing pipe oscillations which may easily occur if synchronism can be picked up from neighbouring machinery and fatigue damage thus result.

Valves may be operated by long extensions to the shafts of the driving gears where these are fitted (as with the British Engines design described in Section 13.4.3). The valve body and connections can then be conveniently shielded by steel plate.

Reactors, whether in the form of vessels or of long coils of tubing, can with advantage be placed so that a heavily reinforced wall separates them from the control centre and other points which must be manned. Usually, the reactors are the greatest concentrations of stored energy, and the locations where the consequences of failure would be most serious. In some cases this protection would no doubt have to take the form of a complete "bomb-proof" cell, (possibly produced by locating underground) and the design considerations for this are discussed below (Section 17.5.2). On the other hand, long experience may cause us to revise our ideas on such matters; for instance, when the synthetic ammonia process was first operated on a large commercial scale the general practice was to enclose the reactors within concrete stalls (which might or might not have really protected the neighbourhood had a vessel burst), but today—when the size of the vessels is much greater and the stored energy correspondingly increased—this is seldom considered necessary, and accidents with them are almost unknown. By contrast it would appear that high pressure ethylene polymerising reactors are still mostly located within fully shielding cells, as well as having pressure-relieving devices fitted. The main reason for this is doubtless the much higher pressure, usually now of the order of 2,000 bars, and also the tendency for the decomposition reaction to come into play if the polymerisation control becomes unsteady; on the other hand, as more experience is gained, and as the reaction control improves, these precautions may well come to be regarded as unnecessary, in spite of the tendency for the working pressure to be increased for reasons of product quality.

Compression and pumping equipment is more difficult to shield because it usually requires manual attention from time to time while running. On the other hand, it is a potential source of danger because the working substance can be subjected to sudden variations of temperature and pressure within the cylinders. This has, for instance, caused ethylene decompositions in a compressor cylinder. However, the usual practice in polyethylene manufacture—so far as can be judged from the scanty information published about such plants and processes—is to mount the compressors without any protection of this kind. High pressure delivery pipes are, however, often placed in trenches below floor level.

An additional source of danger here is the possible heating by friction of the working substance if for any reason the lubricating system fails. With large modern plants the compression units are often so large and powerful that incipient seizure may go unnoticed; this could lead to a serious breakdown and the possible liberation of large quantities of the working fluid if it were not detected and put right in good time, and the only way to get notice of such troubles is by suitably placed temperature measuring points with alarm signals. Breaking links are often fitted by the manufacturers in the drive mechanism of small reciprocating machines and these may be of great value if foreign matter gets into the system. They are less likely, however, to be able to deal with process irregularities.

17.4 Pressure Testing

Conventional practice has changed considerably in recent years as the working pressures have risen. The old idea, still occasionally met with amongst boiler manufacturers, that pressure vessels should always be hydraulically tested to twice their designed working pressure has had to be revised for the simple reason that many such units would be seriously overstrained if they were. On the other hand, tests that carry the stresses in the material close to its elastic limit are often valuable since they tend to cause localised overstrain at stress concentrations, thereby spreading the load over the neighbouring metal and reducing the peak values.

Another important change of attitude relates to hydraulic testing. The old idea that water tests were perfectly safe may have been true with boilers and riveted vessels. In fact, men used often to clamber over them while the pressure was on, looking for leaks and stopping them with caulking tools. On the other hand, hydraulic testing in the kilobar pressure range is an operation calling for careful planning and control, and where possible it should be carried out in a properly constructed safety chamber. Fig. 17.1 is a photograph of the pit and equipment used at the River Don works of the English Steel Corporation. The vessel to be tested is placed in a pit which is cleared of all personnel and covered with heavy steel plates before the pressure is applied, and all measurements are made by

Figure 17.1 Safety pit for testing high pressure vessels *(English Steel Corporation Ltd, Sheffield)*

transducers and distant reading instruments so that those in charge can see exactly what is happening. In addition, close-circuit television is available for investigating unexpected effects. The inset at the top left-hand corner of this illustration shows the arrangement inside the control cubicle.

It should be noted that, at these pressures, the velocity of a water jet through a small hole or crack may be sufficient to penetrate a man's body, which emphasises the need for remote control in such operations.

The actual test pressures to be used can only be decided on the basis of many considerations, e.g. the type of construction, the material used, the conditions of service (especially the temperature), and the working shear stress in relation to the limiting elastic shear stress – to cite only a few. Generally speaking, however, for pressures between 500 and 1,000 bars, a 50% excess over the working pressure is likely to be about what is required, and from 1,000 to 2,000 bars this can be reduced to 33%. Above 2,000 bars for monobloc construction it is likely that some overstrain, i.e. autofrettage, will be needed, and the test pressure must then be related to the initial autofrettage pressure. This will mean also that the design will be based on the Ultimate Bursting Pressure and the margin available must also be taken into account in deciding the test pressure.

There is a tendency in recent years to swing rather to the other extreme and conclude that pressure testing is unnecessary. The authors cannot however subscribe to such a view and, in support of this, would point out the number of recorded incidents in which serious failures have occurred on test which would otherwise almost certainly have failed catastrophically in service. Saibel[17.2] quotes the case of a forged vessel which actually resisted some 200 applications of its designed working pressure before failing catastrophically, while the report on the Cockenzie boiler drum[17.3] is an example of failure on test at nearly 50% overpressure, again with evidence pointing to its likely failure in service.

17.5 Protection

17.5.1 *General Considerations*

The degree of protection which is required for high pressure operations depends on four main factors, namely (i) the size and amount of energy contained in the plant, (ii) the location, i.e. the proximity to other plants, residential areas, etc., (iii) the presence of weakening effects such as fatigue, corrosion, etc. and (iv) the degree of uncertainty which may still exist in regard to the process and the proposed plant. So far as the last is concerned, it is obvious that for a project with major uncertainties unresolved the selection of the site would be of great importance and, unless an isolated place could be found, it would be essential to surround the plant with shielding designed on the most conservative basis.

In many instances there are no more dangers in operation than exist in other types of plant of comparable size, and there is a real possibility that the inexperienced designers will pay too much attention to the high pressure units and overlook more serious hazards elsewhere. Apart from process uncertainties the most important considerations from a safety point of view are undoubtedly those covered by (iii) above which can only be controlled by very careful maintenance.

17.5.2 *Design of Complete Shields*

The problem here is first to ascertain the maximum energy release that could take place and the form in which it is most likely to appear. With very rapid and brittle failure of a vessel or container, as with a detonation, the energy will be chiefly in the form of a shock wave, which is essentially a pressure wave which travels at a speed greater than that of sound in the medium. This is probably the most destructive from the point of view of external damage, although the simultaneous shooting out of high speed fragments can be a serious extra problem for the designer; his protecting walls must also withstand penetration by these. It is useful to leave some blank panels for venting the pressure waves, but these may then allow some of the fragments to escape by ricochet from the walls and these can be even more difficult to contain. So far as the pressure wave is concerned, it should be noted that it is the impulse rather than the pressure intensity which is the real measure of its destructive power, and the rate of pressure rise is thus a vital factor as well as being one that is difficult to determine and even more difficult to estimate in advance.

Several investigations of the strength of defences of this kind have been undertaken and reported in the literature. High[*loc. cit*], for instance, describes a study made in collaboration with the British Atomic Weapons Research Establishment at Foulness in Essex, and H. C. Browne et al. of the Monsanto Chemical Company in the U.S.A.[17.4] describe a special type of cell which their company devised, and J. P. Weber et al. of the Armour Research Foundation in Chicago[17.5] give the results of tests of these cells when explosives were fired inside and the shock characteristics measured outside.

The estimation of the amount of energy that can be released under various conditions is discussed in the Appendix to this chapter, in which as an example the bursting of an ethylene polymerisation reactor of 0·5 m^3 capacity, working at 2,000 bars and at 250°C, in a completely brittle manner is considered. Here it is also assumed that a chemical decomposition occurs, and that 10% of the ethylene reacts in this way during the expansion. The total release of energy is reckoned to be of the order of $1{\cdot}9 \times 10^8$ joules, which is equivalent to exploding no less than 43 kg of trinitrotoluene (T.N.T). The resistance of heavily reinforced concrete to sudden eruptions of energy of this kind is remarkable, and Weber and his colleagues[17.5] fired increasing amounts within their experimental cell—which apparently had an internal capacity of only 820 ft^3—ending with a final shot of 50 lb (22·7 kg) of T.N.T. Even this did not destroy it, nor did it produce the sort of shock wave that would do much damage outside; on the other hand, it did do a lot of damage to the cell and was evidently about the limit that it could cope with.

The ethylene reactor considered in the Appendix would certainly be of considerable external dimensions (whether it was of the form of a cylindrical vessel or a coil of high pressure tubing) and it is reckoned that a space of something like 65 m^3 (say 2,300 ft^3) would be needed. High quotes various authorities who conclude that the pressure effects of firing explosives in confined spaces will be substantially the same if the ratio of the weight of the charge to the surrounding volume is the same. Thus assuming that the wall thickness and reinforcement is increased pro rata, this larger cell might be expected to resist the firing of:

$$\frac{2{,}300}{820} \times 50 = 140 \text{ lb approximately}$$

which is considerably more than the 43 kg charge equivalent to the failing reactor. Weber's cell had 12 in thick walls, so these would presumably need to be about:

$$12 \times \left(\frac{2{,}300}{820}\right)^{\frac{1}{3}} = 17 \text{ in thick in this example.}$$

The Monsanto experimental cell was of a rather special shape with an arched roof designed to force any flying fragments down into a prepared sand pit. A more conventional shape is described by High and shown diagrammatically in Fig. 17.2. The actual tests were carried out with small models one sixth the size of an actual cell that was already in service; owing to the reduction in scale the explosive charges were small—never more than about $\frac{1}{2}$ oz—but the results were considered satisfactory.

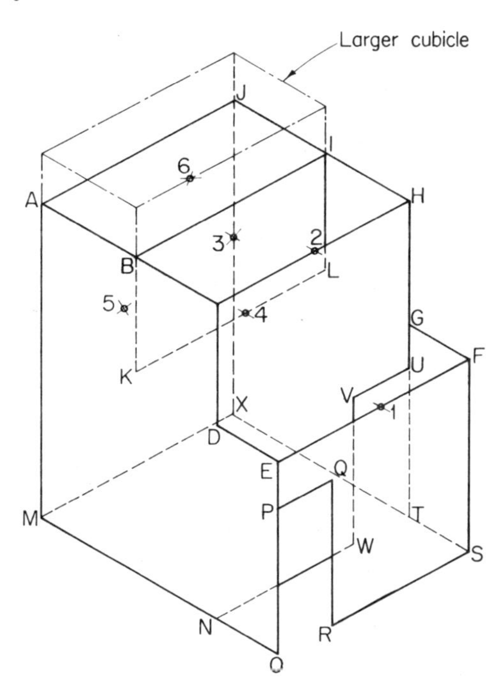

Figure 17.2 Sketch of model protective cell used in experiments with explosives *(described by High (Ref 17.1))*

In these experiments arrangements were also made to determine the pressure-time relations, from which the value of the impulses could be calculated. Fig. 17.3, which is reproduced from this paper, shows some of these traces and the calculated impulses. The effects of reflected waves are also shown and illustrate the complexity of the problem. This work suggests that, in any particular problem, the safest procedure would be to adopt High's policy and arrange for a

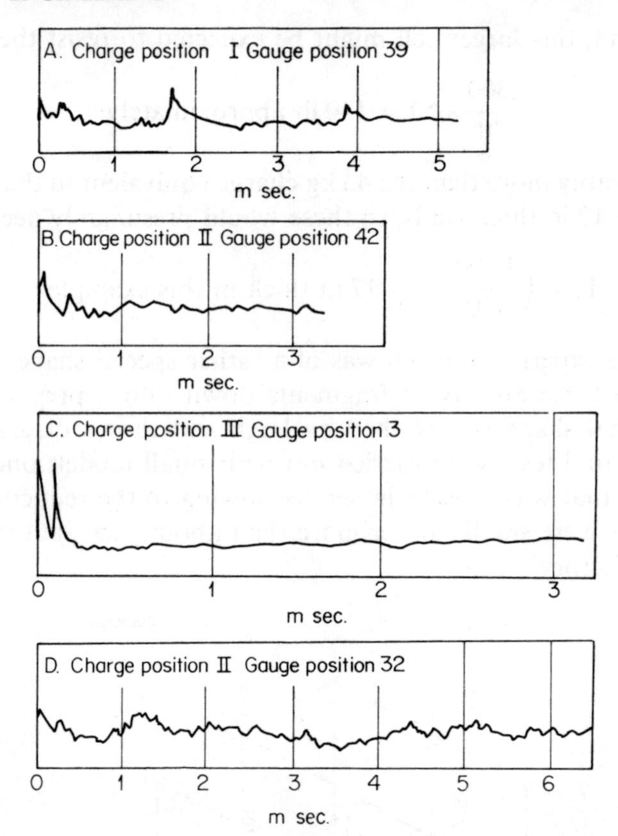

Figure 17.3 Typical pressure-time records measured in experiments *(described by High (Ref 17.1))*

model of the cell or barricade to be made and subjected to fully controlled tests with explosives.

There remains the problem of containing flying fragments. Here there is the difficulty that we have no definite means of ascertaining how much of the energy released will reappear in the form of kinetic energy in these, nor how it will be distributed between pieces of different size. Christopherson[17.6] examined a great mass of data which had been collected by the British Government between 1939 and 1945, partly obtained from descriptions of war incidents in various belligerent countries, and partly from ad hoc tests of air raid shelters, etc. From this he concluded that the depth of penetration d could be related to the weight of the fragment W_f and its velocity at impact v_f by an equation of the type:

$$d = \mathrm{k}W_f^{0.4}v_f^{1.5} \tag{17.1}$$

where k is a constant depending on the material resisting the penetration; and for concrete k has the value of about $5{\cdot}5 \times 10^{-6}$ in m.k.s. units, i.e. d in metres, W_f in kg and v_f in m/s.

The penetration resistance can be increased by covering the inner surface of the concrete by sheets of soft steel, although this may increase the energy of the ricochet. High reckons that the proportion of the total released energy appearing in the fragments is not likely to exceed about 20%, although cases have been reported where it was as much as 80%; but these only occurred when one end of a vessel was driven out like a projectile from a gun. It would seem too that in such circumstances the failure must have been fairly ductile and that an appreciable time lag would ensue, probably leading to a smaller total energy evolution.

In the above considerations the worst conditions have been assumed, and it will usually be prudent to begin by thinking in terms of protection that would cope with these extremes. The probability is however that failure, if it does take place, will be less severe, and that less elaborate structures will suffice to guard the neighbourhood. In fact, in many instances the pressure rise may be slow enough for relieving devices of the kind considered in the next section to deal with the situation.

A useful and cheap protection for all but the most severe energy releases can be obtained by casting a weak concrete mix between shuttering of steel plate which is then left in position. The concrete provides mass to resist the impulse, and the plate—which may conveniently be $\frac{1}{2}$ in thick for a 12 in wall—helps to prevent penetration by fragments and the spalling effects at the outer surface resulting from reflected stress waves. By using weak concrete, new holes can easily be drilled and old ones filled in, a useful facility with pilot-plant operations.

17.5.3 *Pressure-Relieving Devices*

Many pressure vessels such as chemical autoclaves are normally fitted with devices which will allow any excess pressure to be blown away. Where the pressures are low (e.g. less than 20 bars) ordinary safety valves such as those conventionally fitted on steam boilers can be used, but for really high pressure service more sophisticated devices are called for.

These can be grouped into two main classes, namely automatic valves and rupturing elements. Both have been tried in a number of different forms and in essence the main advantages of the valves is that their blow-off pressure is not affected by temperature and that they can sometimes be made to re-seat when the pressure excess has been removed. Rupturing elements have the advantage that, generally speaking, their operation is quicker, and therefore they tend to give a greater degree of safety, but they have the inherent disadvantage that they can never re-seat, and consequently their operation must inevitably result in losing the entire contents of the system they are protecting.

The problems relating to these devices are much influenced by the causes of the rises in pressure and by their rates. In the simplest case—as in a steam boiler, for instance—the demand may fall quicker than the output can be checked, and all that is then needed is some means of blowing off the excess until the system is again brought under control. At the other extreme is the reaction vessel whose contents, for chemical reasons, suddenly detonate; the resulting build-up of pressure may then be so rapid as to destroy the vessel before even the quickest device can function. Thus the pressure-time gradient is very important, especially with vessels worked at different pressures, either for the purpose of reaction control, or to vary the product properties. Fig. 17.4 illustrates this point. The line OA indicates steady working conditions at a pressure represented by O, a small amount below P, which is the pressure at

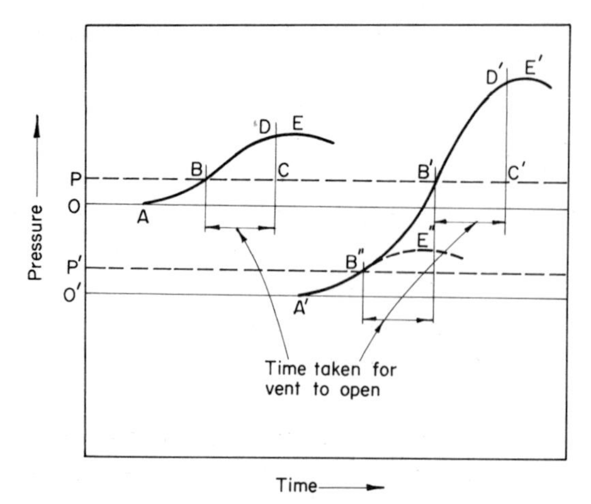

Figure 17.4 Diagram to show possible rates of pressure rise within vessel

which the relieving device is set to start opening. At *A*, for some reason, e.g. an ethylene decomposition, the pressure starts to rise and when it reaches *B* the device starts opening; it is not fully open until the point *C* is reached, but by that time the pressure will have reached *D* and may rise still further to a maximum at *E*.

Now let us consider the same process operating at a lower pressure O', but with the same setting of the device. Here at A' the decomposition begins in what are virtually constant volume conditions, with the result that the rate of rise of pressure rapidly increases. However, relief cannot start to come into operation until the pressure has reached *P*, i.e. at the point B', but by now the rate of pressure rise is very high, and – as the time to open fully will be much the same ($BC = B'C'$) – the pressure will now have got up to D' before the device is fully open and may rise further to a maximum at E'. This could be avoided if there was some simple control by which the setting of the relieving device could be altered when the working pressure was altered, i.e. for operation at a working pressure O' the setting should be at P' and the maximum pressure would not then go above E'', well below any danger to the plant.

It is an obvious objection to rupturing elements that they cannot be quickly adjusted for changed operating conditions, although the contingency envisaged here is probably not encountered very often. Devices of this kind can therefore save much trouble and damage in many forms of high pressure plant; they should not, however, necessarily be regarded as doing away with the need for protective wall. or barricades.

There is the further advantage for the rupturing element that it is much less likely to leak than most types of valve, although – from the safety point of view – this may be less serious, since a vent pipe to carry away any escaping gas will usually be provided for any such device, and explosive mixtures or toxic concentrations are thus unlikely to be formed.

17.5.4 *Rupturing Elements*

A great variety of different types have been tried, and many have given excellent service in industrial plants. A considerable range of well tested rupturing elements are marketed by Marston Excelsior Ltd of Wolverhampton, a subsidiary of Imperial Metal Industries Ltd., most of which are based on experience gained by I.C.I. in their Billingham and other factories.

Most of these devices, however, involve the dilemma that, if they are allowed to get hot, creep and weakening will occur, and they may then fail at the normal working pressure: if, on the other hand, they are kept cool, they may become blocked by the solidification of the process material. This can, to some extent, be avoided by the use of tubular elements, as will be described later.

Fig. 17.5 shows some typical elements. At (a) we see the domed membrane type, formed from metal sheet. These are very reliable in the lower pressure

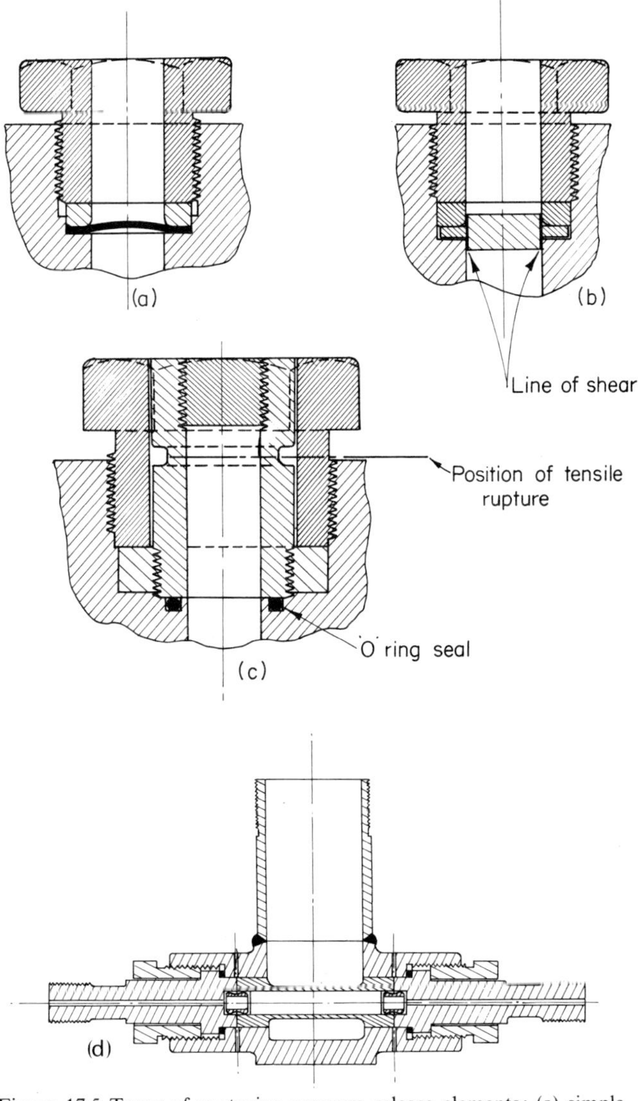

Figure 17.5 Types of rupturing pressure release elements: (a) simple dome (b) shearing element (c) tension element (d) through flow tube element

ranges, but they need periodic replacement if worked near their intended bursting pressure. They can, however, be used up to 200 bars, and to within 10% or so of their bursting pressure, and with care in selection and fitting they have been found to fail within ±3% of the mean value. Lake and Inglis(17.7) give the preferred thickness to diameter ratio as $p/2{\cdot}6$, where p is in lb force/in^2 for pressures up to 1,500 in those units (about 100 bars). Austenitic stainless steel is their preferred material for the higher pressures, and with suitable precautions this may even be raised to 1,000 bars. Tongue(17.8) quotes a case where, on a laboratory scale, membranes of this type, made from S80 steel (which normally contains some 18% chromium and 2% nickel and has an ultimate tensile strength of about 9,000 bars) had a consistent bursting pressure of 1,700 bars.

For pressures higher than about 200 bars, however, rupturing elements that fail by shear or tension are generally preferred. Fig. 17.5(b) shows an example of the former and 17·5(c) of the latter. Both have the advantage that, after blowing out, they leave a full-bore path for the issuing gases, but against this must be set the need to provide means of catching the plug which blows out, usually with very considerable speed.

The tensile type is relatively cheap and easy to make because it can be produced in quantity from lengths of solid-drawn tube, and its quality can be controlled merely by pulling an occasional piece in tension; the bursting pressure will be substantially proportional to the tensile strength with such elements. They require a certain amount of skill and experience to fit if leakage is to be avoided, since most of the designs depend on small O-rings for sealing, but this ought not to give much trouble amongst people used to precision high pressure work.

The shear types are usually easier to fit because it is generally possible to arrange them so that they can be sealed on the unsupported area principle, as shown in Fig. 17.5(b), where the ring shown above the element's fin is made of softer material and seals by reason of the higher pressure which it carries. They are also quite inexpensive to make, but some sort of gauging jig is usually desirable if the dimensions are to be kept within close limits, and this is obviously necessary if the rupturing pressures are to be consistent. Materials with good chemical resistance are always to be preferred, since the slightest attack must weaken any such element and this is likely to lead to premature failure.

Finally, there is the tube type, of which Fig. 17.5(d) (taken from Ref. 17.9) illustrates an example. The great advantage here is that the element can be made a part of the actual process plant system, for instance by fitting in the inlet pipe to a reactor, thereby ensuring that the approaches to it are kept clear and that the element can never be prevented from functioning by blockages. For consistent reproducibility it is best to machine these elements from solid bar, but this is a simple operation and the control of the dimensions is easy. Manufacture from solid-drawn tube is also possible, but in that case it is advisable to obtain a considerable length and to check its dimensions carefully, since the tube manufacturers cannot usually hold the bore to as close a tolerance as is desirable for this purpose.

The example shown has wave-ring seals at each end, but for most purposes

O-rings can be used. For instance, the connecting plugs shown in Fig. 17.5(d) could be made with projecting spigots so that they could each carry an O-ring groove to seal against the bore of the element at its thickened ends.

17.5.5 *Relief Valves*

The main problems here are to obtain a reliably pressure-tight seal and to reduce the inertia of the parts that have to move so that the valve will open quickly. The first requirement can usually be met by using a self-tightening seal, for instance a Bridgman unsupported area type, or a wave-ring, although this may mean that a bigger travel is needed before the valve can open. On the other hand, sealing on a flat or conical seat will require a greater force. The low inertia condition obviously precludes the use of any dead-weight device, and springs become very cumbersome for any but small units and moderate pressures. Thus a 2·5 cm bore valve designed to open at 1,000 bars would need a force of more than 5 tonnes to hold it shut, evidently requiring a very massive spring. Fig. 17.6 shows a possible design for a 1 cm bore valve for 600 bars relief, using a Bridgman seal and a medium length travel.

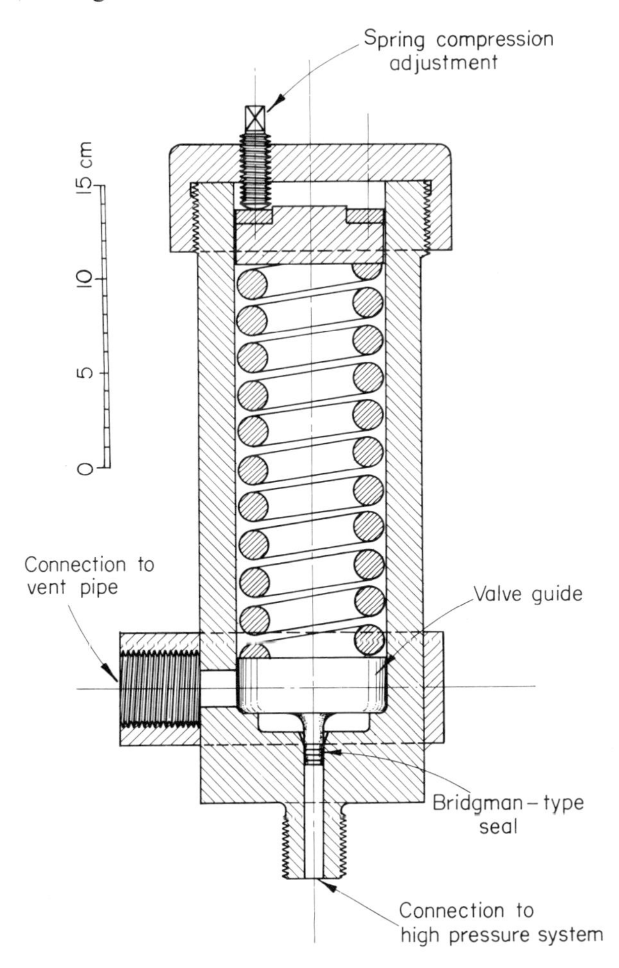

Figure 17.6 Typical spring-loaded relief valve for 600 bars

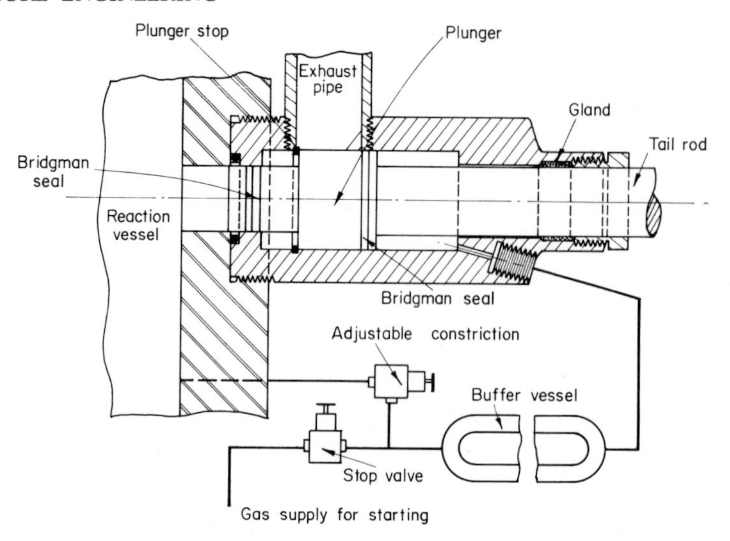

Figure 17.7 Proposed hydrostatically balanced relief valve with automatic compensation for variation in operating pressure

An alternative method of balancing the force of the working pressure is by means of a lower fluid pressure acting over a larger area, as in an intensifier. Unfortunately this also introduces enough inertia to make the action too sluggish for most requirements, especially with chemical reactors. Gas-filled accumulators provide reasonably constant pressure supplies with fairly low inertial resistance, but the ideal from this point of view would seem to be a valve in which the working substance provided the resisting pressure. Fig. 17.7 shows schematically how this might be done. The area of the annulus to the right of the larger diameter part would be slightly greater than the area of the valve port, so that when both these spaces were connected to the reaction pressure the plunger would be forced to the left and the valve would remain shut. A stop would be needed if the seal was of the radial pressure type, as with the Bridgman device shown in the figure, but there would seem to be no difficulty in arranging this, so that under steady pressure conditions the valve would hold indefinitely without leak. However, the connection to the annulus would pass through a buffer vessel and a constricting valve, so that changes of pressure would be felt much sooner at the valve port than in the annulus. Thus any sudden rise of pressure in the reaction space would force the plunger to the right and open the valve, but a slower rise—such as might occur for instance if compressors were being changed by personnel not fully experienced—would not immediately open the valve and could be made to give warning and enable the situation to be brought under control before the contents of the system were lost, as they would be with any kind of rupturing element. This idea was put forward in the 1963 Bulleid Memorial Lectures[(17.10)], but the authors do not know if it has ever been tried. If it would work, however, it would automatically change the blow-off pressure with the working pressure, thus meeting the requirement stated in Section 17.5.3.

17.5.6 *Vent Pipe Reaction*

The release of the contents of a high pressure vessel through an open-ended pipe will result in a large reaction in the form of a compressive force on the pipe, which can – if proper care is not taken to give it adequate support – cause it to fail as a strut, i.e. by buckling. The fact that the duration of such a force would be very short would not prevent it from being very destructive.

The problem is often complicated in practice by the need to change the direction of the issuing gas stream, either at the valve or along the pipe, and this introduces lateral reactions which must also be properly resisted.

The estimation of the magnitude of these forces is often a matter of some conjecture. For instance, with a chemical reactor the pressure rise may be due to some unexpected reaction or decomposition, so that we cannot decide the composition of the exit gases with any certainty. Normally in such a release the gas will leave at the speed of sound in that gas at the temperature and pressure of the issuing jet, and this speed can only be calculated if we know these conditions as well as the composition. By way of illustration, some idea of these magnitudes can be obtained by considering the ethylene reactor mentioned in Section 17.5.2 and assuming that a relieving device opened when the contents were at 2,000 bars pressure and at 250°C. We will also assume that the initially issuing gas was undecomposed ethylene, for which adequate thermodynamic data are available. Using a certain amount of trial and error with an enthalpy–pressure chart, it would appear that the exit velocity would be about 950 m/s and the pressure of the order of 500 bars and the density 330 kg/m^3. Now, if we neglect friction, the force P of the reaction on a straight vent pipe will be given by:

$$P = \frac{\rho A v^2}{g} \tag{17.2}$$

where ρ is the density, A the area of the pipe, and v the issuing velocity. If the pipe had a diameter of 10 cm this would mean a force of no less than 238,000 kgf or 238 tonnesf, so that it would have to be made strong enough to resist an axial force of that magnitude, as well as an internal pressure of 500 bars. In practice frictional effects would probably reduce these figures quite considerably, but the design would certainly have to assume a pressure of that order. That would require a diameter ratio for the pipe of about 1·33, i.e. a wall thickness of some 17 mm, and the free length would then be limited to about 5 metres if the danger of buckling was to be avoided.

Although these figures are only used to illustrate orders of magnitude, it is clear that we are dealing here with large forces, and the design of a venting system of this kind requires careful attention.

Appendix to Chapter 17

Energy Release

This will generally consist mainly of the energy given out when the fluid contents expand rapidly down to atmospheric pressure, or somewhere near it. In some cases also an exothermic reaction may be actually proceeding as the material leaves the vessel and some at least of this chemically released energy must be added in. There is also the elastic strain energy of the metal of the container walls, but normally this is very small.

Considering first the expanding gas, we must assume for the worst case that the containing vessel has disintegrated instantaneously, so that there are no friction losses by "wire-drawing" through long cracks, or frequent changes of direction through zigzag or twisting escape paths. It seems that there is some difference of opinion as to which of the thermodynamic functions will most satisfactorily represent this energy release; both the Gibbs and the Helmholz free energy functions have been suggested, as well as the simple integral:

$$\int p\, dV$$

which under conditions where no heat either enters or leaves the system has advantages, and this we shall consider here. For an ideal gas this would be a comparatively simple calculation, since it would follow the adiabatic law:

$$pV^{\gamma} = \text{constant}$$

Unfortunately, however, γ may vary considerably with real gases, and the only satisfactory procedure is to make use of a thermodynamic chart such as a Mollier diagram with constant pressure and constant volume lines on it, so that the pressure and volume can be read off as expansion proceeds along a line of constant entropy. Then, by dividing the process into a number of smaller intervals of pressure, say for ratios of pressure of between 2 and 3, it is usually possible to fit into each a relation of the type:

$$pV^{n} = \text{constant}$$

This leads to:

$$\int_{1}^{2} p\, dV = \frac{1}{n-1}(p_1 V_1 - p_2 V_2) \qquad (17A.1)$$

This method is reasonably satisfactory for most cases, at any rate for estimating the order of magnitude of this release, although under certain circumstances the fluid state will be represented as partially liquefied owing to the great cooling, effect of adiabatic expansion through a very wide pressure range. In such circumstances it is probable that it will remain as a super-cooled vapour, but the thermodynamic properties will then be largely a matter of guesswork.

To illustrate the above method, and to show how large the energy release can be, an example will be given. Suppose that an ethylene polymerising reactor has an internal volume of 0·5 m^3 and that it is operating at 2,000 bars and at 250°C when it suddenly bursts in a completely brittle manner. Using the pressure-enthalpy chart prepared by the Kältetechnischen Institut der Technischen Hochschule Karlsruhe in 1954 (by H. Benzler and A. v. Koch) it is possible to calculate the energy in pressure ratio steps of from about 2 to 2·5 down to about 30 bars, at which point the equilibrium condition would involve the appearance of some liquid.

In practice, such a pressure release in this process would most likely be caused by the decomposition of the ethylene, and in such circumstances the issuing gas will actually be decomposing as it escapes. The reaction is thought to be somewhat complex, but for our present purpose it will suffice to assume that it takes place according to the equation:

$$C_2H_4 \rightarrow CH_4 + C + 30 \text{ kcal per g mol.},$$

and also that only 10% of it will have so reacted. Since the number of molecules in the system will not be changed, we can add the chemical energy to the pressure energy, and we can, as a first approximation, assume that the final stage of the expansion takes place under ideal gas conditions, with γ = (say) 1·4.

By the above means the energy of the expanding gas is estimated to be about 37,300 m kgf/kg or 365,000 J/kg. The chemical energy, if the decomposition is complete, will be:

$$\frac{1{,}000}{28} \times 30 = 1{,}070 \text{ kcal per kg}$$

or $4{\cdot}48 \times 10^6$ joules. In this case, however, it would be impossible for the whole of the gas to decompose because this takes time, and the drop in pressure and temperature as it escapes would certainly quench the reaction; thus we shall assume that 10% decomposes, and the energy release on this account is therefore 448,000 J/kg.

Now the density of ethylene at 2,000 bars and 250°C is 0·475, according to the Karlsruhe chart already mentioned, and therefore the quantity of gas originally inside the half metre cube reactor is 238 kg.

The elastic strain energy of the container is never likely to be more than a small fraction of the energy in the gas, although with liquid systems it may occasionally be of some significance and it will therefore be briefly mentioned here for the sake of completeness. The easiest way to consider this is to imagine the container being expanded by forcing into it a completely incompressible fluid. Since the whole system is elastic, the resulting increase in volume ΔV will be directly proportional to the pressure, and consequently the energy U_c will be given by:

$$U_c = \tfrac{1}{2} p \Delta V \qquad (17A.2)$$

If the container is assumed to be cylindrical in shape, and if we neglect the energy which may be due to distortion of the end covers—which would be small in most designs—we can assess ΔV as follows:

$$\Delta V = 2\pi r_i u_i L + \pi r_i^2 \varepsilon_z L$$

where r_i and L are the internal radius and length, u_i is the radial shift at the bore, and ε_z is the longitudinal strain. Then substituting for u_i in terms of ε_θ, the tangential strain:

$$\Delta V = \pi r_i^2 L(2\varepsilon_\theta + \varepsilon_z) = V(2\varepsilon_\theta + \varepsilon_z)$$

The values of the strains can then be substituted by means of the relations given in Chapter 2 to give:

$$U_c = \frac{p_i^2 V}{2E(K^2-1)} \{3(1-2\nu) + 2K^2(1+\nu)\} \qquad (17A.3)$$

where E and ν are respectively Young's Modulus and Poisson's Ratio, and K is the radius or diameter ratio.

With the ethylene polymerisation reactor we have been considering, the value of K would be about 2, which would give for U_c in its 0·5 m^3 volume a value of $1{\cdot}93 \times 10^4$ m kgf or $1{\cdot}89 \times 10^5$ joules.

The energy evolution in this example can now be summarised as in Table 17.1 for the whole reactor, i.e. for 238 kg of original ethylene.

Table 17.1

	Energy evolved	
	joules	*ft lbf*
Adiabatic expansion of gas	$8{\cdot}65 \times 10^7$	$6{\cdot}38 \times 10^7$
Decomposition of 10% of gas	$10{\cdot}65 \times 10^7$	$7{\cdot}86 \times 10^7$
Elastic strain energy of container	$0{\cdot}019 \times 10^7$	$0{\cdot}014 \times 10^7$
Total energy	$19{\cdot}32 \times 10^7$	$14{\cdot}25 \times 10^7$

As will be appreciated, this is a very large amount of energy, which would be very difficult to contain if it were liberated in a very short interval of time. It should be noted also that the elastic strain energy is less than $\frac{1}{4}$% of the expansion energy of the gas.

Correlation with Conventional Explosives. This is usually based on a similarity of released energy. T.N.T. gives out approximately $1{\cdot}50 \times 10^6$ ft lbf/lb, or $4{\cdot}48 \times 10^6$ J/kg. From this we see that the reactor in the example is equivalent to 43·1 kg of T.N.T., so far as the energy release is concerned.

REFERENCES

17.1. HIGH, W. G., *Chem. Ind.*, June 1967, 899.
17.2. SAIBEL, E., *Ind. Eng. Chem.*, **53,** 56A, July 1961.
17.3. Report on the Brittle Fracture of a High-Pressure Boiler Drum at the Cockenzie Power Station, South of Scotland Electricity Board, January 1967.
17.4. BROWNE, H. C., HILEMAN, H. and WEGER, L. C., *Ind. Eng. Chem.*, **53,** 82A, Oct. 1961.
17.5. WEBER, J. P., SAVITT, J., KRE, J. and BROWNE, H. C., *Ind. Eng. Chem.*, **53,** 52A, Nov. 1961.
17.6. CHRISTOPHERSON, D. G., *Structural Defence,* H.M. Stationery Office, London 1945.
17.7. LAKE, G. F. and INGLIS, N. P., *Proc. Inst. Mech. Eng.*, **142,** 365, 1940.
17.8. TONGUE, H., *The Design and Construction of High Pressure Chemical Plant,* 2nd Edition, Chapman & Hall 1959, 20.
17.9. MANNING, W. R. D., *Trans. Inst. Chem. Eng.*, **31,** 138, 1953.
17.10. MANNING, W. R. D., Bulleid Memorial Lectures, Nottingham University, 1963, Lecture No. IV, 14.

Name Index

Subject Index